SOIL AND WATER ENGINEERING

Principles and Applications of Modeling

SOIL AND WATER ENGINEERING

Principles and Applications of Modeling

Edited by
Balram Panigrahi, PhD
Megh R. Goyal, PhD, PE

APPLE ACADEMIC PRESS

Apple Academic Press Inc.	Apple Academic Press Inc.
3333 Mistwell Crescent	9 Spinnaker Way
Oakville, ON L6L 0A2	Waretown, NJ 08758
Canada	USA

© 2016 by Apple Academic Press, Inc.
First issued in paperback 2021
Exclusive worldwide distribution by CRC Press, a member of Taylor & Francis Group
No claim to original U.S. Government works

ISBN-13: 978-1-77463-706-7 (pbk)
ISBN-13: 978-1-77188-392-4 (hbk)

Library and Archives Canada Cataloguing in Publication

Soil and water engineering : principles and applications of modeling/edited by
Balram Panigrahi, PhD, Megh R. Goyal, PhD, PE.

(Innovations in agricultural and biological engineering)
Includes bibliographical references and index.
Issued in print and electronic formats.
ISBN 978-1-77188-392-4 (hardcover).--ISBN 978-1-77188-393-1 (pdf)
1. Soil conservation. 2. Hydraulic engineering. 3. Water conservation. 4. Crops and soils. 5.
Crops and water. I. Panigrahi, Balram, author, editor II. Goyal, Megh Raj, editor III. Series:
Innovations in agricultural and biological engineering

S623.S63 2016	631.4'5	C2016-901605-6	C2016-901606-4

Library of Congress Cataloging-in-Publication Data

Names: Panigrahi, Balram, editor. | Goyal, Megh Raj, editor.
Title: Soil and water engineering : principles and applications of modeling/ editors: Balram Panigrahi, Megh R. Goyal.
Description: Oakville, ON ; Waretown, NJ : Apple Academic Press, [2016] |
Includes bibliographical references and index.
Identifiers: LCCN 2016010358 (print) | LCCN 2016012253 (ebook) | ISBN 9781771883924 (hardcover : alk. paper) | ISBN 9781771883931 ()
Subjects: LCSH: Soil mechanics--Mathematical models. | Water resources development--Math-ematical models. | Crops and water--Mathematical models.
Classification: LCC TA710 .S587 2016 (print) | LCC TA710 (ebook) | DDC 333.91--dc23
LC record available at http://lccn.loc.gov/2016010358

Apple Academic Press also publishes its books in a variety of electronic formats. Some content that appears in print may not be available in electronic format. For information about Apple Academic Press products, visit our website at **www.appleacademicpress.com** and the CRC Press website at **www.crcpress.com**

CONTENTS

LIST OF CONTRIBUTORS

M. S. Behera
ICAR – Central Research Institute for Jute and Allied Fibers, Barrackpore, Kolkata-700120, India.
E-mail: kundu_crijaf@yahoo.com, behera_ms@rediffmail.com

S. K. Behera
Senior Scientist (SWCE), All India Coordinated Research Project on Dry Land Agriculture (OUAT),
AT/PO: Phulbani, Dist.: Kandhamal, Odisha-762001, India, Mobile: +91-9437619398,
E-mail: subrat_behera@rediffmail.com

P. R. Choudhury
Ex-Scientist (ICAR) and Development Researcher and Consultant, Plot No 786/1720, Sampur,
PO: Ghatikia, Bhubaneswar – 751003, Odisha, India, E-mail: prchoudhury@rediffmail.com

V. M. Chowdary
Scientist, Regional Remote Sensing Centre, National Remote Sensing Centre, Kolkata–700156, West
Bengal, India; E-mail: muthayya.chowdary@gmail.com

S. Roy Chowdhury
ICAR – Indian Institute of Water Management, Opposite Rail Vihar, Chandrasekharpur,
Bhubaneswar-751023, India. Mobile: +91-9437221616, E-mail: skjena_icar@yahoo.co.in, somnath_
rc@yahoo.com, rajeebm@yahoo.com

Anchal Dass
Senior Scientist, Division of Agronomy, ICAR-Indian Agricultural Research Institute, Pusa Campus,
New Delhi - 110012, India. E-mail: anchal_d@rediffmail.com

Devi Dayal
Central Arid Zone Research Institute, Regional Research Station, Kukma – 370105, Bhuj, Gujarat,
India. Tel.: +91-2832-271238; Fax: +91-2832-271238, E-mail: dmachiwal@rediffmail.com, devidayal.
cazri@yahoo.co.in

Proloy Deb
Center for Water, Climate and Land Use (CWCL), Faculty of Science and Information Technology,
School of Environmental and Life Sciences, University of Newcastle, Callaghan, Newcastle, Australia,
2308 E-mail: debproloy@gmail.com; Proloy.Deb@uon.edu.au; Phone: +61-478744219

M. L. Gaur
Department of Soil and Water Engineering, College of Agricultural Engineering and Technology,
Anand Agricultural University, Godhra–389001, Gujarat, India; Mobile: +91 9428152757,
E-mail: sanjaykraul@gmail.com, mlgaur@yahoo.com

Sampad Ghosh
Department of Chemistry, Indian Institute of Technology, Kharagpur, West Bengal, Kharagpur 721302,
India. Tel.: +91 9434238986, E-mail: justsampad@gmail.com

Susanta Kumar Jena
ICAR – Indian Institute of Water Management, Opposite Rail Vihar, Chandrasekharpur,
Bhubaneswar-751023, India. Mobile: +91-9437221616, E-mail: skjena_icar@yahoo.co.in, somnath_
rc@yahoo.com, rajeebm@yahoo.com

Madan Kumar Jha
Agricultural and Food Engineering Department, Indian Institute of Technology, Kharagpur – 721302, West Bengal, India, E-mail: madan@agfe.iitkgp.ernet.in

K. K. Khatua
Department of Civil Engineering, National Institute of Technology, Rourkela, India, E-mail: kcpatra@nitrkl.ac.in, kkkhatua@nitrkl.ac.in

D. K. Kundu
ICAR – Central Research Institute for Jute and Allied Fibers, Barrackpore, Kolkata-700120, India. E-mail: kundu_crijaf@yahoo.com, behera_ms@rediffmail.com

Deepesh Machiwal
Central Arid Zone Research Institute, Regional Research Station, Kukma – 370105, Bhuj, Gujarat, India. Tel.: +91-2832-271238; Fax: +91-2832-271238, E-mail: dmachiwal@rediffmail.com, devidayal.cazri@yahoo.co.in

Soumen Maji
Department of Civil Engineering, Central Institute of Technology, Kokrajhar, Assam-783370, India. E-mail: s.maji@cit.ac.in

R. K. Mohanty
ICAR – Indian Institute of Water Management, Opposite Rail Vihar, Chandrasekharpur, Bhubaneswar-751023, India. Mobile: +91-9437221616, E-mail: skjena_icar@yahoo.co.in, somnath_rc@yahoo.com, rajeebm@yahoo.com

S. Mohanty
ICAR-Indian Institute of Water Management, Chandrasekharpur, Bhubaneswar-751 023, India, Mobile: +919438008253; E-mail: smohanty.wtcer@gmail.com

G. Chandra Mouli
Principal Investigator, Precision Farming Development Centre, Institute of Agricultural Engineering and Technology, PJTS Agricultural University, Rajendra Nagar, Hyderabad–500030, India; Tel.: +91-9437667879; E-mail: gaddamchandramouli@gmail.com

Trushnamayee Nanda
Research Scholar, Department of Agricultural and Food Engineering, Indian Institute of Technology, Kharagpur, West Bengal, India, 721302, Tel.: +91-8895705688, E-mail: nanda.trushnamayee@yahoo.com

Sudhindra N. Panda
Professor (Land & Water Resources Engineering), Department of Agricultural & Food Engineering, Indian Institute of Technology, Kharagpur-721302, West Bengal, India. Phone: +91-3222-283140 (Work); E-mail: snp@iitkgp.ac.in, snp@agfe.iitkgp.ernet.in, sudhindra.n.panda@gmail.com

R. K. Panda
Dean (R&D), Indian Institute of Technology Bhubaneswar, Samantapuri, Bhubaneswar, Odisha-751013, India, E-mail: rkpanda@iitbbs.ac.in

Balram Panigrahi
Professor and Head, Department of Soil and Water Conservation Engineering, College of Agricultural Engineering and Technology, OUAT, Bhubaneswar, Odisha, India. Mobile: +91-9437882699; E-mail: kajal_bp@yahoo.co.in

Banamali Panigrahi
Research Associate, Department of Regulatory Sciences, GVK Informatics Pvt. Ltd., Hyderabad, Telangana, India, Tel.: +91-9440942065, E-mail: banamali.panigrahi25@gmail.com

S. K. Pattanaik
Assistant Professor (Senior Scale), Department of Agricultural Engineering, College of Horticulture and Forestry, Central Agricultural University, Pasighat, Arunachal Pradesh–791102, India; Mobile: +91-9436630596; E-mail: saroj_swce@rediffmail.com

K. C. Patra
Department of Civil Engineering, National Institute of Technology, Rourkela, India, E-mail: kcpatra@nitrkl.ac.in, kkkhatua@nitrkl.ac.in

J. C. Paul
Associate Professor, Department of Soil and Water Conservation Engineering, CAET, OUAT, Bhubaneswar–751003, Odisha, India. Mobile: +91-9437762584, E-mail: jcpaul66@gmail.com

L. P. Pholane
Project Officer, Chemical Engineering Department, Indian Institute of Technology, Kharagpur-721 302, West Bengal, India; E-mail: pholane@gmail.com

Nirakar Pradhan
Environmental Engineering and Management Program, Asian Institute of Technology, Pathumthani, 12120, Thailand

S. K. Raul
Department of Soil and Water Engineering, College of Agricultural Engineering and Technology, Anand Agricultural University, Godhra–389001, Gujarat, India; Mobile: +91 9428152757, E-mail: sanjaykraul@gmail.com, mlgaur@yahoo.com

Anirban Ray
Agricultural and Food Engineering Department, Indian Institute of Technology, Kharagpur, West Bengal, Kharagpur 721302, India. Tel.: +91 9231695435, E-mail: anirbanrayiitkgp@gmail.com, anirban@agfe.iitkgp.ernet.in

N. Sahoo
Retired Principal Scientist of ICAR-IIWM, Bhubaneswar-751023, India. E-mail: narayansahoo65@yahoo.in

Mrunmayee M. Sahoo
Retired Principal and Scientist of ICAR-IIWM, Bhubaneswar-751023, India. E-mail: narayansahoo65@yahoo.in

Rakesh K. Sahu
Former MTech student, Department of Soil and Water Conservation Engineering, CAET, OUAT, Bhubaneswar–751003, Odisha, India. Mobile: +91-9776356362, E-mail: rakeshkumarsahu57@gmail.com

L. N. Sethi
Associate Professor, Department of Agricultural Engineering, Assam University, Silchar, Assam – 788 011, India, Mobile: +91-9401847943; E-mail: laxmi.narayan.sethi@aus.ac.in

Susama Sudhishri
Principal Scientist, Water Technology Center, ICAR-Indian Agricultural Research Institute (IARI), Pusa Campus, New Delhi-110012, India. Mobile: +91 9971931921, E-mail: susama_s@rediffmail.com

J. B. Swain
Department of Civil Engineering, National Institute of Technology, Rourkela, India, E-mail: Jnkballav.2009@gmail.com

LIST OF ABBREVIATIONS

ADM	above ground dry matter
AET	actual evapotranspiration
AFCR	average feed conversion ratio
AHP	analytic hierarchy process
ANN	artificial neural network
ANOVA	analysis of variance
AR	autoregressive
ARMA	autoregressive moving average
ASW	available soil water
B:C ratio	benefit: cost ratio
BCR	benefit-cost ratio
BD	biodrainage
BF	bottom feeder
BIS	Bureau of Indian Standards
BMPs	best management practices
BOD	biological oxygen demand
BPNN	back propagation neural network
CCA	culturable command area
CCD	central composite design
CCE	crop cutting experiment
CF	column feeder
CGS	critical growth stage
COD	chemical oxygen demand
CRIWAR	crop irrigation water requirement
CWR	crop water requirement
DAP	diammonium phosphate
DAS	days after sowing
DBH	diameter at breast height
DBMS	data base management system
DEA	data envelopment analysis
DM	decision maker
DN	digital number
DO	dissolved oxygen

DOA	Department of Agriculture
DRDA	District Rural Development Agency
dS	decisiemens
DSS	decision support system
DSSAT	Decision Support System for Agrotechnology Transfer
EAS	Employment Assurance Scheme
EC	electrical conductivity
EEC	European Economic Community
EGHL	Eastern Ghat High Land Zone
EP1	experimental plot 1
ERDAS	Earth resource data analysis system
ET	evapotranspiration
FAO	Food and Agriculture Organization
FC	field capacity
FC	Fecal Coli-form
FCC	false color composite
FCR	feed conversion ratio
FLP	fuzzy linear programming
FMEY	finger millet equivalent yield
GCP	ground control point
GDP	gross domestic product
GIR	gross irrigation requirement
GIS	geographical Information System
GNP	gross national product
Govt.	Government
GP	goal programming
GP rank	growth parameter rank
GSM	gram per square meter
GWC	capillary contribution from groundwater
HSJ	hop skip and jump
IARI	Indian Agricultural Research Institute
IBSNAT	International Benchmark Sites Network for Agrotechnology Transfer
ICAR	Indian Council of Agricultural Research
ICMR	Indian Council of Medical Research
IDF	Intensity-Duration-Frequency
IFMOP	interval fuzzy multi objective programming

IFMOPLWS	interval fuzzy multi objective programming for lake watershed system
IIT	Indian Institute of Technology
IIWM	Indian Institute of Water management
IMBW	initial mean body weight
IMC	Indian major carps
IMD	Indian meteorological department
IP	integer programming
IR	irrigation requirement
IRS	Indian remote sensing
ISI	Indian Standard Institution
ITKs	indigenous technical knowledge
IWDP	integrated wasteland development project
IWSM	integrated watershed management system
KSA	Kingdom of Saudi Arabia
Ksf	soil water stress factor
KVK	Krishi Vigyan Kendra
LAI	leaf area index
LBCD	loose boulder check dam
LDP	lexicographic decision programming
LDPE	low density polyethylene
LP	linear programming
LS	lateral seepage
LSD	least significance difference
LULC	land use and land cover
MA	moving average
MAD	maximum allowable depletion
MAE	mean absolute error
MAP	months after planting
MBW	mean body weight
MDPS	Markov decision process
ME	mean error
MM	method of moments
MOLP	multiple objective linear programming
MOP	multiple objective programming
MSE	mean squared errors
MVRA	multivariate variance analysis
NAAS	National Academy of Agricultural Sciences

NC	sum of horizontal and vertical crossings
ND	number of days
NDVI	normalized difference vegetation index
NGO	Non-Government Organization
NH	National Highway
NH4-N	Ammonia as Nitrogen
NH_4^+	ammonium
NIR	near-infrared wave length regions
NO_3^-	nitrate
NP	net present value
NPK	nitrogen, phosphorus and potassium
NPK	total fertilizer applied
NR	reflectance in red
NRM	natural resource management
NSE	nash sutcliffe efficiency
NV	non-vegetable area
O	occurrence
OBC	other backward caste
OC	organic carbon
OFR	on-farm reservoir
OLIC	Orissa Lift Irrigation Corporation
P	phosphorous
P	rainfall
p	soil water depletion factor
PAR	photo synthetically active radiation
PBP	pay-back period
PC	principal component
PCA	principal component analysis
PDF	probability distribution function
PDI	per day increment
Pe	effective precipitation
PE	prediction efficiency
P_{EF}	point performance function relative to efficiency
PET	potential evapotranspiration
ppm	parts per million
PRA	participatory rural appraisal
PSI	pre-sowing irrigation
PVC	poly venyl chloride

RDI	regulated deficit irrigation
RI	recurrence interval
RL	root length
RLD	root length density
RME	reduced mean error
RMSE	root mean square error
Rn	net solar radiation
RN	rainfall
RO	surface runoff
RS	remote sensing
Rs.	Indian rupees
RWH	runoff water harvesting
RWS	relative water supply
SAT	saturation soil water content
SC	stocking composition
SD	stocking density
Seq SS	sequential sum of squares
SF	surface feeder
SI	supplemental irrigation
SI	sub index
SMADA	Storm water Management and Design Aid
SMC	soil moisture contribution
Sn	neutral slope
SOI	Survey of India
SPR	special purpose requirements
SR	survival rate
SSE	summed squared error
SWAT	soil and water assessment tool
SWC	soil and water conservation
SWC	soil water content
SWIR	short wave infrared
TA as $CaCO_3$	total alkali as $CaCO_3$
TC	total coli-form
TCVI	time composited vegetation index
TH as $CaCO_3$	total hardness as $CaCO_3$
TSS	total suspended solid
TW	total storage of water in the on-farm reservoir
UGPL	under ground pipe line

USDA-SCS	United States Department of Agriculture and Soil Conservation Service
USPHS	United States Public Health Services
VP	vertical deep percolation
WD	water depth applied
WHO	World Health Organization
WHP	water harvesting pond
WHS	water harvesting structure
WiFS	wide field sensor
WP	wilting point
WQI	water quality index
WUA	Water User's Association
WUE	water use efficiency

LIST OF SYMBOLS

$	US dollar
A	annual cost
A	area at mid section
a	cross-sectional area of conduit
a	shape factor
A_1	area at top
A_2	area at bottom
B	benefit
C	coefficient of discharge
C1	Bhimpura main canal
C2	Vadhvana Boriad main canal
C3	Dabhoi main canal
C4	Vasai Dangiwada main canal
C5	Simliya canal
C_d	discharge coefficient
CF	column feeder
cm	centimeter
C_o	observed data sample
C_s	simulated data
Cu	copper
CV_S	spatial coefficient of variation
CV_T	temporal coefficient of variation
D	detect ability
D1	depression 1
D_a	depth of irrigation application
dS	decisiemens
E	evapotranspiration
E_t	evapotranspiration during turn-in period
ET_0	reference crop evapotranspiration
ETp	crop evapotranspiration
f	correction factor
Fe	iron
g	acceleration due to gravity

G	length of the grid unit
G	soil heat flux density
H	head causing flow
H	ponding depth
h	pressure head
H1	0 cm weir height
$h1$	hydraulic head at the upper layer of the root-zone
H2	5 cm weir height
$h2$	hydraulic head at the bottom layer of the root-zone
H3	10 cm weir height
H_f	frictional head loss
i	index for type of on-farm reservoir
I_{inv}	initial investment
j	index for type of weir heights
K	potassium
K	unsaturated hydraulic conductivity
K	*kharif*
K^A	anisotropy tensor
K_b	loss coefficients for bends
Kc	crop coefficient
K_c	friction loss coefficient
K_e	entrance loss coefficient
kg	kilogram
km	killometer
K_n	Ponderal index/condition factor = (weight/cube of length)
K_r	relative hydraulic conductivity
K_s	saturated hydraulic conductivity
Ksf	soil water stress factor
L	length of weir's crest
l	pore-connectivity parameter
L1	lined on-farm reservoir
L2	unlined on-farm reservoir
Lakh	10^5
M	million
m	exponent
mg	milligram
Mha	million hectare
Mham	million hectare meter

MM	method of moments
Mm^3	million cubic meter
Mn	manganese
mt	million tons
N	nitrogen
n	pore-size distribution index
N	total number of data samples
N	total number of observation
O	occurrence
P	phosphorous
P	rainfall
p	soil water depletion factor
P1	Platform 1
P_1	pond-1
P_2	pond-2
P_3	pond-3
P_A	point performance function relative to adequacy
Pe	effective precipitation
P_{EF}	point performance function relative to efficiency
PI_A	adequacy
PI_D	dependability
PI_E	equity
PI_{EF}	efficiency
ppm	parts per million
PW_{ac}	present worth of annual cost
PW_b	present worth of benefit
PW_c	present worth of cost
q	quintal (= 100 kg)
Q	discharge
Q	amount of water
Q_D	actual amount of water delivered to the system
Q_E	water lost as evaporation from the on-farm reservoir
Q_P	direct rainfall contribution on the on-farm reservoir
Q_R	actual amount of water required for consumptive use
Q_{RO}	runoff water coming from the field to the on-farm reservoir
Q_{SI}	water used as SI in the cropped field
Q_{SP}	seepage and percolation loss from the unlined on-farm reservoir
r	correlation coefficient

r	interest rate
r	number of rows in the error matrix
R	replication
R	visible red
R	*rabi*
R^2	coefficient of determination
R_e	resilience
Rn	net solar radiation
Rs.	Indian rupees
R_y	reliability
S	severity
S	sink term
S	slope
S	region
S_e	effective water content
Seq SS	sequential sum of squares
Sn	neutral slope
S_p	potential water uptake rate
t	index for time
T	thickness of the effective root-zone
t	time index
T	time period
t.ha^{-1}	ton/hectare
U_2	wind velocity at 2 m height
U_Z	wind velocity at height Z m
V	vegetable area
V	velocity of flow
V	volume
VP	vertical deep percolation
V_y	vulnerability
W	water storage in the on-farm reservoir
x	horizontal space coordinates
Y	rice yield
Z	height above ground, m
z	vertical space coordinates
Zn	zinc

PREFACE 1 BY BALRAM PANIGRAHI

In recent years, modeling aspects have added a new dimension in research innovations in all branches of engineering. In the field of soil and water engineering, they are increasingly used for planning, development, and management of land and water resources, including analysis of quantity and quality parameters of surface and ground water, flood forecasting and control measures, optimum allocation and utilization of irrigation water, increasing water productivity in rainfed and irrigated commands, reclamation of water-logged areas, etc. The application of these models saves considerable time in decision support systems and helps in conservation and proper and optimum allocations of scarce precious natural resources such as soil and water in enhancing agricultural production.

This book *Soil and Water Engineering: Principles and Applications of Modeling*, discusses the development of some useful models and their applications in the field of soil and water engineering. The book is organized into four parts with 15 chapters. Part I titled Modeling Soil Water Engineering Systems in Rainfed Areas contains five chapters; Part II titled Modeling Irrigation systems in Canal Commands contains three chapters; Part III titled Research Advances in Soil and Water Engineering contains four chapters; and the Part IV titled Rainfall Analysis for Crop Planning contains three chapters.

The book addresses the modeling and design of on-farm reservoir irrigation systems, watershed planning and management, water conservation in rainfed areas, and modeling optimum dyke heights to conserve maximum rainfall in rainfed. The volume goes on to examine modeling methods used in canal water delivery systems, the use of remote sensing and geographical information systems in rice irrigated canal commands, and irrigation management using decision support systems. Other topics include the application of aquifer simulation models in study of ground water, ground water quality analysis by artificial neural networks, carbon sequestration study in restoring degraded soils and management of waterlogged areas by bio-drainage, and integrated farming systems.

Rainfall is the main source for supplying water to the crops in both rainfed and irrigated areas. Its analysis is very essential for effective crop planning. Since it is a stochastic variable, analysis of time series of rainfall is very important. A special chapter on a sustainability concept to evaluate performance of rainfall time series is presented, and a study of annual daily extreme precipitation of different rain-gauge stations is presented and will be of immense help in planning and management of water resources.

The book will serve as an invaluable resource for graduate and undergraduate students in the field of agriculture, agricultural, biological and civil engineering and also other branches of natural resources engineering. The book will be helpful for all academicians, research investigators, field engineers, agronomists, soil scientists and extension personnel who directly or indirectly deal with Soil and Water Engineering. The contributions by the authors of different chapters of this book are very valuable which are duly acknowledged. The authors are well experts in their fields and have long years of experience in these areas. It is needless to mention that without their support this book would have not been published successfully. Their names are mentioned in each chapter and also separately in the list of contributors.

I take the opportunity to offer my heartfelt obligations to Prof. Dr. Megh R. Goyal "Father of Irrigation Engineering of 20th Century in Puerto Rico" and editor of this book series who has benevolently given me an opportunity to serve as a lead editor of the book. Through his arduous task of editing various books in agricultural engineering, Dr. Goyal has benefitted educators, planners, decision makers and farmers throughout the world. I feel proud to mention here Prof. Goyal has earlier given me an opportunity to edit a book entitled *Modeling Methods and Practices in Soil and Water Engineering* as lead editor in which he also serves as an author. My special thanks to all the editorial staff of Apple Academic Press Inc. for making every effort to publish both the books.

Readers are requested to offer constructive suggestions that may help to improve the next edition.

I express my deep obligations to my family, friends and colleagues and contributors for their help and moral support during preparation of the book.

—*Balram Panigrahi, PhD*
January 19, 2016

PREFACE 2 BY MEGH R. GOYAL

Due to increased agricultural production, irrigated land has increased in the arid and sub-humid zones around the world. Agriculture has started to compete for water use with industries, municipalities, and other sectors. This increasing demand, along with increments in water and energy costs, have made it necessary to develop new innovative technologies for the adequate management of natural resources. The intelligent use of soil and water for crops requires understanding of evapotranspiration processes and use of efficient irrigation methods under limited resources.

Our planet will not have enough potable water for a population of >10 billion persons in 2115. The situation will be further complicated by multiple factors that will be adversely affected by global warming. The crisis is rampant. I will not be here in 2115, but my great grandchildren will be among the habitants then. We live on our mother planet and not on extra-terrestrial planet. Therefore, it is our ethical and moral responsibility to join hands with the nature, people, and Almighty Creator, to solve this crisis.

I have been involved in soil and water conservation engineering (SWCE) since 1971. I have worked on soil and water conservation measures, irrigation projects, soil crusting, precision farming, acid delinting, pressurized irrigation systems, as well as a professor/researcher/extension specialist/social worker as Pastor throughout my professional career of 51 years. This has helped me to be familiar with innovations and challenges in soil and water conservation engineering. Therefore, I know what the cooperating authors have emphasized in their chapters for this book volume. I am a staunch supporter of preserving our natural resources. The updated seventh edition of *Soil and Water Conservation Engineering* by http://www.asabe.org emphasizes engineering design of soil and water conservation practices and their impact on the environment, primarily air and water quality. Other books on SWCE advocate the same. Importance of the wise use of our natural resources has been taken up seriously by universities, institutes/centers, government agencies and nongovernment agencies. I conclude that the agencies and departments in SWCE have contributed to the ocean of knowledge.

Our book also contributes to the ocean of knowledge on SWCE. Agricultural and biological engineers (ABEs) with expertise in SWCE work to better understand the complex mechanics of natural resources, so that they can be used efficiently and without degradation. ABEs determine crop water requirements and design irrigation systems. They are experts in agricultural hydrology principles, such as controlling drainage, and they implement ways to control soil erosion and study the environmental effects of sediment on stream quality. Natural resources engineers design, build, operate and maintain water control structures for reservoirs, floodways and channels. They also work on water treatment systems, wetlands protection, and other water issues.

While making a call for chapters for a book volume on SWCE, we mentioned to the prospective authors the following focus areas:

- Academia to industry to end user loop in soil and water engineering;
- Aquaculture engineering;
- Biological engineering in SWE;
- Biotechnology applications in SWE;
- Climate change and its impact on SWE;
- Design in irrigation and drainage systems;
- Drainage principles, management, practices;
- Education in SWE: curricula/scope/opportunities;
- Energy potential in SWE;
- Environment engineering;
- Extension methods in SWE;
- Flood damage in crop production;
- Flow through porous media;
- Global warming due ill effects of SWCE;
- Ground water and tube-wells: principles, management, practices;
- Groundwater simulation for sustainable agriculture;
- Human factors engineering in SWE;
- Hydrologic applications in SWE;
- Irrigation principles, management, practices;
- Management of water resources;
- Nanotechnology applications in SWE;
- Natural resources engineering and management;
- Principles of hydraulics in SWE;
- Robot engineering in SWE;
- Simulation, optimization and computer modeling;
- Society and natural resources;
- Soil and water engineering;

- Waste management engineering

Therefore, I conclude that scope of SWCE is very wide, and focus areas may overlap one another. The mission of this book volume is to serve as a reference manual for graduate and undergraduate students of agricultural, biological and civil engineering; and horticulture, soil science, crop science and agronomy. I hope that it will be a valuable reference for professionals who work with soil and water management, for professional training institutes, technical agricultural centers, irrigation centers, Agricultural Extension Service, and other agencies. I cannot guarantee the information in this book series will be enough for all situations.

After my first textbook on *Drip/Trickle or Micro Irrigation Management* by Apple Academic Press Inc., and response from international readers, Apple Academic Press Inc. has published for the world community the ten-volume series on *Research Advances in Sustainable Micro Irrigation*, edited by Megh R. Goyal. I have already published five book volumes under book series *Innovations and Challenges in Micro Irrigation.*

At the 49th annual meeting of Indian Society of Agricultural Engineers at Punjab Agricultural University during February 22–25 of 2015, a group of ABEs convinced me that there is a dire need to publish book volumes on the focus areas of agricultural and biological engineering (ABE). This is how the idea was born on a new book series titled Innovations in Agricultural & Biological Engineering. Here we present the volume titled *Soil and Water Engineering: Principles and Applications of Modeling.*

My long-time colleague, Dr. Balram Panigrahi, joins me as a Lead Editor of this volume. Dr. Panigrahi holds exceptional professional qualities in addition to Professor and Head for Department of Soil and Water Conservation Engineering in College of Agricultural Engineering & Technology (CAET) at Orissa University of Agriculture & Technology, Bhubaneswar, India. His contribution to the contents and quality of this book has been invaluable.

We will like to thank editorial staff, Sandy Jones Sickels, Vice President, and Ashish Kumar, Publisher and President, at Apple Academic Press, Inc., for making every effort to publish the book when the diminishing water resources are a major issue worldwide. Special thanks are due to the AAP Production Staff for the quality production of this book. We request that readers to offers us your constructive suggestions that may help to improve the next edition.

I express my deep admiration to my family for understanding and collaboration during the preparation of this book. Our Almighty God, owner of natural resources, must be very happy on the publication of this book. As an educator, there is a piece of advice to one and all in the world: *"Permit that our almighty God, our Creator and excellent Teacher, irrigate the life with His Grace of rain trickle by trickle, because our life must continue trickling on... and Get married to your profession"*

—*Megh R. Goyal, PhD, PE, Senior Editor-in-Chief*
January 19, 2016

FOREWORD

Ever-increasing global population along with industrialization and urbanization has urged upon us more efficient utilization and conservation of natural resources. Moreover, indiscriminate interference with nature's biological systems by mankind has exacerbated the situation, threatening the livelihood and security through the degradation of land and water resources all over the world. Innovations in research, especially in the development and applications of new models in the field of soil and water engineering, are very much essential for conservation and development of natural resources besides optimally utilizing them for enhancing agricultural production.

In this context, this book, titled *Soil and Water Engineering: Principles and Applications of Modeling* edited by Dr. Balram Panigrahi, Professor and Head, Soil and Water Conservation Engineering, College of Agricultural Engineering and Technology, Orissa University of Agriculture and Technology, Bhubaneswar and distinguished Professor Dr. Megh R. Goyal, under the book series, *Innovations in Agricultural and Biological Engineering*, will definitely serve as an invaluable resource for academicians, practicing engineers, planners, managers, research scientists, and students in the field of soil and water engineering, irrigation engineering and water resources engineering.

I compliment the commendable endeavors of the editors of the book and wish the publication a great success.

—Prof. M. Kar
Vice Chancellor, Orissa University of Agriculture and
Technology, Bhubaneswar
9th October, 2015

WARNING/DISCLAIMER

PLEASE READ CAREFULLY

The goal of this compendium, *Soil and Water Engineering: Principles and Applications of Modeling*, is to guide the world engineering community on how to efficiently design for economical crop production. The reader must be aware that the dedication, commitment, honesty, and sincerity are the most important factors in a dynamic manner for a complete success.

The editor, the contributing authors, the publisher and the printer have made every effort to make this book as complete and as accurate as possible. However, there still may be grammatical errors or mistakes in the content or typography. Therefore, the contents in this book should be considered as a general guide and not a complete solution to address any specific situation in irrigation. For example, one size of irrigation pump does not fit all sizes of agricultural land and to all crops.

The editor, the contributing authors, the publisher, and the printer shall have neither liability nor responsibility to any person, any organization or entity with respect to any loss or damage caused, or alleged to have been caused, directly or indirectly, by information or advice contained in this book. Therefore, the purchaser/reader must assume full responsibility for the use of the book or the information therein.

The mention of commercial brands and trade names is only for technical purposes. It does not mean that a particular product is endorsed over another product or equipment not mentioned. The author, cooperating authors, educational institutions, and the publisher Apple Academic Press Inc. do not have any preference for a particular product.

All weblinks that are mentioned in this book were active on December 31, 2015. The editors, the contributing authors, the publisher, and the printing company shall have neither liability nor responsibility, if any of the weblinks is inactive at the time of reading of this book.

OTHER BOOKS ON AGRICULTURAL & BIOLOGICAL ENGINEERING BY APPLE ACADEMIC PRESS, INC.

Management of Drip/Trickle or Micro Irrigation
Megh R. Goyal, PhD, PE, Senior Editor-in-Chief

Evapotranspiration: Principles and Applications for Water Management
Megh R. Goyal, PhD, PE, and Eric W. Harmsen, Editors

Book Series: Research Advances in Sustainable Micro Irrigation
Senior Editor-in-Chief: Megh R. Goyal, PhD, PE
> Volume 1: Sustainable Micro Irrigation: Principles and Practices
> Volume 2: Sustainable Practices in Surface and Subsurface Micro Irrigation
> Volume 3: Sustainable Micro Irrigation Management for Trees and Vines
> Volume 4: Management, Performance, and Applications of Micro Irrigation Systems
> Volume 5: Applications of Furrow and Micro Irrigation in Arid and Semi-Arid Regions
> Volume 6: Best Management Practices for Drip Irrigated Crops
> Volume 7: Closed Circuit Micro Irrigation Design: Theory and Applications
> Volume 8: Wastewater Management for Irrigation: Principles and Practices
> Volume 9: Water and Fertigation Management in Micro Irrigation
> Volume 10: Innovation in Micro Irrigation Technology

Book Series: Innovations and Challenges in Micro Irrigation
Senior Editor-in-Chief: Megh R. Goyal, PhD, PE
> Volume 1: Principles and Management of Clogging in Micro Irrigation

Volume 2: Sustainable Micro Irrigation Design Systems for
            Agricultural Crops: Methods and Practices
Volume 3: Performance Evaluation of Micro Irrigation Management:
            Principles and Practices
Volume 4: Potential Use of Solar Energy and Emerging Technologies
            in Micro Irrigation
Volume 5: Micro Irrigation Management: Technological Advances
            and Their Applications

Book Series: Innovations in Agricultural and Biological Engineering
Senior Editor-in-Chief: Megh R. Goyal, PhD, PE

Dairy Engineering: Advanced Technologies and Their Applications
Editors: Murlidhar Meghwal, PhD, Megh R. Goyal, PhD, PE, and
Rupesh S. Chavan, PhD

Developing Technologies in Food Science: Status, Applications, and Challenges
Editors: Murlidhar Meghwal, PhD, and Megh R. Goyal, PhD, PE

Emerging Technologies in Agricultural Engineering
Editor: Megh R Goyal, PhD, PE

Engineering Practices for Agricultural Production and Water Conservation: An Interdisciplinary Approach
Editors: Megh R. Goyal, PhD, PE, and R. K. Sivanappan, PhD

Flood Assessment: Modeling and Parameterization
Editors: Eric W. Harmsen, PhD, and Megh R. Goyal, PhD

Food Engineering: Emerging Issues, Modeling, and Applications
Editors: Murlidhar Meghwal, PhD, and Megh R. Goyal, PhD

Food Process Engineering: Emerging Trends in Research and Their Applications
Editors: Murlidhar Meghwal, PhD, and Megh R. Goyal, PhD, PE

Modeling Methods and Practices in Soil and Water Engineering
Editors: Balram Panigrahi, PhD, and Megh R. Goyal, PhD, PE

Soil and Water Engineering: Principles and Applications of Modeling
Editors: Balram Panigrahi, PhD, and Megh R. Goyal, PhD, PE

Soil Salinity Management in Agriculture: Technological Advances and Applications
Editors: S. K. Gupta, PhD, CE, and Megh R. Goyal, PhD, PE

ABOUT THE EDITOR

Balram Panigrahi, PhD
Professor and Head,
Soil and Water Conservation Engineering,
College of Agricultural Engineering and Technology,
Orissa University of Agriculture and Technology,
Bhubaneswar, India

Dr. Balram Panigrahi is an agricultural engineer with specialization in soil and water engineering. Dr. Panigrahi is presently Professor and Head of the Department of Soil and Water Conservation Engineering (SWCE) at the College of Agricultural Engineering and Technology (CAET), Orissa University of Agriculture and Technology (OUAT), in Bhubaneswar, India. He also served as Chief Scientist of the Water Management Project and Associate Director of Research in Regional Research Station of OUAT.

Dr. Panigrahi has published about 180 technical papers in different international and national journals and conference proceedings. He has written several book chapters, practical manuals, and monographs. He has also written two textbooks in the field of irrigation engineering. He has been awarded with 14 gold medals and awards, including the Jawaharlal Nehru Award for best post-graduate research in the field of natural resources management by Indian Council of Agricultural Research, New Delhi; the Samanta Chandra Sekhar Award for best scientist in the state of Odisha, India; and the Gobinda Gupta Award as outstanding engineer of the state of Odisha, given by the Institution of Engineers (India), Odisha state center. He has also received a Japanese Master Fellowship for pursuing a master of engineering study at the Asian Institute of Technology, Thailand. In addition to being the editor and an editorial board member of several journals, Dr. Panigrahi is a reviewer for many journals. He is the member of a number of professional societies at national and international levels. He has chaired several international and national conferences both in India and abroad. With 26 years of of teaching and research experience, he has guided several PhD and many MTech students. Dr. Panigrahi's research interests include irrigation and drainage engineering, water Management in rainfed and irrigated commands, and modeling of irrigation systems.

He obtained his BTech in agricultural engineering from Orissa University of Agriculture and Technology (OUAT), Bhubaneswar, Odisha, India, and his Master of Engineering in water resources engineering from the Asian Institute of Technology, Thailand. He was awarded his PhD in agricultural engineering from the Indian Institute of Technology, Kharagpur, India.

ABOUT SENIOR EDITOR-IN-CHIEF

Megh R. Goyal, PhD, PE
Retired Professor in Agricultural and Biomedical Engineering, University of Puerto Rico, Mayaguez Campus
Senior Acquisitions Editor,
Biomedical Engineering and Agricultural Science, Apple Academic Press, Inc.
E-mail: goyalmegh@gmail.com

Megh R. Goyal, PhD, PE, is a Retired Professor in Agricultural and Biomedical Engineering from the General Engineering Department in the College of Engineering at University of Puerto Rico–Mayaguez Campus; and Senior Acquisitions Editor and Senior Technical Editor-in-Chief in Agriculture and Biomedical Engineering for Apple Academic Press Inc.

He has worked as a Soil Conservation Inspector and as a Research Assistant at Haryana Agricultural University and Ohio State University. He was first agricultural engineer to receive the professional license in Agricultural Engineering in 1986 from College of Engineers and Surveyors of Puerto Rico. On September 16, 2005, he was proclaimed as the "Father of Irrigation Engineering in Puerto Rico for the twentieth century" by the ASABE, Puerto Rico Section, for his pioneer work on micro irrigation, evapotranspiration, agroclimatology, and soil and water engineering.

During his professional career of 45 years, he has received many prestigious awards. A prolific author and editor, he has written more than 200 journal articles and textbooks and has edited over 25 books. He received his BSc degree in engineering from Punjab Agricultural University, Ludhiana, India; his MSc and PhD degrees from Ohio State University, Columbus; and his Master of Divinity degree from Puerto Rico Evangelical Seminary, Hato Rey, Puerto Rico, USA. Readers may contact him at: goyalmegh@gmail.com

BOOK MESSAGES

Conservation of soil and water resources is an urgent need to save our planet from degradation. Agricultural Engineers can help to alleviate these crises. Editors of this book volume have contributed a drop in the ocean. It is our ethical duty to educate our fraternity on this topic.

—Miguel A Muñoz, PhD

In providing these resources in soil and water engineering, Balram Panigrahi and Megh R. Goyal, as well as the Apple Academic Press, are rendering an important service to the conservationists.

—Gajendra Singh, PhD

Water is increasingly scarce and extremely valuable resource, without which sustainable development is impossible. Agriculture is the largest water-using sector worldwide. The gross irrigated area in the world has increased from 94 M-ha in 1950's to about 280 M-ha at present. Most of the areas are in the developing countries, especially in India, the gross irrigated area is more than 100 M-ha. Research in water resources, quantity, quality of water, water management in agriculture including drip irrigation is taken up seriously by the scientists especially for Rice, fruits, vegetables, cotton, banana, sugarcane plantations crops, etc. According to the FAO (1990), 60% of the water supplied for irrigation goes unused and leads to water logging and salinization.

Hence, water requirements for various crops are worked out in surface irrigation, sprinkler and drip irrigation methods to use water efficiently. There are two strategies, which are used to meet challenges: i) Supply management and ii) Demand management. To solve the problems related to water management, water should be considered as an economic asset. The increase in the value of water, demand management will become more important than supply management. This book volume addresses emerging technologies in SWCE.

—R. K. Sivanappan, PhD

The emerging technologies have potential to conserve water that can facilitate timely sowing of crops under the delayed monsoon situation that has occurred in 2014 and provide solutions to monsoon worries. Agricultural Engineers need to provide leadership opportunities for in water resources and water management sector, water resources, irrigation, soil conservation, watersheds, environment and energy for stability of agriculture and in turn stable growth of Indian economy. This book volume is an asset in this path.

—V. M. Mayande, PhD

Visualizing invisible resources using modern tools and modeling approach so as to conserve and manage soil and water resources particularly in the wake of climate change are going to be immensely helpful to the academician and policy makers. I wish the book will be a milestone to deal the important issue of water management.

—Sunil Kumar Ambast, PhD

EDITORIAL

Apple Academic Press Inc., is publishing the AAP book series titled *Innovations in Agricultural and Biological Engineering*. Over a span of 8–10 years, Apple Academic Press Inc., will publish volumes in the specialty areas defined by American Society of Agricultural and Biological Engineers (http://www.asabe.org). The mission of this series is to provide knowledge and techniques for agricultural and biological engineers (ABEs). The series aims to offer high-quality reference and academic content in Agricultural and Biological Engineering (ABE) that is accessible to academicians, researchers, scientists, university faculty, and university-level students and professionals around the world. The following material has been edited/modified and reproduced below [*Goyal, Megh R., 2006. Agricultural and biomedical engineering: Scope and opportunities. Paper Edu_47 at the Fourth LACCEI International Latin American and Caribbean Conference for Engineering and Technology (LACCEI' 2006): Breaking Frontiers and Barriers in Engineering: Education and Research by LACCEI University of Puerto Rico – Mayaguez Campus, Mayaguez, Puerto Rico, June 21 – 23*]:

WHAT IS AGRICULTURAL AND BIOLOGICAL ENGINEERING (ABE)?

"Agricultural Engineering (AE) involves application of engineering to production, processing, preservation and handling of food, fiber, and shelter. It also includes transfer of technology for the development and welfare of rural communities," according to http://www.isae.in. *"ABE is the discipline of engineering that applies engineering principles and the fundamental concepts of biology to agricultural and biological systems and tools, for the safe, efficient and environmentally sensitive production, processing, and management of agricultural, biological, food, and natural resources systems,"* according to http://www.asabe.org. *"AE is the branch of engineering involved with the design of farm machinery, with soil management, land development, and mechanization and automation of livestock farming, and with the efficient planting, harvesting, storage, and processing of*

farm commodities," definition by: http://dictionary.reference.com/browse/agricultural+engineering.

"*AE incorporates many science disciplines and technology practices to the efficient production and processing of food, feed, fiber and fuels. It involves disciplines like mechanical engineering (agricultural machinery and automated machine systems), soil science (crop nutrient and fertilization, etc.), environmental sciences (drainage and irrigation), plant biology (seeding and plant growth management), animal science (farm animals and housing), etc.,*" by: http://www.ABE.ncsu.edu/academic/agricultural-engineering.php.

"According to https://en.wikipedia.org/wiki/Biological_engineering: "*BE (Biological engineering) is a science-based discipline that applies concepts and methods of biology to solve real-world problems related to the life sciences or the application thereof. In this context, while traditional engineering applies physical and mathematical sciences to analyze, design and manufacture inanimate tools, structures and processes, biological engineering uses biology to study and advance applications of living systems.*"

SPECIALTY AREAS OF ABE

Agricultural and Biological Engineers (ABEs) ensure that the world has the necessities of life including safe and plentiful food, clean air and water, renewable fuel and energy, safe working conditions, and a healthy environment by employing knowledge and expertise of sciences, both pure and applied, and engineering principles. Biological engineering applies engineering practices to problems and opportunities presented by living things and the natural environment in agriculture. BA engineers understand the interrelationships between technology and living systems, have available a wide variety of employment options. "*ABE embraces a variety of following specialty areas,*" http://www.asabe.org. As new technology and information emerge, specialty areas are created, and many overlap with one or more other areas.

1. **Aquacultural Engineering**: ABEs help design farm systems for raising fish and shellfish, as well as ornamental and bait fish. They specialize in water quality, biotechnology, machinery, natural resources, feeding and ventilation systems, and sanitation. They seek ways to reduce pollution from aquacultural discharges, to reduce excess

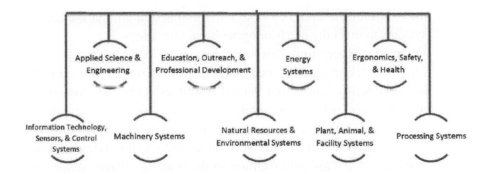

water use, and to improve farm systems. They also work with aquatic animal harvesting, sorting, and processing.

2. **Biological Engineering** applies engineering practices to problems and opportunities presented by living things and the natural environment.

3. **Energy:** ABEs identify and develop viable energy sources – biomass, methane, and vegetable oil, to name a few – and to make these and other systems cleaner and more efficient. These specialists also develop energy conservation strategies to reduce costs and protect

the environment, and they design traditional and alternative energy systems to meet the needs of agricultural operations.

4. **Farm Machinery and Power Engineering**: ABEs in this specialty focus on designing advanced equipment, making it more efficient and less demanding of our natural resources. They develop equipment for food processing, highly precise crop spraying, agricultural commodity and waste transport, and turf and landscape maintenance, as well as equipment for such specialized tasks as removing seaweed from beaches. This is in addition to the tractors, tillage equipment, irrigation equipment, and harvest equipment that have done so much to reduce the drudgery of farming.

5. **Food and Process Engineering:** Food and process engineers combine design expertise with manufacturing methods to develop economical and responsible processing solutions for industry. Also food and process engineers look for ways to reduce waste by devising alternatives for treatment, disposal and utilization.

6. **Forest Engineering**: ABEs apply engineering to solve natural resource and environment problems in forest production systems and related manufacturing industries. Engineering skills and expertise are needed to address problems related to equipment design and manufacturing, forest access systems design and construction; machine-soil interaction and erosion control; forest operations analysis and improvement; decision modeling; and wood product design and manufacturing.

7. **Information and Electrical Technologies Engineering** is one of the most versatile areas of the ABE specialty areas, because it is applied to virtually all the others, from machinery design to soil testing to food quality and safety control. Geographic information systems, global positioning systems, machine instrumentation and controls, electromagnetics, bioinformatics, biorobotics, machine vision, sensors, spectroscopy: These are some of the exciting information and electrical technologies being used today and being developed for the future.

8. **Natural Resources:** ABEs with environmental expertise work to better understand the complex mechanics of these resources, so that they can be used efficiently and without degradation. ABEs determine crop water requirements and design irrigation systems. They are experts in agricultural hydrology principles, such as controlling

drainage, and they implement ways to control soil erosion and study the environmental effects of sediment on stream quality. Natural resources engineers design, build, operate and maintain water control structures for reservoirs, floodways and channels. They also work on water treatment systems, wetlands protection, and other water issues.

9. **Nursery and Greenhouse Engineering**: In many ways, nursery and greenhouse operations are microcosms of large-scale production agriculture, with many similar needs – irrigation, mechanization, disease and pest control, and nutrient application. However, other engineering needs also present themselves in nursery and greenhouse operations: equipment for transplantation; control systems for temperature, humidity, and ventilation; and plant biology issues, such as hydroponics, tissue culture, and seedling propagation methods. And sometimes the challenges are extraterrestrial: ABEs at NASA are designing greenhouse systems to support a manned expedition to Mars!

10. **Safety and Health:** ABEs analyze health and injury data, the use and possible misuse of machines, and equipment compliance with standards and regulation. They constantly look for ways in which the safety of equipment, materials and agricultural practices can be improved and for ways in which safety and health issues can be communicated to the public.

11. **Structures and Environment:** ABEs with expertise in structures and environment design animal housing, storage structures, and greenhouses, with ventilation systems, temperature and humidity controls, and structural strength appropriate for their climate and purpose. They also devise better practices and systems for storing, recovering, reusing, and transporting waste products.

CAREER IN AGRICULTURAL AND BIOLOGICAL ENGINEERING

One will find that university ABE programs have many names, such as biological systems engineering, bioresource engineering, environmental engineering, forest engineering, or food and process engineering. Whatever the title, the typical curriculum begins with courses in writing, social sciences, and economics, along with mathematics (calculus and statistics), chemistry, physics, and biology. Student gains a fundamental knowledge of the life sciences and how biological systems interact with their environment. One also takes engineering courses, such as thermodynamics, mechanics, instrumentation and controls, electronics and electrical circuits, and engineering design. Then student adds courses related to particular interests, perhaps including mechanization, soil and water resource management, food and process engineering, industrial microbiology, biological engineering or pest management. As seniors, engineering students team up to design, build, and test new processes or products.

For more information on this series, readers may contact:

Ashish Kumar, Publisher and President	Megh R. Goyal, PhD, PE
Sandy Sickles, Vice President	Book Series Senior
Apple Academic Press, Inc.,	Editor-in-Chief
Fax: 866-222-9549; E-mail:	*Innovations in Agricultural and*
ashish@appleacademicpress.com	*Biological Engineering*
http://www.appleacademicpress.	E-mail: goyalmegh@gmail.com
com/publishwithus.php	

PART I

MODELING SOIL WATER ENGINEERING SYSTEMS IN RAINFED AREAS

RUNOFF AND ON-FARM RESERVOIR SUPPLEMENTAL IRRIGATION: EFFECTS ON WATER AND NITROGEN DYNAMICS UNDER RAINFED RICE: MUSTARD CROPPING SYSTEM

L. P. PHOLANE,[1] SUDHINDRA N. PANDA,[2] and L. N. SETHI[3]

[1]Project Officer, Chemical Engineering Department, Indian Institute of Technology, Kharagpur-721 302, West Bengal, India; E-mail: pholane@gmail.com

[2]Professor (Land & Water Resources Engineering), Department of Agricultural & Food Engineering, Indian Institute of Technology, Kharagpur-721302, West Bengal, India. Phone: +91-3222-283140 (Work); E-mail: snp@iitkgp.ac.in, snp@agfe.iitkgp.ernet.in, sudhindra.n.panda@gmail.com

[3]Associate Professor, Department of Agricultural Engineering, Assam University, Silchar, Assam – 788 011, India; Mobile: +91-9401847943; E-mail: laxmi.narayan.sethi@aus.ac.in

In this chapter: One US $ = INR Rs. 48.00

Edited version of "*L. P. Pholane, 2011. Effect of runoff and on-farm reservoir irrigation induced soil water regimes on water and nitrogen dynamics under rainfed rice-mustard cropping System. Unpublished PhD Thesis, Agricultural and Food Engineering Department, Indian Institute of Technology, Kharagpur 721302, India.*"

CONTENTS

1.1 INTRODUCTION

The half of annual rainfall in Eastern India (1200–1700 mm) is through storms and about 66%, falls during June to September. The uneven distribution of spatial temporal variability of rainfall causes: surface flooding, runoff, erosion and nutrient losses and water scarcity at the critical crop growth stages [16, 45, 56]. The water management strategy in paddy requires in depth knowledge of soil water dynamics under such monsoon climate.

Water conservation methods in paddy can be through construction of dykes. The Surface runoff, lateral seepage through the dykes, vertical deep percolation and evapotranspiration reduce water application efficiency [82]. All these processes affect nitrogen availability as nitrate-nitrogen (NO_3^--N) that accumulates in aerobic either lost through leaching beyond root zone and contaminate the groundwater [30, 87] or get diffused into the anaerobic (reduced) soil layer below and denitrified to the N_2 and N_2O gaseous forms, which are lost to the atmosphere contributing to global warming/climate change [9, 10, 20]. *In situ* conservation of rainwater in the cropped field, harvesting of excess rainwater in the on-farm reservoirs (OFRs) and utilized for providing supplemental irrigation to rice followed by non-rice crops in dry winter season such as mustard with diversified cropping system are some of the rainwater management strategies for increasing the overall agricultural productivity of rainfed eco-system in a region [13, 28, 31, 76].

The major challenge of the rainfed agriculture is sustainable management of rainwater and nitrogen so that a favorable environment is created for crop growth. Cultivation of non-rice winter crops following rice depends on the availability of residual soil water and water resources. In the absence of groundwater resources, the monsoon rains available in the crop field can be harvested in the OFR for supplying life-saving irrigation to rice in the rainy season and the balance can be usefully utilized for cultivating non-rice crops in winter season. But no efforts has been done yet to investigate the effect of variable ponding depths in rice fields, which leads to different soil water regimes and also generates variable runoff that can be harvested and stored in the lined and unlined OFRs for future use. Much more studies are required on the effect of lining, as the availability of water in the OFR depends on it, for increasing the water use efficiency. Depending on the storage capacity of the OFRs, the crops will be differentially irrigated leading to seasonal variation in soil water regime, which in turn controls the availability and movement of water and NO_3^--N.

In India nitrogen fertilizer is generally applied in the form of urea. When urea is applied in flooded rice fields, it is converted to ammonium-nitrogen (NH_4^+-N) by the process of hydrolysis and finally to NO_3^- by nitrification. With the onset of monsoon and flooding of soil, NO_3^--N, which is soluble in water, either moves with the percolating water and leaches down to groundwater or is lost through denitrification [9, 11]. This results in a net reduction of the mineral nitrogen pool at the beginning of the rice growing period [66]. The two major factors controlling leaching losses of nitrate are concentration of nitrate in the soil profile at the time of leaching and quantity of water leaving the root zone. Ammonium is generally considered immobile, as it gets adsorbed on the soil particles and absorbed by the plants, and is not transported by the percolating water [84] except when fertilizer is applied in very large quantities or soil is coarse textured and has low cation exchange capacity [9]. Thus, one needs a much better understanding of the underlying processes of water extraction and nutrient uptake under the fluctuating conditions of the rainfed rice fields.

Research on the measurement and modeling of water flow and solute transport and transformation processes under transient soil water regime has recently been recognized as an important area of research for paddy production system [77]. Predicting water flow under different soil water regimes is a challenging task because of high variability of soil hydraulic properties with time and space, and the complex nature of underlying flow field. Computer models are becoming increasingly important tools for analyzing complex problems involving water flow and solute transport in the vadose

zone [12]. However, studies simultaneously estimating water flow and solute transport parameters for transient variably-saturated media are less common [37], especially in a layered soil-profile and for field conditions.

The two-dimensional HYDRUS-2D model is one such model that has
been successfully used for simulating soil water transport by various workers
[1, 67].More studies are to be required to evaluate suitability HYDRUS-2D
model to simulate movement of soil water under variable soil water regimes
in rice and mustard fields. The present investigation has, therefore, been
undertaken to assess the influence of runoff and supplemental irrigation
from the OFR on soil water and NO_3^--N dynamics under rainfed upland rice
(*Oryzasativa* L.) and irrigated mustard (*Brassica campestris* L.).

This chapter discusses research results: to simulate soil water under
rice-mustard cropping system using HYDRUS-2D; to evaluate the effect
of runoff induced soil water regimes on water and nitrate-nitrogen dynamics under rainfed upland rice; to study the effect of soil water regimes, as
induced by frequency of irrigation from the lined and unlined on-farm reservoirs, on water and nitrate-nitrogen dynamics under rainfed mustard; and
to assess the effect of runoff and the OFR irrigation induced soil water
regimes on root growth, nutrient uptake, yields, and economics of rice-
mustard cropping system.

Although the impact of soil water regimes on the productivity of rice based
cropping system is well conceived, its variation under changing conditions
of rainfall distribution, runoff and irrigation from the OFR as well as their
impact on the dynamics of water and NO_3^--N are scarcely understood. Not
enough attempts have been made till-date to know the complex dynamics of
the lateritic (*Oxyaquichaplustalf*) tract of eastern India where rice is traditionally adopted as a rainfed crop during monsoon season and a low duty crop like
mustard during the post-monsoon dry winter season. Moreover the impact of
lined and unlined OFRs on volume of storage, water losses and frequency of
supplemental irrigation on yields of crop has been requiring more attention.

1.2   REVIEW OF LITERATURE

1.2.1   EFFECT OF SUBMERGENCE AND WATER STRESS ON NITROGEN DYNAMICS

Rainfed upland rice encounters an environment more complex and unpredictable than other crops. It is grown in diked fields without water control,

and it, therefore, experiences hydrologic conditions fluctuating from complete submergence of the crop to drought situation, often during the same growing season. Such severe changes have marked effects on soil conditions and the availability of nutrients and water [47, 80]. The alternating periods of soil oxidation and reduction resulting from changing hydrologic conditions lead to gaseous loss of N and immobilization of other nutrients, together with changes in soil acidity and the concentrations of toxic iron and aluminum.

In porous soils under rice, continuous flooding cannot be maintained due to high water percolation rates. The high urea application rates that exceed crop demands to the dry season crops [51], resulting in large N losses that range from 34 to 549 kg per cropping sequence [78]. Excessive applied N, not used by the crop, accumulates as NO_3^--N in the soil profile during the dry season. Soil and crop management during both the dry season and dry to wet transition substantially influences soil NO_3^--N [78]. With the onset of monsoon and soil flooding, NO_3^--N either leaches down to groundwater or is lost through denitrification.

Ponding of water in the field was not essential for NO_3^--N losses to occur [69]. When water-filled porosity exceeded 70%, NO_3^--N was lost from all soil layers. Since N did not accumulate in deeper soil layers, losses of NO_3^--N were attributed to denitrification. In most of these studies, the distribution of NO_3^--N in the soil profile has been determined at the end of different field experiments. With the application of nitrogen fertilizer, no studies are available that report changes in NO_3^--N and NH_4^+-N, during and after the harvest of each crop grown in a rotation, occurring under dry and flooded cycles.

Rice is particularly susceptible to soil water deficit [38], and drought affects its growth in about 50% of the world production area [32]. A stress of 12 days during anthesis, adversely affects spikelet fertility with severe reduction in grain yield [24]. Numerous studies conducted on the manipulation of depth and interval of irrigation to save water use without any yield loss have demonstrated that continuous submergence is not essential for obtaining high rice yields [89]. Tabbal [74] reported that maintaining a very thin water layer, saturated soil condition, or alternate wetting and drying could reduce water applied to the field by about 40–70% compared with the traditional practice of continuous shallow submergence, without a significant yield loss. Soil nitrate and ammonium concentrations were similar in continuously shallow-flooded and saturated soil water regimes, implying that plant N availability were not adversely affected when a saturated soil regime was maintained.

In order to predict the concentration of nitrate in the water percolating out of the root zone, it is necessary to determine a nitrate and water budget in the root zone. Since nitrate moves with water, nitrate movement is linked to the output of the soil water balance model developed for rice fields. Quantification of N losses from a rice field requires information of daily percolation rates and daily standing water depths in the fields [89].

1.2.2   EFFECT OF SOIL WATER ON CROP ROOT GROWTH

Soil water availability estimation is critical for assessing crop development and performance. During periods of soil water deficits, the capability of crop roots to extract soil water depends on the distribution and depth of its root system [17]. Most water uptake models assume a relationship between root water extraction and root length density [19, 88].

During early growth stages, root length growth was very sensitive to soil water deficit. Rice is commonly regarded as having a poor root system, often failing to extract water from deeper layers of the soil profile, relative to other crops [38]. Yoshida and Hasegawa [85] reported significant variation in root length density in rice below 30 cm. Consequently, several reviews have concluded that a drought avoidance strategy (maximum rooting depth, a greater root length density at depth, and a greater capacity to conduct water from depth) would be helpful for upland conditions [27].

In puddle rice fields, the bulk density increased with time, due to the settling of particles. Bulk density further increases on drying because of shrinkage. Root growth and grain yield in transplanted rice is negatively correlated with bulk density and soil strength [46]. In contrast, use of deep cultivation or a tap-rooted legume to perforate the hardpan resulted in higher yields of rainfed rice on compaction-prone soils [7].

1.2.3   NUTRIENT UPTAKE AND RICE-BASED CROPPING SYSTEM

For rainfed agriculture, compared to other agricultural systems, it is not clear to what extent rice yields are limited by nutrients, water, and the interactions between them, over the diverse soil types, cultural practices and seasonal conditions. Few of the many experiments have obtained the data essential for thorough interpretation, and none has attempted to develop an understanding of prospects for manipulating nutrient and water interactions across the varied combinations of flood, drought, soil type and cultural practices. More understanding is required of the processes of nutrient release and capture under the fluctuating soil water regimes of the rainfed areas. Furthermore, adoption of proper cropping system that aims to combine maximum capture of water and nutrient resources with minimum use of external inputs is only sustainable if the inputs and outputs of water and nutrients are balanced at the level of a particular target yield [86].

The average recovery efficiency of fertilizer nitrogen in irrigated rice can be as low as 30% [21], but an adequate supply of nitrogen is essential for high yields. By generating wide ranges of dry-season soil NO_3^--N, George [29] demonstrated the significance of N uptake by plants in reducing NO_3^--N loss. If rainfed soils were properly managed, they could conserve up to 130 kg N ha^{-1} for plant uptake, supporting a grain yield of more than 4 tonn ha^{-1} without additional fertilizer-N inputs. Rainfed upland rice is also grown on coarse-textured Alfisols and Ultisols. These soils lose a considerable amount of water and nutrients by deep percolation. The productivity of these soils could be increased if water percolation and leaching of nitrogen were minimized. Under such condition direct seeding may permit a second crop to be grown before or after the main crop of rice, where otherwise it may not be possible. If the additional crop is a legume, this may result in a net input of biologically fixed N into the system [29].

The adoption of legumes into rice-based cropping systems, offers opportunities to increase and sustain productivity and income of small rice farmers in South East Asia. In the irrigated areas, legumes (soybean (*Glycine max* L. Merr.), peanut (*Aracishypogaea* L.) and mungbean (*Vignaradiata* (L.)Wilzek)) are generally grown in rotation of rice-rice-legume or rice-legume-legume with two or more irrigations during the season [46]. Yield increase anged from 42 to 140% as reported by Adisarwanto and Suhartina [4] in experiments conducted in East Java, Indonesia.

1.2.4 MODELING OF WATER AND SOLUTE TRANSPORT

The water status in the rice field, whether rainfed or irrigated, greatly influences nutrient-use efficiency as well as nutrient balances in crop-soil systems. However, owing to the complexity of the soil media, such as the complex fracture system and large heterogeneity of the various hydraulic properties, and limited available field data, flow and solute transport in the complex subsurface environments have not been fully understood. The Richards' equation for water flow and the convection-dispersion equation for solute transport are frequently used to describe water and solute transport experiments in undisturbed soil columns and field soils [26, 61]. Sometimes, however, significant differences are observed between observations and the results of model calculations [40], which is especially the case in structured or macro-porous soils. Jacques [41] evaluated the applicability of Richards' equation for water flow and the convection-dispersion equation for solute transport to model field-scale flow and transport under natural boundary conditions by using detailed experimental data and inverse optimization.

The research on water and solute transport in complex subsurface environments is growing rapidly at present. A number of one-dimensional soil water models have been published over the last decade for predicting the movement of solutes through the soil profile [3, 35, 42, 43]. These models each describe solute transport differently: SLIM is a capacity layer model using mobile/immobile water segmentation, LEACHM is a numerical solution of the Richards' and the Convection-Dispersion equations, CRACK is a model based on the entry of water into soil peds, and MACRO which divides water flow into macropore and micropore domains and applies gravity flow and Richards' equation to micropore water [33].

Chen[18] used a three-dimensional FEM-WATER finite element model developed by Lin [48] to differentiate the lateral seepage and vertical percolation from surface infiltration under varying wet and dry conditions in rice fields, since the one-dimensional SAWAH model cannot simulate both horizontal and vertical water movements. The two-dimensional HYDRUS-2D model is one such model that has been successfully used for simulating soil water transport by various workers [67]. Abbasi [1] calibrated and experimentally validated a two-dimensional numerical flow/transport model (HYDRUS-2D) using data from long furrow irrigation experiments.

1.3 THEORETICAL CONSIDERATIONS

The present study is undertaken with the main objective to evaluate the effect of different soil water regimes due to variable runoff and irrigation from the OFRs on water and nitrogen dynamics. It needs to quantify different inflow and outflow components of the field under different weir heights

(0, 5 and 10 cm) and the OFRs (lined/unlined) to study soil hydraulic parameters and also economic analysis for feasibility study of the OFR system. The present chapter includes the following theoretical procedures.

1.3.1 SIMULATION OF SOIL WATER USING HYDRUS-2D

Model description
HYDRUS-2D [68] model has been used for simulating soil water in the rice and mustard fields under variable runoff and supplemental irrigations from the OFR. Two main inputs that were required for conducting simulations in HYDRUS-2D environment are long-term weather data and soil hydraulic properties.

Governing water flow equations
Variably-saturated water flow in porous media is usually described by the Richards' equation[68]:

$$\frac{\partial \theta}{\partial t} = \frac{\partial}{\partial x_i}[K(K_{ij}^A \frac{\partial h}{\partial x_j} + K_{iz}^A)] - S \tag{1}$$

where, θ = volumetric water content [L^3L^{-3}]; t = time [T]; h = pressure head [L]; $x_i(i =1,2)$ = spatial coordinates [L]; S = sink term [T^{-1}]; K_{ij}^A = components of a dimensionless anisotropy tensor K^A; and K = unsaturated hydraulic conductivity function [LT^{-1}] given by

$$K(h,x,z) = K_s(x,z)K_r(h,x,z) \tag{2}$$

where, K_r= relative hydraulic conductivity; and K_s = saturated hydraulic conductivity [LT^{-1}].

The sink term, S, in eq. 1, represents the volume of water removed per unit time from a unit volume of soil due to plant water uptake. Feddes [25] defined S as:

$$S(h) = a(h)S_p \tag{3}$$

where, $a(h)$ = water tress response function, which is a prescribed dimensionless function of the soil water pressure head ($0 \le a \le 1$); and S_p = potential water uptake rate [T^{-1}].

The Richards equation has been solved numerically using the initial and boundary conditions and two constitutive relations: the soil water retention $\theta(h)$; and hydraulic conductivity, $K(h)$, functions. The soil-hydraulic function of van Genuchten [79], who used the statistical pore size distribution model of Mualem [53] to obtain a predictive equation for the unsaturated hydraulic conductivity function in terms of soil water retention parameters, is given by:

$$\theta(h) = \theta_r + \frac{\theta_s - \theta_r}{[1 + |\alpha h|^n]^m} \qquad h < 0 \tag{4}$$

$$K(h) = K_s S_e^l [1 - (1 - S_e^{1/m})^m]^2 \tag{5}$$

where, S_e = effective water content.

$$S_e = \frac{\theta - \theta_r}{\theta_s - \theta_r} \quad \text{and} \tag{6}$$

$$m = 1 - 1/n, \quad n > 1 \tag{7}$$

where, θ_r = residual water content of soil [L^3 L^{-3}]; θ_s = saturated water content of soil [L^3 L^{-3}]; a = shape factor, approximately equal to the inverse of the air-entry value (or bubbling pressure) [L^{-1}]; n = pore-size distribution index [–]; and l = pore-connectivity parameter [–].

In this chapter, the simulation period was spanned from the date of sowing to date of harvest of the crops and also included the turn-in-period. The data were analyzed using the Richards' equation for water flow under different soil water regimes. Initially, the HYDRUS-2D model was calibrated using 2002 data and validated using 2003 and 2004 data from the rice and mustard fields with three different weir height experiments. Optimization was accomplished by means of the Levenberg-Marquardt optimization method. HYDRUS-2D was also adopted to estimate percolation of water below the root-zone.

1.3.1.1 Initial and Boundary Conditions

Measured soil water contents before the experiments were used as initial conditions within the field. Time-space dependent flow depths (surface ponding) under natural rainfall was specified as the upper boundary condition in the field, while measured pan evaporation rates and estimated reference evapotranspiration rates with the Penman-Monteith method were used as atmospheric boundary conditions during redistribution phase. Free-drainage condition was applied to the lower boundary of the field.

1.3.2 FIELD WATER BALANCE

The field water balance parameters were rainfall (P), runoff (RO) generated at different weir heights, actual evapotranspiration (AET), supplemental irrigation (SI) at critical growth stage, vertical percolation (VP), and lateral seepage (LS) through the dyke (Figures 1.1 and 1.2).

The generalized daily soil water balance model in the effective root-zone of the crops (rice and mustard) ignoring upward flux because of capillary rise from the groundwater is given as:

$$\text{SWC}_{ijt} = \text{SWC}_{ijt\text{-}1} + P_{ijt} + \text{SI}_{ijt} - \text{AET}_{ijt} - \text{RO}_{ijt} - \text{VP}_{ijt} - \text{LS}_{ijt} \qquad (8)$$

where, SWC = soil water content in the effective root-zone of the crop (rice and mustard) (mm); P = rainfall (mm); AET = actual evapotranspiration (mm); VP = vertical deep percolation from effective root-zone (mm); LS = lateral seepage across the boundaries (mm); SI = supplemental irrigation (mm); RO = surface runoff from the cropped field to the OFR (mm); i = index for the field with OFR (i= 1 for polyethylene lined and i= 2 for unlined); j = index for weir heights (j =1 for 0 cm, j = 2 for 5 cm, and j = 3 for 10 cm); and t = index for time in day.

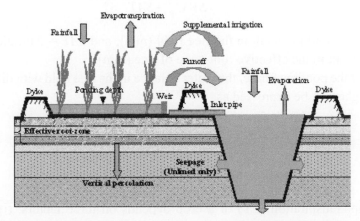

FIGURE 1.1   Schematic presentation of water balance parameters in rice field with the OFR.

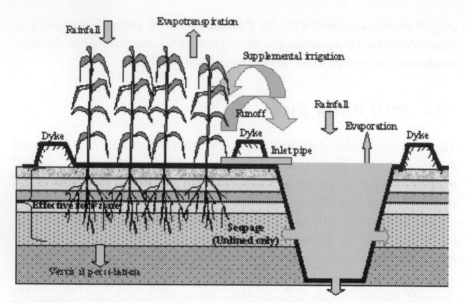

FIGURE 1.2 Schematic presentation of water balance parameters in mustard field with the OFR.

1.3.2.1 Ponding Depth

If soil water content is more than the saturation level, then ponding will occur in the rice field. However, for the mustard field, standing water depth was not considered, i.e., (ponding depth = 0). The ponding depth in the rice field on any day is given as:

$$H_{ijt} = SWC_{ijt} - SAT \tag{9}$$

where, H = ponding depth in the rice field (mm); and SAT = saturation soil water content in the effective root-zone of rice (mm).

Under the ponding phase, the water balance in the rice field with different weir heights can be expressed as

$$H_{ijt} = H_{ijt-1} + P_{ijt} + SI_{ijt} - AET_{ijt} - VP_{ijt} - LS_{ijt} - RO_{ijt} \tag{10}$$

1.3.2.2 Actual Evapotranspiration (AET)

The evapotranspiration (ET) of a crop depends on three factors: (i) Plant characteristics, extent of ground cover, and stage of growth; (ii) Availability of soil water; and (iii) Meteorological parameters or evaporative demand. Under adequate soil water conditions, evapotranspiration of plants occur at potential rate. However, as the ponding water depth from the rice field decreases so that soil in the effective root-zone of rice remains at water stress condition, ET of plants decreases from the potential rate. The AET of rice under soil water stress condition on any day is given [36] as:

$$AET_{ijt} = Kc\,(Ksf)_{ijt}\,(ET_0)_{ijt} \tag{11}$$

where, Kc = crop coefficient; Ksf = soil water stress factor; and ET_0 = reference crop evapotranspiration (mm).

Daily reference crop evapotranspiration (ET_0) was estimated by FAO Penman-Monteith method for the simulation period [8]. The Kc Values of rice for the prevailing climatic conditions of the study area were assumed as 1.05 during the crop establishment, 1.10 during both crop development and mid-season, and 0.95 for late season stage [23]. The value of Ksf in eq. 4 was assumed as 1.0 under no water stress conditions. But as the ponding water disappears from the rice fields, soil water stress occurs that consequently decreases AET, which is governed by Ksf. In this chapter, Ksf was assumed to vary linearly with the relative available SWC (ratio of SWC to saturation soil water content in the crop effective root-zone) in the field under unsaturated condition [5, 44, 59] as:

$$(Ksf)_{ijt} = \frac{SWC_{ijt}}{SAT} \tag{12}$$

Under adequate soil water conditions, potential evapotranspiration on t[th] day (PET_{ijt}) is given by:

$$PET_{ijt} = Kc\,(ET_0)_{ijt} \tag{13}$$

Values of crop coefficient (Kc) of mustard for crop establishment, crop development, mid-season, and late season stages of crop growth were 0.34, 0.61, 0.88, and 0.82, respectively [60]. Under water stress condition, actual evapotranspiration (AET) of mustard is given as:

$$AET_{ijt} = \begin{cases} \dfrac{(SWC_{ijt} - WP)}{(1-p)\,(FC - WP)}\,PET_{ijt} & WP < SWC_{ijt} < (1-p)(FC - WP) \\[2mm] PET_{ijt} & (SWC_{ijt} - WP) \geq (1-p)(FC - WP) \end{cases} \tag{14}$$

where, FC = field capacity (mm); WP = wilting point (mm); and p = soil water depletion factor that was computed with a method by Doorenbos and Kassam [22].

1.3.2.3   Surface Runoff

In this chapter, for 15 days after sowing (DAS) of rice and last 10 days to harvest, no standing water is allowed in the field. During these periods surface runoff (RO) is given as:

$$RO_i = SWC_{ijt-1} + P_{ijt} - SAT \qquad (15)$$

During rest of the periods, except the fields with 0 mm weir height, i.e., $H_{jmax} = 0$, ponding depth of 5 and 10 cm in the field were taken as the maximum limit (H_{jmax}) corresponding to 5 and 10 cm weir heights, respectively; and any excess ponding above H_{jmax} was considered as the surface runoff contribution to the OFR and is given as:

$$RO_{ijt} = SWC_{ijt-1} + P_{ijt} + SI_{ijt} - SAT - H_{jmax} \qquad (16)$$

1.3.2.4   Vertical Percolation

Vertical percolation under variably saturated condition of the cropped field was estimated using HYDRUS-2D model, whereas under ponding condition the vertical percolation was computed using Darcy approach [6, 34, 60, 82], respectively. The vertical percolation (*VP*) was calculated using Darcy's functional relationship

$$VP_{ijt} = \frac{K_s(h1_{ijt} - h2_{ijt})}{T} \qquad (17)$$

where, K_s = saturated hydraulic conductivity (mm day^{-1}); $h1$ and $h2$ = hydraulic head at the upper and bottom layer of the root-zone (mm), respectively; and T = thickness of the effective root-zone (mm).

1.3.2.5   Bare Soil Evaporation During Germination Period of Rice

For computation of different parameters in water balance model during germination period (5 days) of rice, *AET* was replaced by bare soil evaporation (E) in Eq. (8). Bare soil evaporation is estimated from ET_0 subjected to rainfall condition of the day [44, 71] as below:

$$E_{ijt} = \begin{cases} 0.1(ET_0)_{ijt} & P_{ijt} = 0 \\ (ET_0)_{ijt} & P_{ijt} > (ET_0)_{ijt} \\ P_{ijt} & 0 \langle P_{ijt} \langle (ET_0)_{ijt} \end{cases} \tag{18}$$

1.3.2.6 Bare Soil Evaporation During Turn-In and Germination Period of Mustard

Soil water balance of rice field is extended till the end of turn-in period (sowing of mustard). During the said period (15 days), the soil is under bare conditions. During this period soil water is at greater than wilting point. As proposed by Sanchez-Cohen [63] and Abraham and Tiwari [2], when the soil water in the top 150 mm layer is greater than field capacity (*FC*), the bare soil evaporation during the turn-in period (Et_{ijt}) occurs at its potential rate of ET_{0ijt} as:

$$Et_{ijt} = \begin{cases} (ET_0)_{ijt} & SWC_{ijt} \geq FC \\ (ET_0)_{ijt} \dfrac{SWC_{ijt}}{FC} & SWC_{ijt} < FC \end{cases} \tag{19}$$

For computation of soil water balance, AET_{ijt} is replaced by Et_{ijt} using eq. 12 during turn-in period and also in germination period of mustard.

1.3.3 OFR WATER BALANCE

The proper size of the lined and unlined OFR must be designed by considering all the inflows and outflows components of the OFR. The inflow components are the direct rainfall received to the OFR and surface runoff contribution to the OFR from the crop field through different weir heights. The outflow components are SI (supplement irrigation) supplied to crop, evaporation, and seepage losses. The water balance model for the lined and unlined OFR at different weir heights for rice-mustard cropping system with the provision of supplemental and pre-sowing irrigation can be expressed as:

$$W_{ijt} - W_{ijt-1} = \Delta W_{ijt} = Q_{RO_{ijt}} + Q_{P_{ijt}} - Q_{E_{ijt}} - Q_{SI_{ijt}} - Q_{SP_{ijt}} \tag{20}$$

where, W = OFR water storage (m³); Q_{RO} = runoff water coming from the field to the OFR (m³); Q_P = direct rainfall contribution to the OFR (m³); Q_E = water lost as evaporation from the OFR (m³); Q_{SI} = water used as supplemental irrigation (m³); and Q_{SP} = water lost as seepage and percolation from the unlined OFR (m³).

Best quality low density polythene (SILPAULIN Company) of 90 gram per square meter (GSM)was used for the lining of OFR. Therefore, volume of water lost from the lined OFR as seepage was assumed negligible. Total storage of water in the OFR (TW_{ijt}) is:

$$TW_{ijt} = \sum_{t=1}^{ND} \Delta W_{ijt} \qquad (21)$$

where, ND = number of days since simulation started; and TW = total storage of water in the OFR (m³).

1.3.4   ECONOMIC ANALYSIS OF OFR SYSTEM

The total costs of OFR irrigation and the returns from the increased yield of the crops from the OFR irrigation should be evaluated for the lined and unlined OFR system. Net present value (NP), benefit-cost ratio (BCR) and payback periods are the factors considered for economic analysis of the system. Present worth analysis, proposed by Samra [62], is used to evaluate the economics of the OFR system for which following items were considered:

a.   Initial investment for the construction of the OFR.
b.   The OFR maintenance cost.
c.   Irrigation cost.
d.   Land lease cost for the construction of the OFR.
e.   Returns from the OFR system over without OFR system.
f.   Interest rate of 12% and 25 years life span of the lined and unlined OFR are assumed in the present study.

1.3.4.1   Present Worth of Cost (PW_c)

$$PW_c = I_{inv} + PW_{ac} \qquad (22)$$

where, I_{inv} = initial investment; and PW_{ac} = present worth of annual cost.

$$PW_{ac} = \sum_{t=1}^{n} \frac{A_t}{(1+r)^t} \tag{23}$$

where, A = annual cost; r = interest rate; and t = life span of the OFR.

1.3.4.2   Present Worth of Benefit (PW$_b$)

$$PW_b = \sum_{t=1}^{n} \frac{B_t}{(1+r)^t} \tag{24}$$

where, B = benefit.

1.3.4.3   Net Present Value of the OFR System

The net present value (NP) of the OFR is calculated as:

$$NP = PW_b - PW_c \tag{25}$$

1.3.4.4   Benefit–Cost Ratio of the OFR System

Benefit–cost ratio is the present worth of benefits divided by present worth of costs. This is calculated as:

$$BCR = \frac{PW_b}{PW_c} \tag{26}$$

The initial investment of the OFR irrigation system consisted of the construction cost of the OFR and material cost for lining of the OFR. The annual cost comprised of repair and maintenance cost, land lease cost for the construction of the OFR, and irrigation costs. Repair and maintenance cost was assumed to be 2% of I_{inv} [52, 55]. Land lease cost for the construction of the OFR was taken Rs. 3000 ha^{-1} year^{-1}. Irrigation cost depends on the amount of SI applied to the crop, area served by SI as well as the hire charge of pump set. The hiring charge of 5 HP diesel pump set in the region was Rs. 300 for providing 5 cm SI to one hectare area. The minimum government support price per 100 kg of rice grains and mustard seeds is taken as Rs. 450 and 1500, respectively. The price of rice straw and mustard stover in the study area was taken as Rs. 30 and Rs 15 per 100 kg, respectively. The average returns obtained from the increased yields of rice, mustard including its byproduct are used for estimating the net present value.

1.3.4.5   Payback Period

The payback period (PBP) is the period of time that takes to recover its initial investment of the OFR system. The PBP is computed as the reciprocal of the investment and annual cash inflows [52, 58]. The better investment is the one with the shorter payback period. It will evaluate the time period for recovering of the investment in the OFR system.

1.3.5   MODEL EVALUATION

The performance was evaluated by graphical presentation (scattered diagram) between the observed and simulated results, coefficient of determination (R^2) and statistical analysis. The statistical characteristics such as the root mean square error (RMSE) and prediction efficiency (PE) were used for model evaluation [6, 57]. These statistical terms can be expressed as:

$$\text{RMSE} = \left[\frac{1}{N} \sum_{ds=1}^{N} (C_{sds} - C_{ods})^2 \right] \tag{27}$$

$$\text{PE} = 1 - \frac{\sum_{ds=1}^{N} (C_{sds} - C_{ods})^2}{\sum_{ds=1}^{N} (C_{ods} - \overline{C_O})^2} \tag{28}$$

where, N = total number of data samples (ds); C_{sds} = simulated data sample; C_{ods} = observed data sample; and $\overline{C_O}$ = mean of data sample.

1.4   EXPERIMENTAL STUDY

1.4.1   STUDY AREA

The site selected for the present investigation is an agricultural farm of the Department of Agricultural and Food Engineering, Indian Institute of Technology (IIT), Kharagpur, West Bengal State in eastern India located at a latitude of 22° 19' N, longitude of 87° 19' E with an altitude of 48 m above the mean sea level. The site lies in sub-humid, sub-tropical climate zone and receives 1500 mm as mean annual rainfall of which about 80% are received during the rainy season from June to September.

1.4.2 FIELD EXPERIMENTS

Rainfall and hydrograph of ponding in the rainfed upland rice fields are random variables, therefore, substantial damage to crop and yield due to submergence or long dry spell at critical crop growth stages takes place. Keeping these factors and water requirement of rainfed upland rice varieties in view, different weir heights with low-density polyethylene lined and unlined OFRs were considered for the present experiment. Three years (2002, 2003 and 2004) of field experiments were undertaken in the experimental farms of Agricultural and Food Engineering Department, IIT, Kharagpur, West Bengal, India.

The field experiment consists of lined (L1) and unlined (L2) OFRs with three weir heights (H1 = 0 cm, H2 = 5 cm, H3 = 10 cm) replicated thrice (Figure 1.3). In addition, three plots without the OFR were cultivated for economic evaluation of the OFR system over the traditional method. Each plot is 40 × 20 m² size with 30 cm dyke height around. Initially square shaped pyramidal 9 lined and 9 unlined OFRs were constructed at one corner of each plot in an area of 10% of the plot size with 1:1 side slope and 2.4 m depth. The berm width of the OFR at the ground level is 30 cm. Height, top width, bottom width, and side slope of the embankment of the OFR are 30, 30, 90 cm, and 1:1, respectively (Figure 1.4). One inlet pipe fitted with mechanical water meter was placed in each OFR to quantity the runoff water to the OFR. A staff gauge of 3 m long was installed in each OFR to compute the volume of storage. Figures 1.5 and 1.6 present the overview of field experimental setup and data monitoring in the study site.

Different weirs with reference to field level were placed 10 days after germination of rice and removed 10 days before its harvest so as to facilitate quick drain out of any ponding water from the rice field to the OFR.

1.4.3 SOIL

The soil at the site is sandy loam, lateritic (*Oxyaquichaplustalf*) with pH ranging from 4.8 to 5.6. The layer-wise physical properties of soil collected

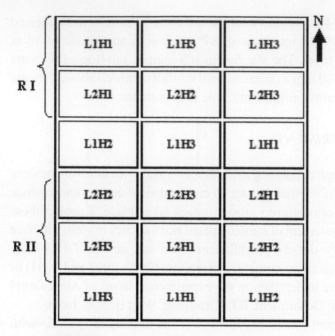

LEGENDS:
L1H1 = Lined OFR with 0 cm weir height L1H2 = Lined OFR with 5 cm weir height
L1H3 = Lined OFR with 10 cm weir height
L2H1 = Unlined OFR with 0 cm weir height
L1H2 = Unlined OFR with 5 cm weir height
L2H3= Unlined OFR with 10 cm weir height
R = Replication

FIGURE 1.3 Layout of experimental setup.

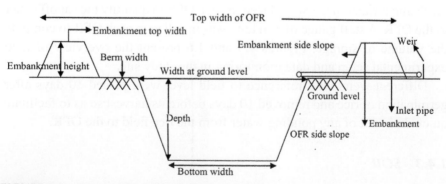

FIGURE 1.4 Cross-section of the on-farm reservoir.

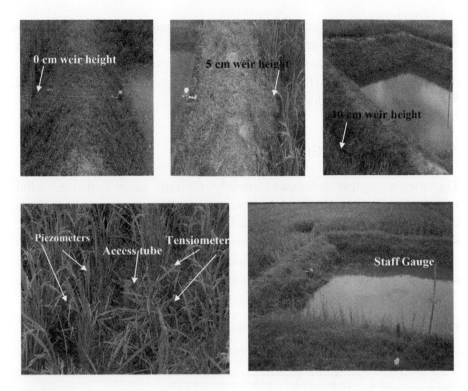

FIGURE 1.5   View of experimental setup.

FIGURE 1.6   View of rice and mustard field with the lined OFR.

from the experimental plot are shown in Table 1.1 and the chemical properties of top 15 cm soil layer are presented in Table 1.2. The soil water retention characteristics of the soil were measured using pressure plate apparatus and depicted in Figure 1.7. The water retention characteristics of soil were used for estimation of its hydraulic parameters.

1.4.4　METEOROLOGICAL PARAMETERS

The meteorological parameters namely rainfall, solar radiation, wind velocity, air temperature, and relative humidity were collected from the meteorological center as well as Automatic Weather Station of IIT, Kharagpur, which is located in the close vicinity of the experimental site.

The site is coming under sub-humid and sub-tropical climate. Total seasonal rainfall during monsoon varies from 787 to 1600 mm. However, the

TABLE 1.1　Soil Physical and Hydraulic Properties of the Experimental Site

Soil layers (cm)	Sand (%)	Silt (%)	Clay (%)	Bulk density (g cm^{-3})	(cm day^{-1})	(cm^3 cm^{-3})	(cm^3 cm^{-3})
0–15	66.4	18.6	15.0	1.65	12.24	0.37	0.0306
15–30	62.5	21.5	16.0	1.60	7.01	0.39	0.0364
30–45	63.0	20.6	16.4	1.58	5.94	0.38	0.0386
45–60	64.2	20.0	15.8	1.60	4.01	0.40	0.0405
60–75	62.8	20.5	16.7	1.62	3.19	0.42	0.0407
75–90	62.7	20.8	16.5	1.61	2.14	0.42	0.0428
90–105	62.5	19.5	18.0	1.68	1.01	0.43	0.0470

TABLE 1.2　Chemical Properties of the Experimental Soil (0–15 cm Profile)

Property	Value	Method, [Ref.]
pH (1 : 2.5 soil : water)	5.20	Glass electrode pH meter
EC (1 : 1 soil : water) (dS m^{-1})	0.60	EC meter
Cation exchange capacity [cmol (p$^+$) kg^{-1}]	7.57	NH$_4$Ac-leaching [39]
Organic carbon (g kg^{-1})	4.2	[83]
Available N (kg ha^{-1})	178	[72]
Available K (kg ha^{-1})	165	[14]
Available P (kg ha^{-1})	12	[15]

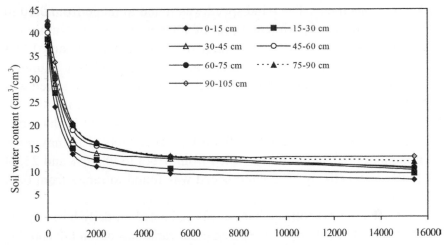

FIGURE 1.7   Soil water retention curves for different soil profile.

annual rainfall varies from 1034 to 2100 mm. The annual and seasonal rainy days varies from 46 to 124 and 34 to 83 days, respectively. Total rainfall during the rice growing season in 2002, 2003, and 2004 were 1275, 972 and 1037 mm, respectively. Total rainfall during the mustard growing season in 2002, 2003, and 2004 were 57, 106, and 4 mm, respectively.

During rice-growing season in the experimental year of 2002, the maximum and minimum temperatures range from 26.5 to 36.3°C and 22.8 to 29.6°C, respectively. The maximum and minimum relative humidity ranged from 85.6 to 99.2% and 33.6 to 93.9%, respectively. Wind speed ranged from 0.58 to 23.62 km hr^{-1}.

In the experimental year of 2003, the maximum and minimum temperatures range from 27.50 to 41.98°C and 22.96 to 27.96°C, respectively. The maximum and minimum relative humidity ranged from 85.57 to 98.50% and 33.63 to 93.98%, respectively. Wind speed ranged from 0.90 to 47.70 km hr^{-1}. During the experimental year of 2004, the maximum and minimum temperatures range from 23.19 to 37.64°C and 21.78 to 31.76°C, respectively. The maximum and minimum relative humidity ranged from 87 to 98% and 40 to 87%, respectively. Wind speed ranged from 0.86 to 39.06 km hr^{-1}.

During mustard growing season in the experimental year of 2002, the maximum and minimum temperatures were recorded in the range of 25.2 to 33.2°C and 10.2 to 25.8°C, respectively. The maximum and minimum relative humidity varied from 79.8 to 99.0% and 25.7

to 84.5%, respectively. The wind speed was found to range from 0.10 to 31.2 km hr^{-1}. For the experimental year of 2003, maximum and minimum temperatures were recorded in the range of 20.55 to 33.87°C and 12.03 to 26.17°C, respectively. The maximum and minimum relative humidity varied from 80 to 98% and 26 to 86%, respectively. The wind speed was found to range from 0.82 to 27.72 km hr^{-1}. For the experimental year of 2004, maximum and minimum temperatures were recorded in the range of 20.20 to 31.36°C and 12.03 to 29.80°C, respectively. The maximum and minimum relative humidity varied from 81.44 to 98.50% and 25.94 to 82.80%, respectively. The wind speed was found to range from 0.90 to 22.86 km hr^{-1}.

Reference crop evapotranspiration was estimated on daily basis by the FAO Penman-Monteith method [8]. Temporal variations of rainfall and ET$_0$ during rice and mustard-growing seasons for three years (2002, 2003, and 2004) are given in Figures 1.8 and 1.9, respectively.

1.4.5 CROP CULTIVATION

During the monsoon season, rice (*Oryzasativa* L., cv. MW 10) of duration 101 days and in winter season, mustard (*Brassica campestris* L. cv. B 54) of 70 days duration were grown in the experimental field. Dry seeding of rice @ 100 kg ha^{-1} with 20 cm row spacing was done on the onset day of monsoon. Farmyard manure was applied @ 5 ton ha^{-1} 15 days before dry seeding of rice. A basal fertilizer dose, 30:45:45 kg ha^{-1} of NPK was applied through urea, single super phosphate and muriate of potash. Remaining 30 kg N was applied 30 days after sowing. Mustard seed @ 5 kg ha^{-1} was sown with 20 cm row spacing 15 days after harvest of rice with basal fertilizer dose of 40:20:20 kg ha^{-1} of NPK. All cultural operations were followed as recommended for both the crops (Table 1.3).

1.4.5.1 Irrigation Strategies

In the present study, a water-saving irrigation technique was considered in which, supplemental irrigation (SI) from the OFR to the rice and mustard crop was applied during the critical growth stages (CGS), if required. The CGS of rice started at the 45th day after germination of seed (booting stage)

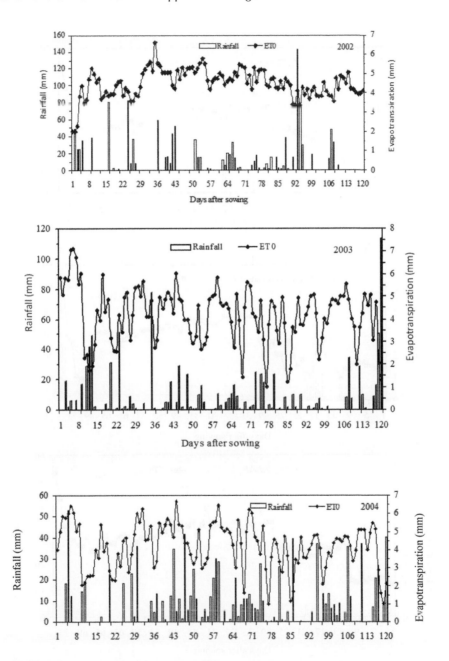

FIGURE 1.8 Temporal variation of rainfall and reference evapotranspiration during rice growing including turn-in period.

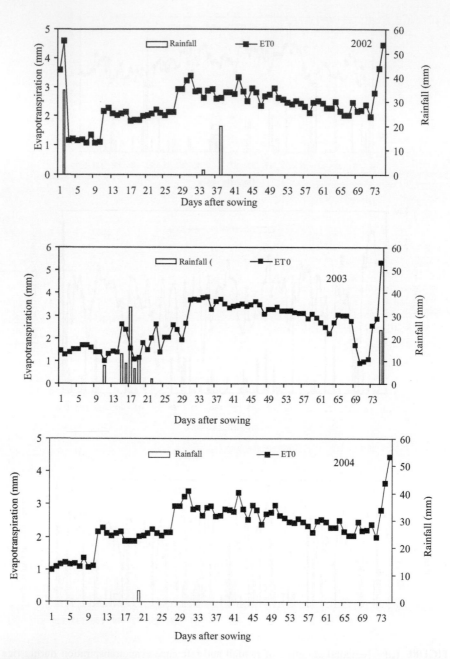

FIGURE 1.9 Temporal variation of rainfall and reference evapotranspiration during mustard growing period.

TABLE 1.3 Schedules of Different Cultural Operations Followed During Field Experiments

Operation	Rice			Mustard		
	2002	2003	2004	2002	2003	2004
Sowing of seeds	June, 9–15	June, 6–12	June, 9–15	October, 8–12	October, 10–15	October, 6–10
Fertilizer application						
Basal dose (N, P, K)	June, 9–15	June, 6–12	June, 9–15	October, 8–14	October, 10–15	October, 6–12
Top dressing (N only)	July, 10–13	July, 7–10	July, 10–13	—	—	—
First SI	—	—	—	November, 15	November, 27	November, 12
Second SI	—	—	—	November, 30	December, 8	November, 27
Harvesting	September 22–26	September 20–24	September 22–26	December 20–23	December 22–26	December 18–22

and continued up to the end of the milking stage. Supplemental irrigation to rice was provided when the soil water content in the 45 cm effective root-zone depth depleted 40% (deficit irrigation) from the saturation soil water content [73].

Supplemental irrigation was applied to rice for raising the soil water content up to field capacity or the effective depth of available water in the OFR (whichever was minimum). Supplemental irrigation (SI) was provided to mustard when the soil water content in the effective root-zone depleted 25% below the available soil water content during the CGS [81]. Depth of irrigation of 5 cm or the actual amount of water available in the OFR (whichever was minimum) will be supplied from the OFRs to the mustard field during CGS (31st to 60th DAS) when the soil water content in the effective root-zone depths depleted to 25% of available soil water content. In this chapter during all the experimental years, SI was not given to rice because the soil water content in the effective root-zone was not depleted 40% of saturation soil water content during the CGS. However, 5 cm depth of SI was applied to the mustard field from the OFRs first at 37, 47 and 37 days after sowing (DAS) from the lined and unlined OFR and second at 52, 58 and 52 DAS from the lined OFR only during the year 2002, 2003 and 2004, respectively. The residual soil water after the harvest of rice was sufficient for the germination of mustard seeds and so no pre-sowing irrigation (PSI) was applied to mustard in all three experimental years.

1.4.6 EVALUATION OF SOIL AND CROP PARAMETERS

1.4.6.1 Soil Properties

Core samples were collected with auger at seven soil depths: 0–15, 15–30, 30–45, 45–60, 60–75, 75–90 and 90–105 cm of the experimental plots before sowing of crops. Soil characteristics like bulk density, particle size distribution, water retention, saturated hydraulic conductivity, etc. were determined. The initial nutrient status in 0–15 cm layer was also determined.

1.4.6.2 Soil Hydraulic Parameters

In rice field, piezometers and tensiometers at 15, 30 and 45 cm depths were installed to monitor daily pressure heads under saturated and unsaturated conditions, respectively; while only tensiometers were installed in mustard field. Access tubes were installed in each experimental plot and daily soil water content was measured using Aqua-pro sensor. Variation in pressure heads and soil water content in space and time were used in monitoring water fluxes and to estimate the soil hydraulic parameters of different soil layers using HYDRUS-2D model.

1.4.6.3 Distribution of Nitrate-N in Soil

Soil water samplers were installed in each plot at 15, 30 and 45 cm depths in rice field while soil core samples were used in mustard field upto 105 cm depth at an interval of 15 cm for first three layers and at 30 cm interval for lower two layers, for monitoring NO_3^--N transport. The extract was analyzed for NO_3^--N by phenoldisulfonic acid method [39].

1.4.6.4 Root Growth and Distribution

Soil samples with root auger were periodically taken from the upper three soil depths (0–15, 15–30 and 30–45 cm) for rice and seven soil depths (0–15, 15–30, 30–45, 45–60, 60–75, 75–90 and 90–105) for mustard.

Sampling depth was 15 cm at first sampling and it was increased at later stage according to rooting depth. The total root length (Eq. 29) of each sample was determined by Newman's [54] method as modified by Tennant [75]. From root length, root density was calculated (Eq. 30).

$$RL = \frac{11}{14} \times NC \times G \qquad (29)$$

$$RLD = \frac{\text{Root length (cm)}}{\text{Volume of soil sample (cm}^3)} \qquad (30)$$

where, RL = root length (mm); NC = sum of horizontal and vertical crossings; G = length of the grid unit (1 cm, in this case); and RLD = root length density(cm cm^{-3}).

1.4.6.5   Crop Yield and Nutrient Uptake

At harvest, yields of rice and mustard (grain and straw) were determined. Crop samples were also taken and oven dried at 70°C for 72 hours. Concentration of N, P and K in grain and straw of both rice and mustard samples were determined. Nutrient uptake was calculated by multiplying with respective yields.

1.5   RESULTS AND DISCUSSION

1.5.1   *WATER DYNAMICS IN THE CROPPED FIELD*

In order to assess the depth and time variation of soil water content (SWC) under different weir heights, which led to variable soil water regimes, SWC was determined periodically upto 45 cm depth for rice and up to 105 cm for mustard. Vertical percolation (VP) is the important parameter in the cropped field water balance, in addition to actual evapotranspiration (AET) that influence the soil water status in the effective root-zone. Hence, there is need to simulation soil hydraulic parameters against the independently observed data. The details of HYDRUS-2D model calibration and validation for the prediction of soil water content (SWC) and vertical percolation (VP) in the field with different weir heights are discussed below.

1.5.1.1 Simulation of Water Movement Using HYDRUS-2D

Inverse modeling technique was used to calibrate HYDRUS-2D model. The calibration was carried out considering the observed soil hydraulic parameters and boundary conditions. It is assumed that fields are level and field scale spatial variability for each soil layer is uniform during the experiments. It is assumed that the observed data at a particular depth of soil layer are uniform for the entire layer. The initial calibrated parameters, coefficient in soil water retention function (α) and pore size distribution index (n), for each layer of soils were estimated using neural network prediction and van Genuchten and Mualem hydraulic models. The inputs used for neural network predictions were measured soil texture, bulk density, field capacity, and wilting point values for each soil layer. Simulation for soil water content of 7 soil layers for mustard and 3 for rice (15 cm increment) was carried out with respect to observation nodes in the model domain and field observation points. The model simulated parameters for rice and mustard field were tested with the field observations of the year 2002–2004. The optimized model calibrated parameters for each layer of soils were determined using inverse modeling with 95% confidence intervals (Table 1.4).

1.5.1.2 Soil Water Content

Using optimized calibrated parameters, HYDRUS-2D model was used to simulate soil water content for rice and mustard fields under variably saturated conditions. The regression analysis of observed and simulated

TABLE 1.4   Initial and Optimized Calibrated Parameters for Different Soil Layers

Soil depth (cm)	α (cm⁻¹)		n	
	Initial	Optimized	Initial	Optimized
0–15	0.0070	0.0072	1.4634	1.4500
15–30	0.0077	0.0057	1.4893	1.4749
30–45	0.0122	0.0039	1.5000	1.5619
45–60	0.0094	0.0036	1.5000	1.5611
60–75	0.0070	0.0035	1.5000	1.5669
75–90	0.0070	0.0034	1.2000	1.5597
90–105	0.0070	0.0018	1.2000	1.6945

SWC for different soil layers of mustard field for the experimental year 2002 are shown in Figure 1.10. The slope of the regression line is close to unity, which further indicates good agreement between the observed and simulated SWC. The clustering of observed and simulated SWC around the 1:1 line and the high value of R^2 (more than 0.72) indicate that HYDRUS-2D model is efficient in predicting daily variation of SWC in the root-zone of mustard in the rainfed ecosystem. In addition to scattered and regression presentation, the performance of HYDRUS-2D model was also

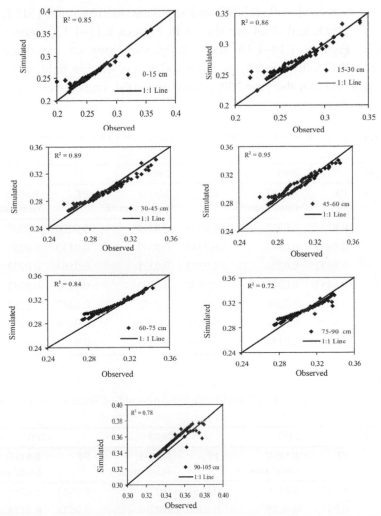

FIGURE 1.10 Regression analysis of observed and simulated soil water content (mm³/mm³) for different soil layers of mustard field for 2002.

evaluated using error statistics such as prediction efficiency (PE) and root mean square error (RMSE) for simulation of SWC in different soil layers of mustard and rice fields for the year 2002, 2003, and 2004. The error statistics revealed that the PE and RMSE values are within the acceptable limit (Tables 1.5 and 1.6). The high value of R^2, PE (more than 0.90), and low value of RMSE (less than 0.05) indicates that the HYDRUS-2D model is quite efficient in predicting daily variation of SWC in the cropped field with variably saturated condition.

Daily variation of simulated and observed SWC for seven soil layers of mustard field with rainfall and supplemental irrigation (SI) for the year 2002, 2003, and 2004 are shown in Figures 1.11–1.13, respectively. Similarly, Figures 1.14–1.16 show daily variations of simulated and observed SWC with rainfall for three soil layers of rice field with 0 cm weir height (H1) for the year 2002, 2003, and 2004, respectively. However,

TABLE 1.5 Error Statistics for HYDRUS-2D Simulation of Soil Water Content in Mustard Field

Soil layer (cm)	2002		2003		2004	
	PE	RMSE (mm^3 mm^{-3})	PE	RMSE (mm^3 mm^{-3})	PE	RMSE (mm^3 mm^{-3})
0–15	0.9337	0.0317	0.9335	0.0318	0.9620	0.0239
15–30	0.9342	0.0315	0.9345	0.0313	0.9530	0.0266
30–45	0.9432	0.0291	0.9353	0.0311	0.9503	0.0273
45–60	0.9457	0.0283	0.9518	0.0266	0.9588	0.0247
60–75	0.9612	0.0240	0.9525	0.0263	0.9442	0.0289
75–90	0.9700	0.0211	0.9616	0.0236	0.9408	0.0297
90–105	0.9796	0.0174	0.9708	0.0263	0.9357	0.0309

TABLE 1.6 Error Statistics for HYDRUS-2D Simulation of Soil Water Content in Rice Field

Soil layer (cm)	2002		2003		2004	
	PE	RMSE (mm^3 mm^{-3})	PE	RMSE (mm^3 mm^{-3})	PE	RMSE (mm^3 mm^{-3})
0–15	0.9326	0.0230	0.9132	0.0263	0.9523	0.0207
15–30	0.9475	0.0225	0.9360	0.0209	0.9623	0.0278
30–45	0.9531	0.0221	0.9525	0.0294	0.9709	0.0191

rice field with 5 (H2) and 10 cm weir height (H3) remains under both ponding and variably saturated condition. So, the soil water status and/ponding were simulated by the field water balance model and presented in the subsequent sections.

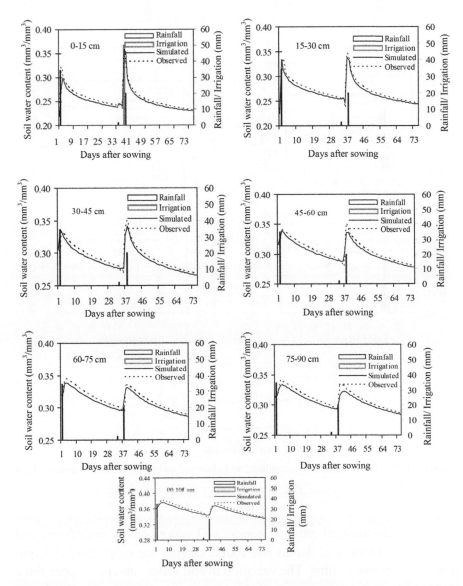

FIGURE 1.11 Daily variation of simulated and observed soil water content in mustard field during 2002.

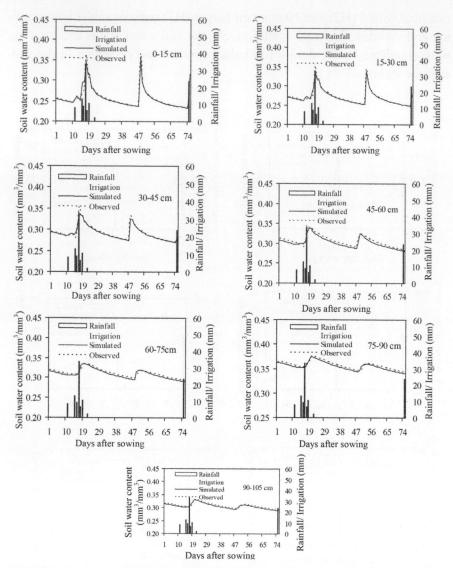

FIGURE 1.12   Daily variation of simulated and observed soil water content in mustard field during 2003.

During the mustard growing period, soil water content in different soil layers of crop root-zone depths was found varying due supplemental irrigation over a period of time. The variation in soil water content in upper three layers of crop root-zone depths (0–45 cm) was found higher than the lower depths (45–105 cm) during three years of experiment (Figures 1.11–1.13).

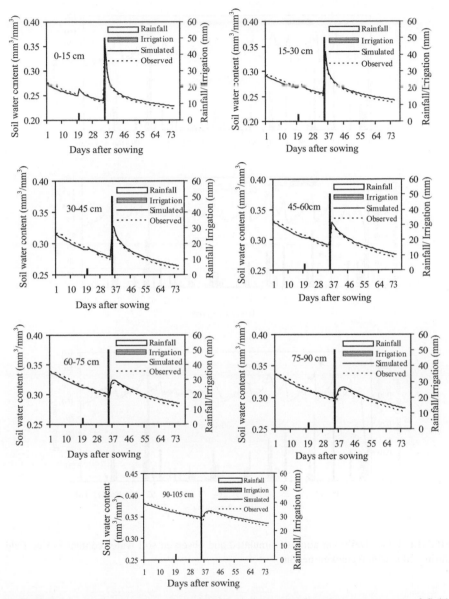

FIGURE 1.13 Daily variation of simulated and observed soil water content mustard field during 2004.

The variation in soil water content in upper layers of crop root-zone depths affects the availability of water and nutrient to mustard and ultimately affects crop yields.

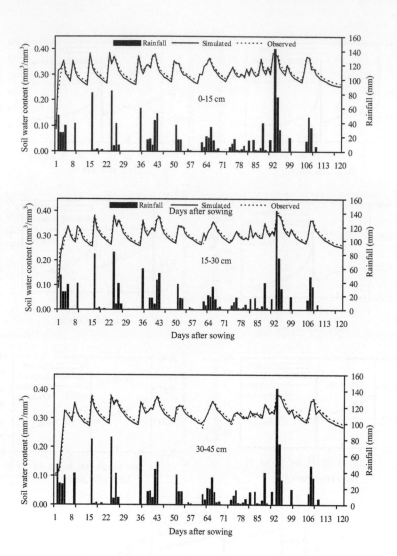

FIGURE 1.14 Daily variation of simulated and observed soil water content in rice field during 2002 with 0 cm weir height.

Variation of soil water content in different soil layers of crop root-zone depths was found to have cyclic experience in rice fields due to variation in rainfall. However, during the period when there was no rainfall or no SI is applied, SWC was found to decline gradually because of uptake of water by the plant roots and VP from the root-zone (Figures 1.14–1.16). The average value of PE and RMSE for simulated soil water content in rice field

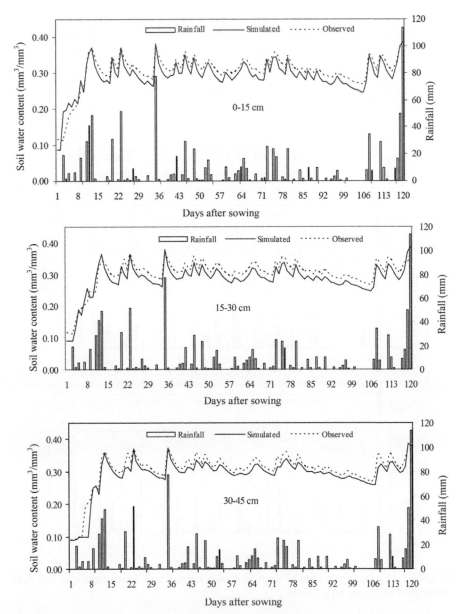

FIGURE 1.15 Daily variation of simulated and observed soil water content in rice field during 2003 with 0 cm weir height.

was found as 0.9467 and 0.0235, respectively (Table 1.6). So, the calibrated parameters can be used for the simulation of soil water content in the rice field with variably saturated condition.

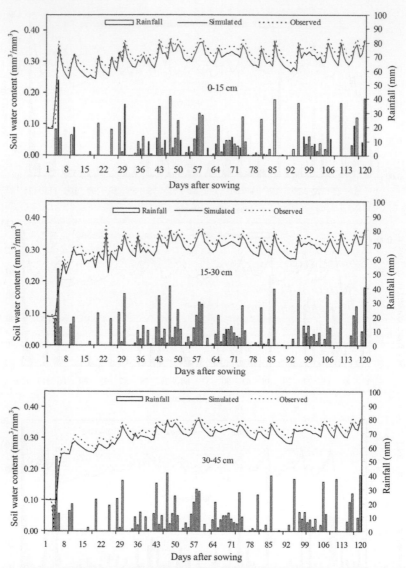

FIGURE 1.16 Daily variation of simulated and observed soil water content in rice field during 2004 with 0 cm weir height.

1.5.1.3 Vertical Percolation in the Cropped Field

Vertical percolation (VP) losses in the rice field with 0 cm weir height were simulated using HYDRUS-2D model under variably saturated condition, whereas in case of 5 and 10 cm weir heights (ponding condition) Darcy's approach was used. The regression analysis of observed and simulated VP losses in the root-zone of rice at different weir heights and mustard field for the year 2002 are shown in Figures 1.17 and 1.18, respectively. The regression analysis revealed that the simulated fluxes were matched reasonably well with the observed values with acceptable R^2 (more than 0.90). The daily variation of simulated and observed VP from the rice field with different weir heights for the year 2002, 2003 and 2004 are presented in Figures 1.19, 1.20, and 1.21, respectively. The performance of the model was also evaluated using error statistics such as PE and RMSE to simulate VP from the rice field at different weir heights for the year 2002, 2003 and 2004 (Table 1.7).

The statistical analysis revealed that the PE, and RMSE values are also within acceptable limit. The high value of R^2, PE (more than 0.90), and low value of RMSE indicate that the HYDRUS-2D model is quite efficient for simulation from the cropped field under variable saturated

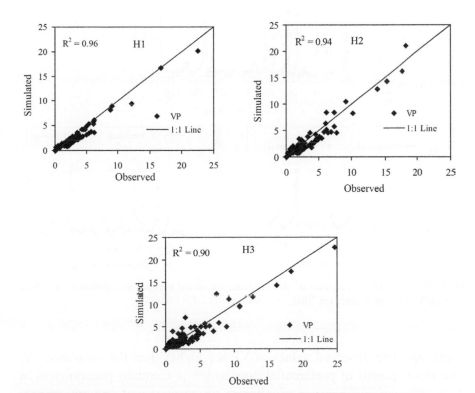

FIGURE 1.17 Regression analysis of simulated and observed vertical percolation (mm) in the rice field with different weir heights for 2002.

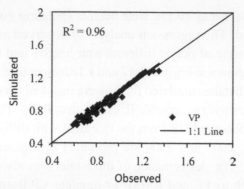

FIGURE 1.18 Regression analysis of simulated and observed vertical percolation (mm) in the mustard field during 2002.

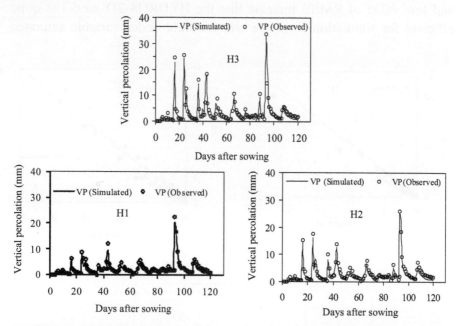

FIGURE 1.19 Daily variation of observed and simulated vertical percolation in rice field with different weir heights for 2002.

condition. The observed values of VP are higher than the simulated VP. The development of preferential flow which is common phenomenon in near-saturated soil because of inherent structure of soil and macropores created by soil fauna, decayed root channels, and shrinking clay materials

TABLE 1.7 Error Statistics of Vertical Percolation in the Rice Field with Different Weir Heights

Parameter	2002		2003		2004	
	PE	**RMSE (mm)**	**PE**	**RMSE (mm)**	**PE**	**RMSE (mm)**
0 cm weir height (H1)						
VP	0.921	0.636	0.984	0.102	0.979	0.159
5 cm weir height (H2)						
VP	0.948	0.856	0.908	0.266	0.949	0.582
10 cm weir height (H3)						
VP	0.903	0.563	0.867	0.955	0.891	0.754

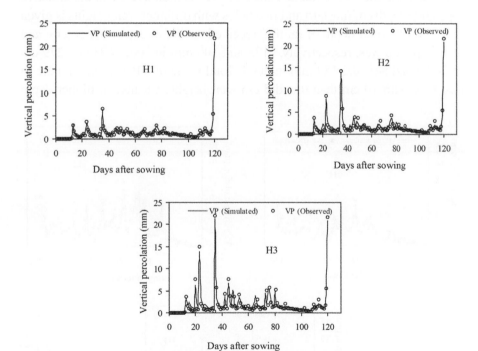

FIGURE 1.20 Daily variation of observed and simulated vertical percolation in rice field with different weir heights for 2003.

which leads to observed vertical percolation higher compared with that of simulated one [40, 64, 65, 70].

The observed VP under 0, 5, and 10 cm weir heights in the rice growing and turn-in period was ranging, respectively, from 0 to 22.59, 0 to 25.90, and

0 to 33.80 mm day⁻¹ for 2002; 0 to 20.90, 0 to 21.70, and 0 to 22.00 mm day⁻¹ for 2003; and 0 to 5.60, 0 to 8.35, and 0 to 11.59 mm day⁻¹ for 2004. The average observed values of VP under 0, 5, and 10 cm weir heights in the rice growing and turn-in period was found, respectively, as 2.68, 3.14, and 3.62 mm day⁻¹ for 2002; 1.36, 1.67, and 1.90 mm day⁻¹ for 2003; and 1.61, 2.20, and 2.56 mm day⁻¹ for 2004.

The VP was found same in the rice field with 0, 5, and 10 cm weir heights during initial stage (up to 15 days after sowing), 10 days to harvesting and turn-in period because no ponding water was allowed in the field.

The variation of VP was found because of deterministic and stochastic input variables such as supplemental irrigation, available soil water in the crop root-zone, and rainfall. The simulated and observed cumulative VP in the effective root-zone depths of mustard and rice fields with different weir heights is given in Table 1.8. Total depth of rainfall received during rice growing season and turn-in period was, respectively, 1175 and 100 mm in 2002; 693 and 279 cm in 2003; and 848 and 189 mm in 2004. Total value of VP was found more in rice field with 10 cm than 0 and 5 cm weir heights because of higher depth

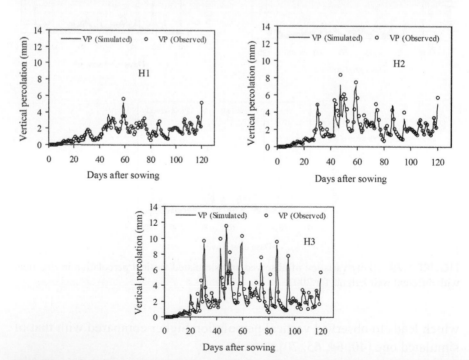

FIGURE 1.21 Daily variation of observed and simulated vertical percolation in rice field with different weir heights for 2004.

TABLE 1.8 Total Simulated and Observed Vertical Percolation (mm) in the Rice with Different Weir Heights and Mustard Field

Year					
2002		2003		2004	
Simulated	Observed	Simulated	Observed	Simulated	Observed
Rice Field with 0 cm weir height					
241.15	277.67	122.82	119.40	149.95	156.26
Rice Field with 5 cm weir height					
303.72	330.10	157.66	141.25	194.34	225.47
Rice Field with 10 cm weir height					
374.62	389.78	204.06	184.01	253.85	270.50
Mustard Field					
66.45	65.43	66.00	64.43	41.57	40.22

of ponding. Out of total rainfall received in 2002 during rice growth period including turn-in period, the percentage observed total VP was 21.78 for 0 cm; 25.90 for 5 cm; 30.58 for 10 cm weir heights. Under similar rice growth period as stated above, the percentage of VP in 2003 was found as 12.29 for 0 cm; 14.54 for 5 cm; and 18.94 for 10 cm weir heights, respectively. Similarly for 2004, the percentage of VP out of total rainfall received during rice growing season including turn-in period was found as 15.07 for 0 cm; 21.74 for 5 cm; and 26.08 for 10 cm weir heights, respectively.

Three years of observations revealed that minimum loss of the VP in the rice fields with 0 cm weir height (184.44 mm) than the field with 5 cm weir height (232.27 mm) and 10 cm weir height (281.43 mm), which indicates that 0 cm weir height in the rice field can minimize 47.83 and 96.99 mm loss of VP than the 5 and 10 cm weir height, respectively. So, the rice field without standing water (0 cm weir height) may be adopted to minimize these major losses and in other words fields with standing water may enhance groundwater recharge processes.

However, out of three years (2002, 2003 and 2004) of average rainfall (1095 mm) during the rice growing and turn-in period, the percentage of average total VP loss found from the field with 0, 5, and 10 cm weir height are 16.38, 20.72 and 25.20, respectively, which indicates that the fields with 0 cm weir height can conserve 4.34 and 8.82% of excess rainfall in the OFR for sustainable integrated farming in rainfed ecosystem.

Figure 1.22 shows the daily variation of observed and simulated VP losses in the mustard field for the year 2002, 2003, and 2004. In mustard field VP was observed ranging from 0 to 1.28 mm day^{-1} for 2002; 0 to 1.13 mm day^{-1} for 2003; and 0 to 1.61 mm day^{-1} for 2004. Three years average observed VP from the mustard field is found as 57 mm. The observed and simulated VP in three years of experiment is varied with SI, rainfall and available soil water in the crop root-zone.

1.5.2 WATER BALANCE IN THE CROPPED FIELDS

Actual evapotranspiration (AET), surface runoff (RO) and supplemental irrigation (SI)/pre-sowing irrigation (PSI) simulated by the field water balance simulation model were validated using field experimental data of three years (2002–04). The parameters such vertical percolation (VP) from the effective root-zone of rice field with different weir heights and mustard field were

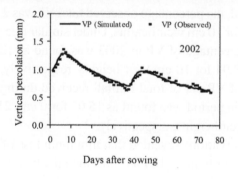

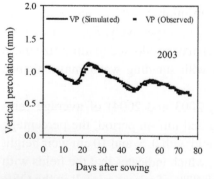
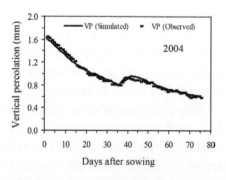

FIGURE 1.22   Daily variation of observed and simulated vertical percolation in the mustard field.

simulated by using HYDRUS-2D model as explained in earlier sections and used as inputs to the field water balance model. However, the simulated and observed findings of soil water content/ponding depth, AET, RO and SI/PSI in the cropped field were explained in the following sections.

1.5.2.1 Soil Water Content and Ponding Depth

Prediction of soil water content (SWC) in the root-zone and ponding depth in the rice field with different weir heights is used to reveal the period when there is supplemental irrigation requirement and need of drainage to the OFR.

The simulated soil water content and ponding depth in the rice field with different weir heights was compared with its corresponding daily measured values for all the three experimental years. The observed ponding depth in the rice field with 5 and 10 cm weir heights for 2002 is shown in Figure 1.23.

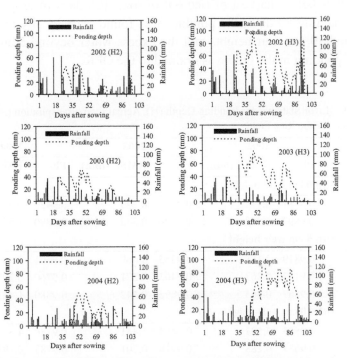

FIGURE 1.23 Daily variation of ponding depth in the rice field with 5 and 10 cm weir heights.

Values of R^2 between the observed and simulated values of ponding depth in the rice field were found more than 0.92. The performance of the field water balance simulation model was also evaluated using error statistics such as PE and RMSE to predict ponding depth in the rice field with different weir heights for the year 2002, 2003, and 2004 (Table 1.9). Since the values of R^2 for the prediction of ponding depth is high and RMSE is less, the field water balance simulation model can be safely used to simulate variably ponding conditions.

Occurrence of rainfall and/or application of irrigation during crop growing season controls daily variation in SWC in the effective root-zone of rice and mustard fields. This is because of the VP and root water uptake from different soil layers. Daily variation of SWC with rainfall/irrigation during three years of field experiments are shown in Figures 1.11–1.13, for mustard; and Figures 1.14–1.16 for rice.

Daily variation of simulated and observed ponding depth in the rice field with 5 and 10 cm weir heights for the year 2002, 2003, and 2004 are shown in Figure 1.23. The weir heights were introduced 15 days after sowing and 10 days before harvest of rice out of total 105 days of crop duration. So in remaining 80 days, the ponding was found occasionally in the field with 5 and 10 cm weir heights and the excess water beyond 5 and 10 cm depth of ponding in the field was drained to the OFR.

TABLE 1.9   Error Statistics for Ponding Depth (H), Actual Evapotranspiration (AET) and Runoff (RO) for the Rice Field

Parameter	2002		2003		2004	
	PE	RMSE (mm)	PE	RMSE (mm)	PE	RMSE (mm)
Rice field with 0 cm weir height						
AET	0.8781	0.414	0.9542	0.243	0.9420	0.306
RO	0.9897	1.052	0.9896	0.332	0.9890	1.545
Rice field with 5 cm weir height						
H	0.9819	2.111	0.9355	3.660	0.9601	2.946
AET	0.9147	0.363	0.9465	0.271	0.9760	0.209
RO	0.9867	0.865	0.9892	0.342	0.9880	1.427
Rice field with 10 cm weir height						
H	0.9805	3.597	0.9454	5.451	0.9786	4.582
AET	0.9633	0.288	0.9718	0.201	0.9783	0.198
RO	0.9876	0.856	0.9900	0.592	0.9627	2.587

Total days of ponding in the rice field with 5 and 10 cm weir heights were observed, respectively, as 30 and 71 in 2002; 46 and 68 in 2003; and 42 and 52 in 2004. The variation in ponding depth has shown cyclic trend depending on the rainfall intensity and duration during the rice-cropping season. During three experimental years of 2002, 2003, and 2004, sufficient quantity of rainfall of 217, 174, and 290 mm occurred during CGS of rice (49 to 78 days after sowing), respectively, that created ponding in fields for which SI was not required. More number of days of ponding was observed in 10 cm in comparison to 5 cm weir heights. During three years of experiment on winter mustard, it was observed that PSI was not required for the germination of mustard because residual soil water content was not depleted below 75% of available water. However, SI was applied during mustard growing period.

1.5.2.2   Actual Crop Evapotranspiration

Actual crop evapotranspiration (AET) is one of the important parameters of water balance that influences the soil water status of rainfed crop fields. Total seasonal observed and simulated AET for rice and mustard for 2002, 2003, and 2004 are presented in Table 1.10, which demonstrates higher AET in rice with 10 cm in comparison to 0 and 5 cm weir heights because of relatively more days of ponding. Out of total rainfall received in a cropping season, percentage total observed AET of rice under 0, 5, and 10 cm weir heights were, respectively, 31.06, 37.62, and 39.19 for 2002; 49.89, 57.48, and 58.59 for 2003; and 44.21, 48.16, and 48.20 for 2004. Three years average observed total AET of rice under 0, 5, and 10 cm weir heights was found, respectively, as 44.21, 48.16, and 48.20% of average total rainfall (905 mm). Total AET of rice at 0 cm weir height was 6 and 7% less than the AET at 5 and 10 cm weir heights because of variably saturated condition.

1.5.2.3   Surface Runoff

Total surface runoff (RO) generated from the rice fields with 0, 5, and 10 cm weir heights were observed, respectively, as 324.00, 298.79, and 285.97 mm for 2002; 91.14, 40.86, and 15.46 mm for 2003; and 81.84, 44.21, and 4.96 mm for 2004. However, the total rainfall measured during rice cropping season of 2002, 2003, and 2004 were 1175, 693 and 848 mm, respectively. While rainfall and runoff observed during turn-in period was

TABLE 1.10   Total Seasonal Values of Simulated and Observed Runoff and Actual
Evapotranspiration in the Rice and Mustard Fields During Three Years of Study

Parameter	Year					
	2002		2003		2004	
	Simulated	Observed	Simulated	Observed	Simulated	Observed
Rice Field with 0 cm weir height						
Rainfall (mm)	1175		693		848	
Runoff (mm)	340.03	324.00	95.33	91.14	106.25	81.84
AET (mm)	363.02	364.97	336.22	345.68	374.79	374.79
Rice Field with 5 cm weir height						
Runoff (mm)	290.03	298.79	45.27	40.86	56.25	44.21
AET (mm)	449.85	442.01	394.58	398.34	400.03	408.32
Rice Field with 10 cm weir height						
Runoff (mm)	240.03	285.97	17.21	15.46	6.25	4.96
AET (mm)	471.06	460.48	401.82	405.98	402.63	408.63
Mustard Field						
Rainfall (mm)	57		106		4	
Runoff (mm)	0.00	0.00	0.00	0.00	0.00	0.00
AET (mm)	133.50	130.45	142.92	138.64	129.66	133.94

100 and 0 mm for 2002; 279 and 136 mm for 2003; and 189 and 84 mm
for 2004, respectively. Values of RO are found maximum from the rice
fields with 0 cm weir height and the values decreased with the increasing weir heights (Table 1.10). The variations are found only due to rainfall
intensity and weir heights. It is further observed that percentage seasonal
(120 days) RO generated from rainfall in the rice field with 0, 5, and 10 cm
weir heights are 27.57, 25.43, and 24.33 for 2002; 13.15, 5.90, and 2.23 for
2003; 9.65, 5.21, and 0.06 for 2004, respectively. Thus there is more scope
for harvesting RO in the OFR by generating from the rice field with 0 cm
weir height in comparison to the field with 5 and 10 cm weir heights and
reuse the harvested water for integrated farming. However, there was no
RO observed from the mustard field in all the three years of experiments.

1.5.2.4 Supplemental Irrigation Requirement

Strategies for supplemental irrigation (SI) from the OFR to the rice field
were kept at 40% depletion of SWC from saturation soil water content

(170 mm) in the effective root-zone depth during the reproductive stage. During rest of the periods, rice was grown under purely rainfed condition. During three years of field experiments SI was not applied to the rice due to uniform temporal distribution of rainfall.

However, PSI and SI to mustard was applied at 25% depletion of available soil water content. Field experimental study from 2002 to 2004 revealed that PSI to mustard was not required due to availability of sufficient residual soil water contents after the harvest of rice. But first SI of 5 cm was applied to mustard field using sprinkler irrigation on 37, 47 and 37 days after sowing (DAS) from the lined and unlined OFR and second at 52, 58 and 52 DAS from the lined OFR only during the year 2002, 2003 and 2004, respectively.

1.5.3 WATER BALANCE IN THE ON-FARM RESERVOIR (OFR)

For sustainable farming, it is imperative to observe the inflow and outflow components of the OFR system using cropped field. The actual capacity of the OFR (120 m³) used for field experiment is not the same as the simulated capacity. Daily variation of water storage in the lined and unlined OFRs with different weir heights for the experimental year 2002, 2003, and 2004 are shown in Figures 1.24–1.27.

Average observed volume of water storage in the OFR was found less in comparison to the simulated value, which may be due to the seepage through the lining material as a result of natural and/or manmade damage. Simulated maximum storage in the lined OFR with 0, 5, and 10 cm weir heights was found, respectively, as 300.42, 268.55 and 236.43 m³ for 2002; 211.40, 181.36 and 164.57 m³ for 2003; and 189.48, 158.14, and 126.51 m³ for 2004. However, the simulated maximum storage in the unlined OFR at 0, 5, and 10 cm weir heights was found respectively, as 185.94, 173.57, and 159.16 m³ for 2002; 138.45, 130.48, and 125.42 m³ for 2003, and 110.51, 99.38, and 86.45 m³ for 2004.

1.5.3.1 The OFR Inflows

The components of observed and simulated OFR inflows and outflows for the years 2002, 2003 and 2004 are presented in Table 1.11, which indicates higher total inflow to the OFR during the entire experimental period of rice-mustard cropping season that requires proper sizing.

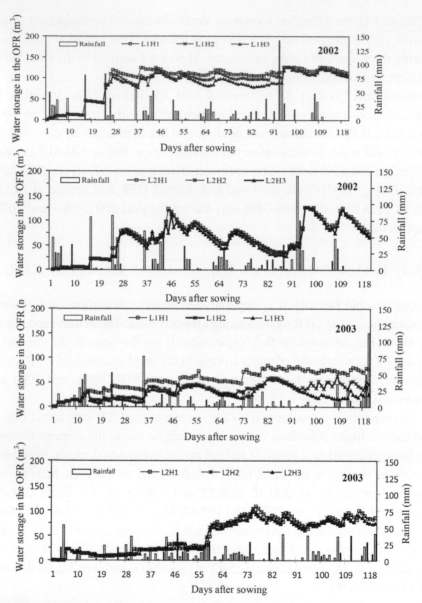

FIGURE 1.24   Daily variation in water storage in the lined and unlined OFRs during rice growing season.

1.5.3.2   Rainfall Contribution

Rainfall constituted the major components of total inflows. Rainfall contribution to the OFRs during rice and mustard growing seasons are,

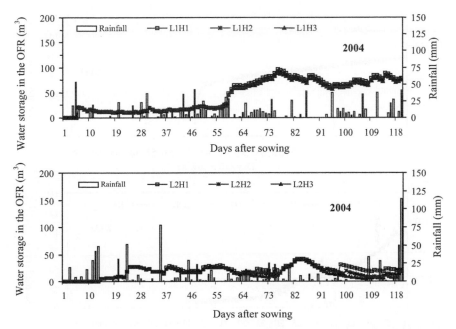

FIGURE 1.25 Daily variation in water storage in the lined and unlined OFRs during rice growing season.

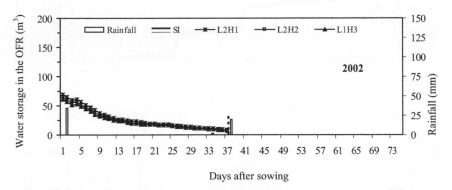

FIGURE 1.26 Daily variation in water storage in the lined and unlined OFRs during mustard growing season: 2002.

respectively, 101.98 and 4.53 m³ in 2002; 77.71 and 8.48 m³ in 2003; and 82.95 and 0.32 m³ in 2004 (Tables 1.11 and 1.12).

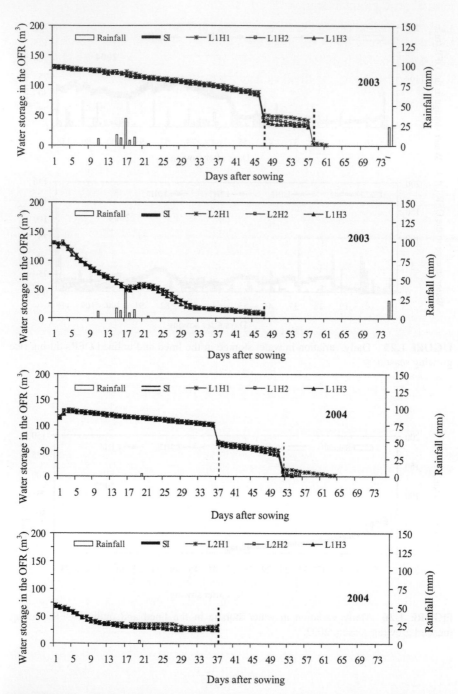

FIGURE 1.27 Daily variation in water storage in the lined and unlined OFRs during mustard growing season: 2003 and 2004.

TABLE 1.11   Seasonal Water Balance Parameters of the OFR During Rice Growing and Turn-In Periods

Parameters	Year					
	2002		2003		2004	
	Lined	Unlined	Lined	Unlined	Lined	Unlined
0 cm weir height						
Rainfall[a] (m³)	101.98		77.71		82.95	
Runoff[b] (m³)	244.82		166.47		137.19	
Total Inflow (m³)	346.80		244.18		220.14	
Evaporation (m³)	53.04	41.99	32.78	24.37	30.66	24.29
SP (m³)	0.00	173.93	0.00	81.49	0.00	85.59
Total Outflow (m³)	53.04	215.92	32.78	105.86	30.66	109.88
5 cm weir height						
Rainfall[a] (m³)	101.98		77.71		82.95	
Runoff[b] (m³)	208.82		130.43		101.19	
Total Inflow (m³)	310.80		208.14		184.14	
Evaporation (m³)	48.53	38.56	26.78	19.67	26.01	20.36
SP (m³)	0.00	150.03	0.00	58.12	0.00	64.65
Total Outflow (m³)	48.53	188.59	26.78	77.79	26.01	85.01
10 cm weir height						
Rainfall[a] (m³)	101.98		77.71		82.95	
Runoff[b] (m³)	172.82		110.22		65.19	
Total Inflow (m³)	274.80		187.93		148.14	
Evaporation (m³)	44.24	35.04	23.37	16.73	21.63	16.40
SP (m³)	0.00	127.62	0.00	45.91	0.00	45.54
Total Outflow (m³)	44.24	162.66	23.37	62.64	21.63	61.94

[a]Direct rainfall contribution to the OFR, [b]Runoff contribution from the cropped field.

1.5.3.3   Surface Runoff Contribution

Out of total volume of inflows to the OFR during rice-growing season, the percentage of surface runoff (RO) contribution to the OFRs with 0, 5, and 10 cm weir heights are, respectively, 70.59, 67.19, and 62.89 in 2002; 68.18, 62.66, and 58.65 in 2003; and 62.32, 54.95, and 44.01 in 2004. Highest percentage of RO contribution to the OFR with 0, 5, and 10 cm weir heights

TABLE 1.12   Seasonal Water Balance Parameters of the OFR During Mustard Growing Periods

Parameters	Year					
	2002		**2003**		**2004**	
	Lined	**Unlined**	**Lined**	**Unlined**	**Lined**	**Unlined**
0 cm weir height						
Rainfall[a] (m³)	4.53		8.48		0.32	
Total Inflow (m³)	4.53		8.48		0.32	
Evaporation (m³)	23.35	10.55	20.83	11.68	17.57	8.47
SP (m³)	0.00	68.69	0.00	77.63	0.00	54.75
SI/PSI (m³)	72.00	36.00	72.00	36.00	72.00	36.00
Total Outflow (m³)	95.35	115.24	92.83	125.31	89.57	99.22
5 cm weir height						
Rainfall[a] (m³)	4.53		8.48		0.32	
Total Inflow (m³)	4.53		8.48		0.32	
Evaporation (m³)	21.98	10.02	19.20	11.21	15.82	7.63
SP (m³)	0.00	63.64	36.00	72.96	0.00	48.29
SI/PSI (m³)	72.00	36.00	72.00	36.00	72.00	36.00
Total Outflow (m³)	93.98	109.66	91.20	120.17	87.82	91.22
10 cm weir height						
Rainfall[a] (m³)	4.53		8.48		0.32	
Total Inflow (m³)	4.53		8.48		0.32	
Evaporation (m³)	20.50	9.36	18.22	10.90	13.82	6.41
SP (m³)	0.00	57.77	0.00	70.00	0.00	40.83
SI/PSI (m³)	72.00	36.00	72.00	36.00	72.00	36.00
Total Outflow (m³)	92.50	103.13	90.22	116.90	85.82	83.24

[a]Direct rainfall contribution to the OFR.

was due to relatively more seasonal rainfall in 2002 (1175 mm) in comparison to 2003 and 2004. Three years (2002, 2003, and 2004) average percentage of RO contributed to the OFR with 0, 5, and 10 cm weir heights was 67.03, 61.60, and 55.18, respectively. During mustard growing season, there was negligible quantity of rainfall for three years, which did not generate any runoff.

1.5.3.4 The OFR Outflows

The OFR outflows (evaporation, pre-sowing/supplemental irrigation, and seepage) during rice and mustard growing seasons for all the experimental years are shown in Table 1.11 and 1.12, respectively.

1.5.3.5 Evaporation Loss

The outflows from the lined OFR during rice growing season are evaporation loss and SI. Since, SI for rice was not required during the experimental years, so only outflow from the lined OFRs was evaporation loss. Evaporation loss from the lined OFR with 0, 5, and 10 cm weir heights during rice-growing season was, respectively, as 53.04, 48.53, and 44.24 m^3 in 2002; 32.78, 26.78, and 23.37 m^3 in 2003; and 30.06, 26.01, and 21.63 m^3 in 2004. Evaporation loss from the lined OFR was observed to be more in 2002, than that of 2003 and 2004 because of more volume of water storage that leads to larger water surface area. However, in the unlined OFRs with 0, 5, and 10 cm weir heights, the outflow components are evaporation, supplemental irrigation, and seepage. Out of total outflows in the rice-growing season, the percentage of evaporation loss from the unlined OFRs with 0, 5, and 10 cm weir heights was, respectively, 21.54, 20.45, and 19.45 in 2002; 26.71, 25.29, and 23.02 in 2003; and 26.48, 23.95, and 22.11 in 2004. Average evaporation loss from the unlined OFR with 0, 5, and 10 cm weir heights during the rice-growing season was 21.52, 23.23, and 24.91% of the total outflow, respectively. The percentage of evaporation loss from the OFR was decreased with the increase in weir heights due to the decrease in OFR water storage that leads to reduction in water surface area.

Whereas during the mustard growing season, total evaporation loss from the carry-over water volume of the lined OFR with 0, 5, and 10 cm weir heights was, respectively, 23.35, 21.98, and 20.50 m^3 in 2002; 20.83, 19.20, and 18.22 m^3 in 2003 and 17.57, 15.82, and 13.82 m^3 in 2004. Similarly, evaporation loss from the unlined OFRs with 0, 5, and 10 cm weir heights was, respectively, 10.55, 10.02, and 9.36 m^3 in 2002; 11.68, 11.21, and 10.90 m^3 in 2003; and 8.47, 7.63, and 6.41 m^3, in 2004.

1.5.3.6 Pre-Sowing Supplemental Irrigation

During three years of experiments, SI to rice and PSI to mustard was not applied because SWC not depleted below 75% of ASW. However, first SI of 5 cm (equivalent volume of the OFR water = 36 m³) was applied to mustard on 37, 47 and 37 days after sowing (DAS) from the lined and unlined OFR and second at 52, 58 and 52 DAS from the lined OFR only during the year 2002, 2003 and 2004, respectively, of all the weir heights. It was observed that there was negligible quantity of water available in the unlined OFR after giving one SI to mustard.

1.5.3.7   Seepage

In the lined OFRs, the seepage was considered negligible due to the application of a better quality LDPE lining material of 90 GSM. However, out of the total water outflow during rice-growing season, percentage of seepage (includes seepage and vertical percolation) loss from the unlined OFRs with 0, 5, and 10 cm weir heights was, respectively, 80.55, 79.55, and 78.46 in 2002; 76.98, 74.71, and 73.29 in 2003; and 77.89, 76.05, and 73.52 in 2004. Average SP loss from the unlined OFRs with 0, 5, and 10 cm weir heights during the rice-growing season was 78.48, 76.77 and 75.09% of total water outflow, respectively. Whereas, during the mustard growing season, the SP loss from the total carry-over water volume of the unlined OFR with 0, 5, and 10 cm weir heights was 58.91, 57.23 and 54.98% of the total water outflow.

1.5.4   ROOT GROWTH UNDER VARIABLE SOIL WATER REGIMES

For nutrient-use efficiency in rainfed ecosystems, the special challenge is the dynamic of the water regime. If water is available but nutrients are limiting, the crop may not be able to utilize the water efficiently. Root development may be critical in this regard. Figures 1.28 and 1.29 showed the root growth of rice and mustard, respectively.

From Figure 1.28 it is evidenced that different weir heights does not affect the root growth in rice but in mustard weir heights affect the root growth and its density (Figure 1.29). The SI from the lined OFR with 0 cm weir height contributes more soil water at upper layer of the soil depicting more root length density in 0 cm as compared to 10 cm weir height. It is evidence that with less irrigation the root tends to grow deeper in the soil profile.

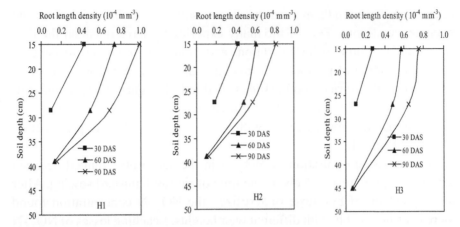

FIGURE 1.28 Effect of different weir heights on root growth of rice.

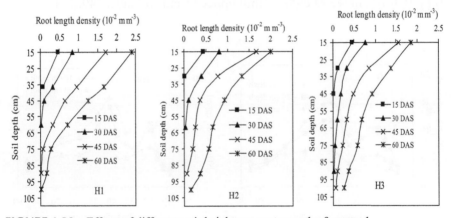

FIGURE 1.29 Effects of different weir heights on root growth of mustard.

During periods of soil water deficits, the capability of crop roots to extract soil water depends on the distribution and depth of its root system. Most water uptake models assume a relationship between root water extraction and root length density [19].

1.5.5 NITROGEN DYNAMICS IN THE CROPPED FIELD

For rainfed upland rice as compared to other agricultural systems, it is not clear to what extent rice yields are limited by nutrients, water, and the interactions between them, over the diverse soil types, cultural practices and seasonal conditions. Few of the many experiments have obtained the data essential for thorough interpretation, and none has attempted to develop an understanding of prospects for manipulating nutrient and water interactions across the varied combinations of flood, drought, soil type and cultural practices. The average temporal changes of NO_3^--N at different soil depths in rice and mustard field with various weir heights are presented in Figures 30 and 31.

The initial concentration of NO_3^--N in the soil profile ranged between 3.5 and 4.5 ppm, respectively, at the time of rice and mustard sowing. After the application of first dose of fertilizer, the NO_3^--N concentration found increasing noticeably with different weir heights. Leaching losses of NO_3^--N in the rice field is more due to more water storage in 10 than 0 cm weir heights. The NO_3^--N concentration during the experiment found as high as 20 ppm in 10 cm weir height within upper 15 cm soil layer [49, 77].

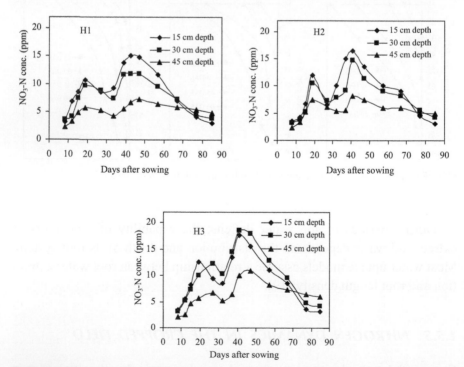

FIGURE 1.30 Temporal variation of NO_3^--N at different depth in the rice fields at various weir heights.

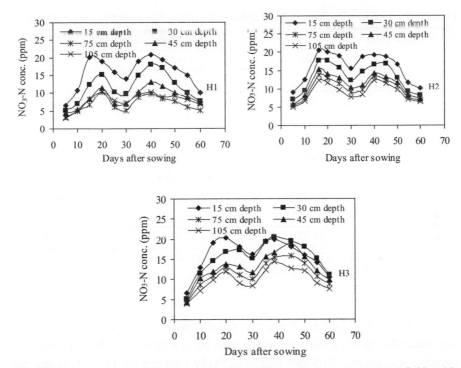

FIGURE 1.31 Temporal variation of NO_3-N at different depths in the mustard fields with the lined OFR at various weir heights.

In mustard, NO_3^-- N concentration was found increasing at a slow rate as compared to rice fields. Variation in NO_3^--N concentration in mustard is governed by the application if SI from the OFR [9]. The NO_3^--N concentration during the experiment was found as high as 20 ppm in all the treatments within upper 15 cm soil layer while it was 10 ppm at lower soil layer (Figure 1.31)

1.5.5.1   Nitrate Leaching

The leaching loss of NO_3^--N in the rice field was affected by distribution and amount of rainfall occurred during growing period. The leaching loss of NO_3^--N was 32.34, 41.41, 55.01 kg ha^{-1} in 2002; 13.91, 17.72, 25.97 kg ha^{-1} in 2003; and 18.20, 28.28, 38.17 kg ha^{-1} in 2004 with 0, 5, and 10 cm weir heights, respectively (Table 1.13). The leaching rate of NO_3^--N in the rice field was found higher in 10 cm than 0 cm weir height field as VP is higher

TABLE 1.13   Effect of Different Weir Heights on Leaching Losses of Nitrogen (kg ha^{-1})

Weir height (cm)	Year		
	2002	**2003**	**2004**
Rice			
0	30.04 d	12.61 d	16.72 e
Lined 5	39.10 cd	16.21 cd	26.03 cd
10	53.74 ab	24.01 ab	36.02 ab
0	34.64 d	15.21 d	19.68 e
Unlined 5	43.72 c	19.23 c	30.53 bc
10	56.28 a	27.93 a	40.32 a
Mustard			
—	15.17	13.08	8.87

Note: Within a column for each year, means (average of three replications) followed by the same letters are not significantly different according to LSD (0.05).

in 10 cm weir height field Nitrate is soluble and moves with the percolating water through the soil. Nitrogen loss by leaching is mainly as NO_3^--N because it is soluble in water and negatively charged [9, 11, 78]. From the mustard it was found to be 15.17, 13.08, 8.87 kg ha^{-1} with 0, 5, and 10 cm weir heights, respectively. The leaching loss of NO_3^--N was reduced with 0 cm weir height in the rice field.

1.5.6 NUTRIENT UPTAKE AND CROP YIELDS

Uniform rice yield was observed during the course of three years of experimentation when no SI was applied, but affected by the amount and distribution of rainfall (Table 1.14). The rice yields were found to be 4748–5455 kg ha^{-1} during 2002, which was higher over traditional method by 1000 -1500 kg ha^{-1}. This was due to better soil environment for root growth and nutrient uptake. In 2003, the rice yields were found to be 3867–4065 kg ha^{-1}, which was lower than 2002 mainly due to low and uneven distribution of rainfall.

Mustard was greatly affected due to the frequency of SI applied from the lined and unlined OFRs. The mustard yields were found to be 560–917 kg ha^{-1} in the field with the lined OFR while these were recorded 318–633 kg ha^{-1} in the field with the unlined OFRs.

TABLE 1.14 Effect of Different Weir Heights in Lined and Unlined OFR on Yield and Nitrogen Uptake by Rice and Mustard

Weir height (cm)	Yield (kg ha⁻¹)		Nitrogen uptake (kg ha⁻¹)	
	Rice	Mustard	Rice	Mustard
2002				
0	5455 a	832 a	122.74 a	30.78 a
Lined 5	5114 a	787 a	115.07 a	29.12 a
10	5191 a	772 ab	116.80 a	28.56 ab
0	5434 a	448 c	122.27 a	15.24 c
Unlined 5	4945 a	440 c	111.26 a	16.58 c
10	4748 a	412 c	106.83 a	16.28 c
2003				
0	4016 a	917 a	90.36 a	33.93 a
Lined 5	3977 a	858 a	89.48 a	31.75 a
10	4030 a	793 ab	90.68 a	29.34 ab
0	3859 a	633 c	86.83 a	23.42 bc
Unlined 5	3867 a	564 c	87.01 a	20.87 c
10	4065 a	552 c	91.46 a	20.42 c
2004				
0	5299 a	732 a	119.23 a	27.08 a
Lined 5	5574 a	577 b	125.42 a	21.35 b
10	5457 a	560 bc	122.78 a	20.72 bc
0	4811 a	441 d	108.25 a	16.32 d
Unlined 5	5108 a	402 de	114.93 a	14.87 de
10	5381 a	318 e	121.07 a	11.77 e

The variation in crop yield among different years can be explained by erratic distribution of rainfall. The yields of rice were not affected much by weir heights and lined/unlined system but affected significantly from year-to-year due to magnitude and distribution of rain [50]. Mustard yield obtained more (1.5 times) in lined than the unlined field. The variation in yield during different treatment in a particular year can be attributed due to more frequency of supplemental irrigation.

Nitrogen uptake is calculated from the yield and N concentration in plant shoot. Nitrogen uptake was found to be more in the lined field compared to unlined plots for both the crops. Total nitrogen removal from the rice-mustard cropping system was highest in 2002 followed by 2004 and 2003. Crop-wise rice removed more nitrogen as compared to mustard, which was due to its high internal requirement of nitrogen.

The nitrogen uptake in rice field was influenced by variation in crop yields obtained as a result of distribution pattern of rainfall. While in mustard field, it was affected to a great extent by different yields as influenced by various weir heights of the lined and unlined OFR. The minimum and maximum nitrogen uptake by rice during three growing seasons was observed as 90 and 122 kg ha^{-1}, respectively. The nitrogen uptake by mustard was found as 20–30 kg ha^{-1} and 11–23 kg ha^{-1} with the lined and unlined OFRs, respectively. The variation in uptake of nitrogen by mustard was due to amount of water available in the OFR and the magnitude and frequency of supplemental irrigation applied.

1.5.7   ECONOMIC ANALYSIS OF OFR SYSTEM

The factors considered for the economic evaluation were initial investment, the OFR maintenance cost, land lease cost, irrigation cost, and annual returns from the OFR system. Initial investment in the lined and unlined OFR was Rs. 5788 (1 US $ = INR Rs. 48) (earthwork Rs. 2108, lining material Rs. 3680 and labor cost for lining Rs. 200) and Rs. 2108 (only earthwork), respectively.

Considering 12% bank interest rate and 25 years life span of the lined and unlined OFR, present worth value of total annual cost for the lined and unlined OFR was calculated. From the increased production of rice and mustard against the traditional rainfed condition (without the OFR), present worth value of the total return was estimated and net profit (NP), benefit cost ratio (BCR) and payback period (PBP) was calculated. Three years of economic analysis of the OFR system over traditional rainfed system reveal that the OFR systems are economically viable as BCR is more than one (Tables 1.15–1.17).

Average BCR for the lined and unlined OFR systems was found as 1.1 and 1.7, respectively, revealing the fact that rice yields are almost same in both the system, whereas mustard yields were higher in lined than the unlined OFR system. Lower BCR in the lined system is due to the higher initial cost incurred for lining material. The PBP was found around two years for unlined and 8 years for the lined OFR system.

TABLE 1.15 Economic Viability Study of OFR Systems for Farm Area of 800 m² in 2002

Particulars	Lined OFR			Unlined OFR		
	0 (cm)	5 (cm)	10 (cm)	0 (cm)	5 (cm)	10 (cm)
Returns obtained – A						
Increase of rice grain (Rs.)	462.35	351.86	376.81	455.54	297.11	233.28
Increase of mustard seeds (Rs.)	744.75	694.13	677.25	272.25	312.75	303.75
Increase of rice straw (Rs.)	46.23	35.19	37.68	45.55	29.71	23.33
Increase of mustard stover (Rs.)	14.30	13.33	13.00	5.23	6.00	5.83
Total returns (Rs.)	1267.63	1094.50	1104.75	778.58	645.57	566.19
Present worth of returns (Rs.)	8527.97	7363.25	7432.16	5237.85	4343.09	3809.03
Cost incurred – B						
Construction cost (Rs.)	2108	2108	2108	2108	2108	2108
Lining material cost (Rs.)	3680	3680	3680	0	0	0
Initial investment (Rs.)	5788	5788	5788	2108	2108	2108
Maintenance cost (Rs.)	115.76	115.76	115.76	42.16	42.16	42.16
SI to mustard applied (cm)	10	7	7	5	5	5
Cost of SI to mustard (Rs.)	43.2	30.24	30.24	21.6	21.6	21.6
Irrigation cost (Rs.)	43.2	30.24	30.24	21.6	21.6	21.6
Land lease cost (Rs.)	24	24	24	24	24	24
Total annual cost (Rs.)	182.96	170	170	87.76	87.76	87.76
Present worth of annual cost (Rs.)	1231	1144	1144	590	590	590
Total cost (Rs.)	7019	6932	6932	2698	2698	2698
Net present value (Rs.)	1509	432	500	2539	1645	1111
Benefit-cost ratio	1.22	1.06	1.07	1.94	1.61	1.41
Payback period (years)	4.65	16.06	13.85	1.06	1.64	2.43

TABLE 1.16 Economic Viability Study of OFR Systems for Farm Area of 800 m² in 2003

Particulars	Lined OFR			Unlined OFR		
	0 (cm)	5 (cm)	10 (cm)	0 (cm)	5 (cm)	10 (cm)
Returns obtained – A						
Increase of rice grain (Rs.)	318.49	305.86	323.03	267.62	270.22	334.37
Increase of mustard seeds (Rs.)	817.88	751.50	678.38	498.38	420.75	407.25
Increase of rice straw (Rs.)	31.86	30.60	32.31	26.77	27.03	33.45

TABLE 1.16   Continued

Particulars	Lined OFR			Unlined OFR		
	0 (cm)	5 (cm)	10 (cm)	0 (cm)	5 (cm)	10 (cm)
Increase of mustard stover (Rs.)	15.70	14.43	13.02	9.57	8.08	7.82
Total returns (Rs.)	1183.93	1102.38	1046.74	802.34	726.08	782.88
Present worth of returns (Rs.)	7964.87	7416.25	7041.94	5397.74	4884.67	5266.84
Cost incurred – B						
Construction cost (Rs.)	2108	2108	2108	2108	2108	2108
Lining material cost (Rs.)	3680	3680	3680	0	0	0
Initial investment (Rs.)	5788	5788	5788	2108	2108	2108
Maintenance cost (Rs.)	115.76	115.76	115.76	42.16	42.16	42.16
SI to mustard (cm)	10	10	10	5	5	5
Cost of SI to mustard (Rs.)	43.2	43.2	43.2	21.6	21.6	21.6
Irrigation cost (Rs.)	43.2	43.2	43.2	21.6	21.6	21.6
Land lease cost (Rs.)	24	24	24	24	24	24
Total annual cost (Rs.)	182.96	182.96	182.96	87.76	87.76	87.76
Present worth of annual cost (Rs.)	1231	1231	1231	590	590	590
Total cost (Rs.)	7019	7019	7019	2698	2698	2698
Net present value (Rs.)	946	397	23	2699	2186	2568
Benefit-cost ratio	1.13	1.06	1.00	2.00	1.81	1.95
Payback period (years)	7.42	17.66	304.18	1.00	1.23	1.05

TABLE 1.17   Economic Viability Study of OFR Systems for Farm Area of 800 m² in 2004

Particulars	Lined OFR			Unlined OFR		
	0 (cm)	5 (cm)	10 (cm)	0 (cm)	5 (cm)	10 (cm)
Returns obtained – A						
Increase of rice grain (Rs.)	420.88	509.98	462.35	327.56	423.79	512.24
Increase of mustard seeds (Rs.)	609.75	435.38	382.50	282.38	238.50	144.00
Increase of rice straw (Rs.)	42.09	51.00	46.23	32.76	42.38	51.22
Increase of mustard stover (Rs.)	11.71	8.36	7.34	5.42	4.58	2.76
Total returns (Rs.)	1084.42	1004.71	898.43	648.12	709.25	710.23
Present worth of returns (Rs.)	7295.42	6759.15	6044.15	4360.20	4771.47	4778.08

TABLE 1.17   Continued

Particulars	Lined OFR			Unlined OFR		
	0 (cm)	5 (cm)	10 (cm)	0 (cm)	5 (cm)	10 (cm)
Cost incurred – B						
Construction cost (Rs.)	2108	2108	2108	2108	2108	2108
Lining material cost (Rs.)	3680	3680	3680	0	0	0
Initial investment (Rs.)	5788	5788	5788	2108	2108	2108
Maintenance cost (Rs.)	115.76	115.76	115.76	42.16	42.16	42.16
SI to mustard (cm)	10	7	7	5	5	5
Cost of SI to mustard (Rs.)	43.2	30.24	30.24	21.6	21.6	21.6
Irrigation cost (Rs.)	43.2	30.24	30.24	21.6	21.6	21.6
Land lease cost (Rs.)	24	24	24	24	24	24
Total annual cost (Rs.)	182.96	170	170	87.76	87.76	87.76
Present worth of annual cost (Rs.)	1231	1144	1144	590	590	590
Total cost (Rs.)	7019	6932	6932	2698	2698	2698
Net present value (Rs.)	277	-173	-888	1662	2073	2080
Benefit-cost ratio	1.04	0.98	0.87	1.62	1.77	1.77
Payback period (years)	25.38	--	--	1.62	1.30	1.30

1.6   CONCLUSIONS

The present investigation was undertaken to assess the influence of runoff and the OFR induced irrigation on soil water regimes and NO_3^--N dynamics under rainfed upland rice (*Oryzasativa* L.) and irrigated mustard (*Brassica campestris* L.). The movement of soil water under rice and mustard crops was simulated using HYDRUS-2D model. Brief description about the results obtained from the study is discussed in this section.

The impact of the on-farm reservoir with different weir heights on the dynamics of water and NO_3^--Nitrogen under variable soil water regime has been studied for rainfed upland rice based cropping system in eastern India. Based on three years of field experimental studies (2002–2004), the following conclusions have been drawn:

   1.   Calibration and validation of HYDRUS-2D model based on three years of field experiments resulted high coefficient of determination (more than 0.72), and low root mean square error (less than 0.032) revealing

its suitability to predict soil water dynamics in the effective root-zone of rainfed rice and mustard under varying saturated condition.

2. In rice fields, soil water content was found cyclic due to the variation of rainfall. However, no supplemental irrigation was applied to rice as soil water content was not depleted 40% below saturation during critical growth stage while in mustard; the variation in soil water content in the root-zone affects the availability of water and nutrient and ultimately impact on crop yields.

3. Vertical percolation (VP) loss in the rice field with 0 cm weir height (without allowing any standing water) is found to be 4.34 and 8.82% less than the fields with 5 and 10 cm weir heights, respectively. The rice field without standing water may be adopted to minimize VP losses.

4. Seepage loss from the unlined OFR is found to be 2 and 4% more for 0 cm weir height field than that of 5 and 10 cm weir heights, respectively. This indicates the need to adopt technology to minimize seepage losses from the unlined OFR. In other words, the unlined OFR enhances the scope for groundwater recharge to the tune of 79, 78 and 76% of total outflow with 0, 5 and 10 cm weir heights, respectively.

5. The leaching loss of NO_3^--N in the rice field was reduced to 50% with 0 cm weir height as compared to 10 cm weir height without affecting yields. The leaching loss of NO_3^--N in mustard field with the lined OFR was 50% higher than the unlined system due to more frequency of irrigation.

6. The nitrogen uptake by rice during its growth period was affected by rainfall amount and its distribution pattern. The variation in uptake of nitrogen by mustard was due to amount of water available in the OFR and the frequency of supplemental irrigation applied.

7. The rice yields were not affected by weir heights so rice cultivation without weir height is a feasible option. The mustard yields obtained were 1.5 times more in lined than the unlined OFR system as governed by quantity and frequency of supplemental irrigation.

8. Cost benefit ratio (BCR) for the unlined OFR system was found to be 1.7 with payback period (PBP) of around two years while for the lined OFR system the BCR is 1.1 with PBP of 8 years knowing the fact that rice yields are almost same and mustard yields were 1.5 times higher than the unlined OFR system. The difference in BCR was due to higher initial cost of investment incurred towards lining material.

1.7   SUMMARY

Calibration and validation of HYDRUS-2D model
Simulation for soil water content of 7 soil layers for mustard and 3 for rice (15 cm increment) fields was carried out with respect to observation nodes in the model domain and field observation points. The model simulated parameters for rice and mustard fields were tested with the field observations of the year 2002–04. The optimized model calibrated parameters for each layer of soils were determined using inverse modeling with 95% confidence intervals.

The slope of the regression line is close to unity, which further indicates good agreement between the observed and simulated soil water content indicate that HYDRUS-2D model is efficient in predicting daily variation of soil water content in the root-zone of mustard in the rainfed ecosystem. In addition to scattered and regression presentation, the performance of HYDRUS-2D model was also evaluated using error statistics such as prediction efficiency (PE) and root mean square error (RMSE). The average value of PE and RMSE for simulated soil water content in rice field was found as 0.9467 and 0.0235, respectively. The high value of R^2, PE (more than 0.90), and low value of RMSE (less than 0.05) indicates that the HYDRUS-2D model is quite efficient in predicting daily variation of soil water content in the cropped field with variably saturated condition.

1.7.1   WATER DYNAMICS IN THE CROP FIELD

In rice fields variation of soil water content in different soil layers of crop root-zone was found to have cyclic experience due to variation in rainfall. However, during the period when there was no rainfall or no supplemental irrigation is applied, soil water content was found to decline gradually because of uptake of water by the plant roots and vertical percolation from the root-zone. During the mustard growing period, soil water content in different soil layers of crop root zone depths was found varying due to supplemental irrigation over a period of time. The variation in soil water content in upper three layers of crop root-zone depths (0–45 cm) was found higher than the lower depths (45–105 cm) during three years of experiment. The variation in soil water content in upper layers of crop root-zone depths affects the availability of water and nutrient to mustard and ultimately affects crop yields.

The variation in ponding depth has shown cyclic trend depending on the rainfall intensity and duration during the rice-cropping season. More number of days of ponding was observed in 10 cm in comparison to 5 cm weir heights. During three years of experiment on winter mustard, it was observed that pre-sowing irrigation (PSI) was not required for the germination of mustard because residual soil water content was not depleted below 75% of available water. However, supplemental irrigations were applied during mustard growing period.

Vertical percolation (VP) losses in the rice field with 0 cm weir height were simulated using HYDRUS-2D model under variably saturated condition, whereas in case of 5 and 10 cm weir heights (ponding condition) Darcy's approach was used. Three years of experiments revealed that minimum loss of the VP in the rice fields with 0 cm weir height (184.44 mm) than the field with 5 cm weir height (232.27 mm) and 10 cm weir height (281.43 mm). It indicates that 0 cm weir height in the rice field can minimize 47.83 and 96.99 mm loss of VP in comparison to 5 and 10 cm weir height, respectively. Out of three years average rainfall (1095 mm) during the rice growing and turn-in period, the percentage of average total VP loss found from the field with 0, 5, and 10 cm weir heights were 16.38, 20.72 and 25.20, respectively. It indicates that the fields with 0 cm weir height can conserve 4.34 and 8.82% of runoff in the OFR, which was supposed to loss as VP. So, the rice field without standing water (0 cm weir height) may be adopted to minimize these major losses and in other words, fields with standing water may enhance groundwater recharge processes. In mustard field vertical percolation was found as 57 mm.

Actual crop evapotranspiration (AET) is one of the important parameters of water balance that influence the soil water status of rainfed crop fields. Higher AET of rice was observed with 10 cm weir height in comparison to 0 and 5 cm weir heights because of relatively more days of ponding. Three years average observed total AET of rice under 0, 5, and 10 cm weir heights was found, respectively, as 44.21, 48.16, and 48.20% of average total rainfall (905 mm). Total AET of rice at 0 cm weir height was 6 and 7% less than the AET at 5 and 10 cm weir heights because of variably saturated condition.

The variations in total surface runoff (RO) are found only due to rainfall intensity and weir heights. The percentage seasonal (120 days) surface runoff generated from rainfall in the rice field with 0, 5, and 10 cm weir heights are 27.57, 25.43, and 24.33 for 2002; 13.15, 5.90, and 2.23 for 2003; 9.65, 5.21, and 0.06 for 2004, respectively. However, there was no surface runoff observed from the mustard field in all the three years of experiments.

1.7.2   WATER BALANCE IN THE OFR

For sustainable farming, it is imperative to observe the inflow and outflow components of the OFR system using cropped field. The actual capacity of the OFR (120 m^3) used for field experiment is not the same as the simulated capacity. Average observed volume of water storage in the OFR was found less in comparison to the simulated value, which may be due to the seepage through the lining material as a result of natural and/or manmade damage.

Rainfall constituted the major components of total inflows. Direct rainfall contribution to the OFRs during rice and mustard growing seasons are, respectively, 101.98 and 4.53 m^3 in 2002; 77.71 and 8.48 m^3 in 2003; and 82.95 and 0.32 m^3 in 2004.

Out of total volume of inflows to the OFR during rice-growing season, the highest percentage of surface runoff (RO) contribution to the OFR with 0, 5, and 10 cm weir heights was due to relatively more seasonal rainfall in 2002 (1175 mm) in comparison to 2003 and 2004. Three years average percentage of RO contributed to the OFR with 0, 5, and 10 cm weir heights was 67.03, 61.60, and 55.18, respectively. During mustard growing season, there was negligible quantity of rainfall for three years, which did not generate any runoff.

The outflows from the lined OFR during rice growing season are evaporation loss and supplemental irrigation. Since, supplemental irrigation for rice was not required during the experimental years, so only outflow from the lined OFRs was evaporation loss. Evaporation loss from the lined OFR was observed to be more in 2002, than that of 2003 and 2004 because of more volume of water storage that leads to larger water surface area. However, in the unlined OFRs with 0, 5, and 10 cm weir heights, the outflow components are evaporation, supplemental irrigation, and seepage. Average evaporation loss from the unlined OFR with 0, 5, and 10 cm weir heights during the rice-growing season was 21.52, 23.23, and 24.91% of the total outflow, respectively. The percentage of evaporation loss from the OFR was decreased with the increase in weir heights due to the decrease in OFR water storage that leads to reduction in water surface area. Whereas during the mustard growing season, total evaporation loss from the carry-over water volume of the lined OFR with 0, 5, and 10 cm weir heights was, respectively, 23.35, 21.98, and 20.50 m^3 in 2002; 20.83, 19.20, and 18.22 m^3 in 2003 and 17.57, 15.82, and 13.82 m^3 in 2004. Similarly, evaporation loss from the unlined OFRs with 0, 5, and 10 cm weir heights was, respectively, 10.55,

10.02, and 9.36 m³ in 2002; 11.68, 11.21, and 10.90 m³ in 2003; and 8.47, 7.63, and 6.41 m³, in 2004.

During three years of experiments, supplemental irrigation to rice and pre-sowing irrigation (PSI) to mustard was not applied because soil water content not depleted below 75% of available soil water content (ASW). However, first supplemental irrigation of 5 cm (equivalent volume of the OFR water = 36 m³)was applied to mustard on 37, 47 and 37 days after sowing (DAS) from the lined and unlined OFR and second at 52, 58 and 52 DAS from the lined OFR only during the year 2002, 2003 and 2004, respectively, of all the weir heights. It was observed that there was negligible quantity of water available in the unlined OFR after giving one supplemental irrigation to mustard.

In the lined OFRs, the seepage was considered negligible. However, average seepage (SP) loss from the unlined OFRs with 0, 5, and 10 cm weir heights during the rice-growing season was 78.48, 76.77 and 75.09% of total water outflow, respectively. Whereas, during the mustard growing season, the SP loss from the total carry-over water volume of the unlined OFR with 0, 5, and 10 cm weir heights was 58.91, 57.23 and 54.98% of the total water outflow.

1.7.3 ROOT GROWTH AND NITROGEN DYNAMICS IN THE FIELD

For nutrient-use efficiency in rainfed ecosystems, the special challenge is the dynamic of the water regime. Root development plays critical role in this regard. The different weir heights does not affect the root growth in rice but in mustard weir heights affect the root growth and its density. The supplemental irrigation from the lined OFR with 0 cm weir height leads more soil water content at upper depth of the soil layer. Due to high water content at upper soil layer root length density was found more in 0 cm as compared to 10 cm weir height. It is evidence that with less irrigation the root tends to grow deeper in the soil profile. During the periods of soil water deficits, the capability of crop roots to extract soil water depends on the distribution and depth of its root system.

For rainfed upland rice as compared to other agricultural systems, it is not clear to what extent rice yields are limited by nutrients, water, and the interactions between them, over the diverse soil types, cultural practices and seasonal conditions. The initial concentration of NO_3^--N in the soil profile ranged between 3.5 and 4.5 ppm, respectively, at the time of rice and mustard sowing. After the application of first dose of fertilizer, the NO_3^--N concentration found increasing noticeably with different weir heights. Leaching losses of NO_3^--N in the rice field is more due to more water storage in 10 than 0 cm weir heights. The NO_3^--N concentration during the experiment found as high as 20 ppm in 10 cm weir height within upper 15 cm soil layer. In mustard, NO_3^-- N concentration was found increasing at a slow rate as compared to rice fields. Variation in NO_3^--N concentration in mustard is governed by the application if supplemental irrigation from the OFR. The NO_3^--N concentration during the experiment was found as high as 20 ppm in all the treatments within upper 15 cm soil layer while it was 10 ppm at lower soil layer.

The leaching loss of NO_3^--N in the rice field was affected by distribution and amount of rainfall occurred during growing period. The leaching rate of NO_3^--N in the rice field was found higher in 10 cm than 0 cm weir height field due to high vertical percolation in 10 cm weir height field. The leaching loss of NO_3^--N was 32.34, 41.41, 55.01 kg ha^{-1} in 2002; 13.91, 17.72, 25.97 kg ha^{-1} in 2003; and 18.20, 28.28, 38.17 kg ha^{-1} in 2004 with 0, 5, and 10 cm weir heights, respectively. From the mustard it was found to be 15.17, 13.08, 8.87 kg ha^{-1} with 0, 5, and 10 cm weir heights, respectively. The leaching loss of NO_3^--N was reduced with 0 cm weir height in the rice field.

1.7.4 CROP YIELDS AND ECONOMIC ANALYSIS

Uniform rice yield was observed during the course of three years of experimentation when no supplemental irrigation was applied, but affected by the amount and distribution of rainfall. The rice yields were found to be 4748–5455 kg ha^{-1} during 2002, which was higher over traditional method by 1000 -1500 kg ha^{-1}. This was due to better soil environment for root growth and nutrient uptake. In 2003, the rice yields were found to be 3867–4065 kg ha^{-1}, which was lower than 2002 mainly due to low and uneven distribution of rainfall.

Mustard yield was greatly affected due to the frequency of supplemental irrigation applied from the lined and unlined OFRs. The mustard yields were found to be 560–917 kg ha^{-1} in the field with the lined OFR while these were recorded 318–633 kg ha^{-1} in the field with the unlined OFRs. Mustard yield obtained more (1.5 times) in lined than the unlined field. The variation in yield during different treatment in a particular year can be attributed due to more frequency of supplemental irrigation.

Total nitrogen removal from the rice-mustard cropping system was highest in 2002 followed by 2004 and 2003. Rice removed more nitrogen as compared to mustard, which was due to its high internal requirement of nitrogen. The nitrogen uptake in rice field was influenced by variation in crop yields obtained as a result of distribution pattern of rainfall. While in mustard field, it was affected to a great extent by different yields as influenced by various weir heights of the lined and unlined OFR. The minimum and maximum nitrogen uptake by rice during three growing seasons was observed as 90 and 122 kg ha^{-1}, respectively. The nitrogen uptake by mustard was found as 20–30 kg ha^{-1} and 11–23 kg ha^{-1} with the lined and unlined OFRs, respectively. The variation in uptake of nitrogen by mustard was due to amount of water available in the OFR and the magnitude and frequency of supplemental irrigation applied.

The factors considered for the economic evaluation were initial investment, the OFR maintenance cost, land lease cost, irrigation cost, and annual returns from the OFR system. Initial investment in the lined and unlined OFR was Rs. 5788 (1 US $ = INR Rs. 48) (earthwork Rs. 2108, lining material Rs. 3680 and labor cost for lining Rs. 200) and Rs. 2108 (only earthwork), respectively. Considering 12% bank interest rate and 25 years life span of the lined and unlined OFR, present worth value of total annual cost for the lined and unlined OFR was calculated. From the increased production of rice and mustard against the traditional rainfed condition (without the OFR), present worth value of the total return was estimated and net profit (NP), benefit cost ratio (BCR) and payback period (PBP) was calculated.

Average BCR for the lined and unlined OFR systems was found as 1.1 and 1.7, respectively, revealing the fact that rice yields are almost same in both the system, whereas mustard yields were higher in lined than the unlined OFR system. Lower BCR in the lined system is due to the higher initial cost incurred for lining material. The PBP was found around two years for unlined and 8 years for the lined OFR system.

ACKNOWLEDGEMENT

The financial assistance provided under National Agriculture Technology Project by Indian council of Agriculture Research, New Delhi, India to carry out research work is highly acknowledged

KEYWORDS

- Alfisols
- Benefit cost ratio
- Critical growth stage
- Crop yield
- Cropping system
- Denitrification
- Eastern India
- Economic analysis
- Evapotranspiration
- Global warming
- Groundwater contamination
- HYDRUS-2D
- Lateritic soil
- Lining material
- Mustard
- Net profit
- Nitrate leaching
- Nitrogen dynamics
- Nitrogen modeling
- Nutrient uptake
- On-farm reservoir
- Pay-back period
- Ponding depth
- Pre-sowing irrigation
- Present worth analysis
- Rainfall
- Rainfed ecosystem
- Rainfed rice
- Root length
- Root length density
- Runoff
- Sandy loam

- Seepage, Percolation
- Soil erosion
- Soil water balance
- Soil water dynamics
- Soil water modeling
- Soil water movement
- Soil water regimes
- Solute transport
- Submergence
- Supplemental irrigation
- Upland crop
- Vadose zone
- Variably-saturated soil
- Water harvesting
- Water stress
- Weir height

REFERENCES

1. Abbasi, F., J. Feyenand, and M.Th. van Genuchten (2004). Two-dimensional simulation of water flow and solute transport below furrows: model calibration and validation. *J. Hydrol.*, 290, 63–79.

2. Abraham, N., and K. N. Tiwari (1999). Modeling hydrological processes in hill slope watershed of humid tropics. *J. Irrig. and Drain. Eng. (ASCE),* 125, 203–211.

3. Addiscott, T. M. and A. P. Whitmore (1991). Simulation of solute leaching in soils of differing permeabilities. *Soil Use Manage.,* 7, 94–102.

4. Adisarwanto, T. and Suhartina (1994). The response of three grain legumes to different soil management practices after wetland rice. In: *Proc. ACIAR 8938, Malang – Indonesia, 15–19 June.* Univ. of Brawijaya, Malang, Indonesia.

5. Agrawal, A. (2000). *Optimal design of on-farm reservoir for paddy-mustard cropping pattern using water balance approach.* MTech. Thesis, Indian Institute of Technology, Kharagpur, India.

6. Agrawal, M. K., S. N. Panda, and B. Panigrahi (2004). Modeling water balance parameters for rainfed rice. *J. Irrig. and Drain. Engrg. (ASCE)*, 130,129–139.

7. Ahmed, H. U., M. L. Ali, S. K. Zaman, M. A. Kabir, and N. M. Miah (1996).Varietal characteristics and soil management to reduce drought stress, 150–167 pages. In: *V. P. Singh et al. (ed.) Physiology of stress tolerance in rice. Proc. Int. Conf. stress physiology of rice, 28 Feb.-5 March,* 1994. Lucknow, India.

8. Allen, R. G., L. S. Pereira, D. Raes and M. Smith (1998). *Crop evapotranspiration. guidelines for computing water requirements*. FAO Irrigation and Drainage Paper 56, FAO, Rome, 300 pages.

9. Aulakh, M. S., and B. Singh (1997). Nitrogen losses and fertilizer N use efficiency in irrigated porous soils. *Nutr. Cycl. Agroecozyst.*, 7, 1–16.

10. Ayala, S (2002). Nitrogen in crop production -Tough to curb liability. *Curr. Sci.*, 82(9),1067–1068.

11. Behera, S. K., and R. K. Panda (2009). Effect of fertilization and irrigation schedule on water and fertilizer solute transport for wheat crop in a sub-humid sub-tropical region. *Agric. Ecozyst. and Environ.*, 130, 141–155.

12. Belder, P., B. A. M. Bouman, and J. H. J. Spiertz (2007). Exploring options for water savings in lowland rice using a modeling approach. *Agric. Syst.*, 92,91–114.

13. Biswas, B., D. C. Ghosh, M. K. Dasgupt, N. Trivedi, J. Timsina, and A. Dobermann (2006). Integrated assessment of cropping systems in the Eastern Indo-Gangetic plain. *Field Crops Res*. 99,35–47.

14. Black, C. A. (1968). *Soil-Plant Relationships*. John Wiley and Sons, Inc., New York.

15. Bray R. H., and L. T. Kurtz (1945). Determination of total, organic and available form of phosphorus in soil. *Soil Sci.*, 59,360–361.

16. Cai, X. L., and B. R. Sharma (2010). Integrating remote sensing, census and weather data for an assessment of rice yield, water consumption and water productivity in the Indo-Gangetic river basin. *Agric. Water Manage.*, 97, 309–316.

17. Cairns, J. E., A. Audebert, C. E. Mullins, A. H. Price (2009). Mapping quantitative trait loci associated with root growth in upland rice (Oryza sativa L.) exposed to soil water-deficit in fields with contrasting soil properties. *Field Crops Res.*, 114, 108–118.

18. Chen, S. K., C. W. Liu, and H. C. Huang (2002). Analysis of water movement in paddy rice fields (II) simulation studies. *J. Hydrol.*, 268,259–271.

19. Dardanelli, J. L., J. T. Ritchie, M. Calmon, J. M. Andriani, and D. J. Collino (2004). An empirical model for root water uptake. *Field Crops Res.*, 87,59–71.

20. Dinnes, D. L., D. L. Karlen, D. B. Jaynes, T. C. Kaspar, J. L. Hatfield, T. S. Colvin, and C. A. Cambardella (2002). Nitrogen management strategies to reduce nitrate leaching in tile-drained Midwestern soils. *Agron. J.* 94,153–171.

21. Dobermann, A., C. Witt, D. Dawe, G. C. Gines, R. Nagarajan, S. Satawathananont, T. T. Son, P. S. Tan, G. H. Wang, N. V. Chien, V. K. T. Thoa, C. V. Phung, P. Stalin, P. Muthukrishnan, V. Ravi, M. Babu, S. Chatuporn, M. Kongchum, Q. Sun, R. Fu, G. C. Simbahan, and M. A. A. Adviento (2002). Site-specific nutrient management for intensive rice cropping systems in Asia. *Field Crops Res.*, 74,37–66.

22. Doorenbos, J., and A. H. Kassam (1979). *Yield Response to Water*. FAO Irrigation and Drainage Paper 33, FAO, Rome.

23. Doorenbos, J., and W. O. Pruitt (1977). *Guidelines for Predicting Crop Water Requirements*. FAO Irrigation and Drainage Paper 24, FAO, Rome.

24. Ekanayake, I. J., P. L. Steponkus, and S. K. De Datta (1989). Spikelet sterility and flowering response of rice to water stress at anthesis. *Ann. Bot.*, 63,257–264.

25. Feddes, R. A., P. J. Kowalik, and H. Zardny (1978). *Simulation of field water use and crop yield*. John Wiley & Sons, New York.

26. Flury, M. (1993). *Transport of bromide and chloride in a sandy and a loamy field soil*, 136 pages. No. 10185. Ph. D. diss., ETH.

27. Fukai, S., and M. Cooper (1995). Review. Development of drought-resistant cultivars using physio-morphological traits in rice. *Field Crops Res.*, 40,67–86.

28. Fusheng Li, Jiangmin Yu, Mengling Nong, Shaozhong Kang, and Jianhua Zhang (2010). Partial root-zone irrigation enhanced soil enzyme activities and water use of maize under different ratios of inorganic to organic nitrogen fertilizers. *Agric. Water Manage.*, 97, 231–239.

29. George, T., J. K. Ladha, D. P. Garrity, and R. J. Buresh (1994). Legumes as nitrate catch crops during the dry-to-wet season transition in lowland rice-based cropping systems. *Agron. J.*, 86, 267–273.

30. Goswami, N. N., S. K. De Datta, and M. V. Rao (1986). Soil fertility and fertilizer management for rainfed lowland rice, 275–283 pages. In: *Progress in Rainfed Lowland Rice.* IRRI, Manila, Philippines.

31. Haefele, S. M., and Y. Konboon (2009). Nutrient management for rainfed lowland rice in northeast Thailand. *Field Crops Res.*, 114, 374–385.

32. Hanson, A. D., W. J. Peacock, L. T. Evans, C. J. Arntzen, and G. S. Khush (1990). Drought resistance in rice. *Nature*, 345, 26–27.

33. Holden, Nicholas M., Andrew J. Rook and David Scholefield (1996). Testing the performance of a one-dimensional solute transport model (LEACHC) using response surface methodology. *Geoderma*, 69, 157–173.

34. Home, P. G., R. K. Panda, and S. Kar (2002). Effect of method and scheduling of irrigation on water and nitrogen use efficiencies of okra. *Agric. Water Manage.*, 55, 159–170.

35. Hutson, J. L. and R. J., Wagenet (1992). *Leaching Estimation And Chemistry Model.* A process based model of water and solute movement, transformations plant uptake and chemical reactions in the unsaturated zone. Water Resources Institute, Cornell University, Ithaca.

36. Idike, F. I., C. L. Larson, and D. C. Slack (1982). Modeling soil moisture and effects of basin tillage. *Trans. ASAE*, 25, 1262–1267.

37. Inoue, M., J. Šimůnek, S. Shiozawa, and J. W. Hopmans (2000). Simultaneous estimation of soil hydraulic and solute transport parameters from transient infiltration experiments. *Adv. Water Resour.*, 23, 677–688.

38. Inthapan, P., and S. Fukai (1988). Growth and yield of rice cultivars under sprinkler irrigation in south-eastern Queensland. 2. Comparison with maize and grain sorghum under wet and dry conditions. *Aust. J. Exp. Agric.*, 28, 243–248.

39. Jackson, M. L. (1973). *Soil Chemical Analysis*, Prentice Hall of India Pvt. Ltd., New Delhi.

40. Jacques, D., D. J. Kim, J. Diels, J. Vanderborght, H. Vereecken, and J. Feyen (1998). Analysis of steady-state chloride transport through two heterogeneous field soils. *Water Resour. Res.*, 34, 2539–2550.

41. Jacques, D., J. Šimůnek, A. Timmerman, and J. Feyen (2002). Calibration of Richards' and convection-dispersion equations to field-scale water flow and solute transport under rainfall conditions. *J. Hydrol.*, 259, 15–31.

42. Jarvis, N. J. (1989). *CRACK: a model of water and solute movement in cracking clay soils.* Report 159, Swedish University of Agricultural Sciences, Department of Soil Science, Uppsala, Sweden.

43. Jarvis, N. J. (1991). *MACRO – a model of water movement and solute transport in macroporous soils.* Swedish University of Agricultural Sciences, Department of Soil Science, Reports and Dissertations No. 9, Uppsala, Sweden.

44. Jensen, J. R., S. M. A. Mannan, and S. M. N. Uddin (1993). Irrigation requirement of transplanted monsoon rice in Bangladesh. *Agric. Water Manage.*, 23, 199–212.

45. Kar, G., and H. N. Verma (2005). Climatic water balance, probable rainfall, rice crop water requirements and cold periods in AER 12.0 in India. *Agric. Water Manage.* 72, 15–32.

46. Kirchhof, G., H. B. Soa, T. Adisarwantob, W. H. Utomoc, S. Priyonoc, B. Prastowod, M. Basird, T. M. Landod, Subandid, E. V. Dacanaye, D. Tan-Elicanoe, and W. B. Sani-dade (2000). Growth and yield response of grain legumes to different soil management practices after rainfed lowland rice. *Soil Till. Res.*, 56, 51–66.

47. Kumar, Arvind, Satish Verulkar, Shalabh Dixit, Bhagirath Chauhan, Jerome Bernier, Ramaiah Venuprasad, Dule Zhao, and M. N. Shrivastava (2009). Yield and yield-attributing traits of rice (Oryza sativa L.) under lowland drought and suitability of early vigor as a selection criterion. *Field Crops Res.*, 114, 99–107.

48. Lin, H. C., D. R. Richards, G. T. Yeh, J. R. Cheng, H. P. Cheng, and N. L. Jones (1996). *FEMWATER: a three-dimensional finite element computer model for simulating density dependent flow and transport.* Technical Report, HL-96.

49. Liu, C. W., S. K. Chen, S. W. Jou, and S. F. Kuo (2001). Estimation of the infiltration rate of a paddy field in Yun-Lin, Taiwan. *Agric. Sys.* 68, 41–54.

50. Lu, G., R. Cabangon, T. P. Tuong, P. Belder, B. A. M. Bouman, and E. Castillo (2002). The effects of irrigation management on yield and water productivity of inbred, hybrid, and aerobic rice varieties. *Proceedings of a thematic workshop on water wise rice production, 8–11 April 2002* at IRRI headquarters in Los Baños, Philippines. In: *Water-wise rice production*, Edited by B. A. M. Bouman, H. Hengsdijk, B. Hardy, P. S. Indraban, T. P. Tuong, and J. K. Ladha, 22–35 pages.

51. Lucas, M. L., S. Pandey, E. O. Agustin, R. Villeno, T. F. Marcos, D. R. Culannay, P. C. Sta Cruz, and S. R. Obien (1999). Characterization and economic analysis of intensive cropping systems in rainfed lowlands of Ilocos Norte, Philippines. *Exp. Agric.*, 35, 211–224.

52. Mishra, P. K., C. A. Rama Rao, and S. Siva Prasad (1998). Economic evaluation of farm pond in a micro-watershed in semi-arid alfisoldeccan plateau. *Indian J. Soil Conserv.,* 26, 59–60.

53. Mualem Y. (1976). A new model for predicting the hydraulic conductivity of unsaturated porous media. *Water Resour. Res.*, 12, 513–22.

54. Newman, J. (1966). A method of estimating the total length of root in a sample. *J. Appl. Ecol.*, 3, 139–145.

55. Palmer, W. L., B. J. Barfield, and C. T. Haan (1982). Sizing farm reservoirs for supplemental irrigation of corn. Part II. economic analysis. *Trans. ASAE*, 15, 377–387.

56. Panigrahi, B., and S. N. Panda (2003a). Field test of a soil water balance simulation model. *Agric. Water Manage.*, 58, 223–240.

57. Panigrahi, B., and S. N. Panda (2003b). Optimal sizing of on-farm reservoirs for supplemental irrigation. *J. Irrig. and Drain. Engrg. (ASCE)*, 129, 117–128.

58. Panigrahi, B., S. N. Panda, and A. Agrawal (2005). Water balance simulation and economic analysis for optimal size of on-farm reservoir. *Water Resour. Manage.*, 19, 233–250.

59. Panigrahi, B., S. N. Panda, and R. Mull (2001). Simulation of water harvesting potential in rainfed rice lands using water balance model. *Agric. Syst.*, 69, 165–182.

60. Paulo, A. M., L. A. Pereira, J. L. Teixeira, and L. S. Pereira (1995). Modeling paddy rice irrigation. *Crop-water-simulation models in practice*, L. S. Pereira, B. J. van Broek, P. Kabat, and R. G. Allen, eds., WageningenPers, Wageningen, The Netherlands, 287–302 pages.

61. Radcliffe, D. E., P. M. Tillotson, P. F. Hendrix, L. T. West, J. E. Box, and E. W. Tollner (1996). Anion transport in a piedmont ultisol: I. Field-scale parameters. *Soil Sci. Soc. Am. J.*, 60, 755–761.

62. Samra, J. S., V. N. Sharda, and A. K. Sikka (2002). *Water harvesting and recycling Indian experiences*. ICAR, New Delhi, Indian.

63. Sanchez-Cohen, I., V. L. Lopes, D. C. Slack, and M. M. Fagel (1997). Water balance model for small-scale water harvesting systems. *J. Irrig. and Drain. Engrg. (ASCE)*, 123, 123–128.

64. Sander, T., and H. H. Gerke (2007). Preferential Flow Patterns in Paddy Fields Using a Dye Tracer. *Vadose Zone J.* 6, 105–115.

65. Schaap M. G., and M.Th. Van Genuchten (2005). A modified Mualem-van Genuchten formulation for improved description of the hydraulic conductivity near saturation. *Vadose Zone J.* 5, 27–34.

66. Shrestha, R. K., and J. K. Ladha (2000). Recycling of residual soil nitrogen in a lowland rice-sweet pepper cropping system. *Soil Sci. Soc. Am. J.*, 64, 1689–1698.

67. Šimůnek, J., and J. W. Hopmans (2002). Parameter estimation and nonlinear fitting. 139–157 pages. In: J. H. Dane and G. C. Topp (ed.) *Methods of Soil Analysis. Part 4. Physical Methods*, SSSA Book Series 5, SSSA, Madison, WI.

68. Šimůnek, J., M. Šejna, and M.Th. van Genuchten (1999). *The HYDRUS-2D Software Package for Simulating the Two dimensional Movement of Water, Heat, and Multiple Solutes in Variably Saturated Media, Version 2.0*, IGWMC-TPS-70, International Ground Water Modeling Center, Colorado School of Mines, Golden Colorado.

69. Singh B., Y. Singh, and G. S. Sekhon (1995). Fertilizer-N use efficiency and nitrate pollution of groundnut in developing countries. *J. Contam. Hydrol.*, 20, 167–184.

70. Snow, V. O., B. E., Clothier, D. R., Scotter, and R. E., White (1994). Solute transport in a layered soil: experiments and modeling using the convection-disperion approach. *J. Cont. Hydrol.* 16, 339–358.

71. Srivastava, R. C. (1996). Design of runoff recycling irrigation system for rice cultivation. *J. Irrig. and Drain. Engrg. (ASCE)*, 122, 331–335.

72. Subbiah B. V., and G. L. Asija (1956). A rapid procedure for determination of available nitrogen in soils. *Curr. Sci.*, 25, 259–260.

73. SWIM Mission Report (1997). *Water Saving Techniques in Rice Irrigation.* SWIM Mission to Guilin Prefecture, Guangxi Region, China.

74. Tabbal, D. F., R. M. Lampayan, and S. I. Bhuiyan (1992). Water-efficient irrigation technique for rice. 146–159 pages. In: *V. V. N. Murty and K. Koga (ed.) Soil and water engineering for paddy field management. Proc. Int. Worksh. soil and water engineering for paddy field management. 28–30 Jan. 1992.* AIT, Bangkok, Thailand.

75. Tennant, D. (1975). A test of a modified line intersect method of estimating root length. *J. Ecol.*, 63, 995–1001.

76. Toure, Amadou, Mathias Becker, David E. Johnson, BrahimaKone, Dansou K. Kossou, and Paul Kiepe (2009). Response of lowland rice to agronomic management under different hydrological regimes in an inland valley of Ivory Coast. *Field Crops Res.*, 114, 304–310.

77. Tournebize, J., H. Watanabe, K. Takagi, and T. Nishimura (2006). The development of a coupled model (PCPF-SWMS) to simulate water flow and pollutant transport in Japanese paddy fields. *Paddy Water Environ.*, 4, 39–51.

78. Tripathi, B. P., J. K. Ladha, J. Timsina, and S. R. Pascua (1997). Nitrogen dynamics and balance in intensified rainfed lowland rice-based cropping systems. *Soil Sci. Soc. Am. J.*, 61, 812–821.
79. vanGenuchten, M. Th. (1980). A closed-form equation for predicting the hydraulic conductivity of unsaturated soil. *Soil Sci. Soc. Am. J.*, 44, 892–898.
80. van Oosterom, E. J., A. K. Borrell, S. C. Chapman, I. J. Broad and G. T. Hammera (2010). Functional dynamics of the nitrogen balance of sorghum: I. N demand of vegetative plant parts. *Field Crops Res.*, 115, 19–28.
81. Verma, H. N., and P. B. S. Sarma (1990). Criteria for identifying effective monsoon for sowing in rainfed agriculture. *J. Irrig. Power,* 35, 177–184.
82. Walker, S. H., and K. R. Rushton (1984). Verification of lateral percolation losses from irrigated rice fields by a numerical model. *J. Hydrol.*, 71, 335–351.
83. Walkley, A., and C. A. Black (1934). Estimation of soil organic carbon by the chromic acid titration method. *Soil Sci.*, 37, 29–38.
84. Wiklander, L. (1977). Leaching of plant nutrient in soils. IV. Contents in drainage water and groundwater. *Acta Agric. Scand.*, 27, 175–189.
85. Yoshida, S., and S. Hasegawa (1982). The rice root system, its development and function. *Drought Resistance in Crops with Emphasis on Rice.* 97–114 pages. IRRI, Los Baños, Philippines.
86. Youzun Xu, LixiaoNie, Roland J. Buresh, Jianliang Huang, Kehui Cui, Bo Xu, Weihua Gong, and Shaobing Peng (2010). Agronomic performance of late-season rice under different tillage, straw, and nitrogen management. *Field Crops Res.* 115, 79–84.
87. Zhang, Limeng, Shan Lin, B. A. M. Bouman, Changying Xue, Fengtong Wei, Hongbin Tao, Xiaoguang Yang, Huaqi Wang, Dule Zhao, and Klaus Dittert (2009). Response of aerobic rice growth and grain yield to N fertilizer at two contrasting sites near Beijing, China. *Field Crops Res.*, 114, 45–53.
88. Zhang, Xiying, Suying Chen, Hongyong Sun, Yanmei Wang, and Liwei Shao (2009). Root size, distribution and soil water depletion as affected by cultivars and environmental factors. *Field Crops Res.*, 114, 75–83.
89. Zotarelli, L., M. D. Dukes, J. M. S. Scholberg, R. Munoz-Carpena, and J. Icerman (2009). Tomato nitrogen accumulation and fertilizer use efficiency on a sandy soil, as affected by nitrogen rate and irrigation scheduling. *Agric. Water Manage.*, 96, 1247–1258.

INNOVATIONS IN WATER CONSERVATION FOR RAINFED AREAS OF EASTERN INDIA

SUSAMA SUDHISHRI,[1] ANCHAL DASS,[2] and P. R. CHOUDHURY[3]

[1]Principal Scientist, Water Technology Center, ICAR-Indian Agricultural Research Institute (IARI), Pusa Campus, New Delhi-110012, India. Mobile: +91 9971931921, E-mail: susama_s@rediffmail.com

[2]Senior Scientist, Division of Agronomy, ICAR-Indian Agricultural Research Institute, Pusa Campus, New Delhi - 110012, India. E-mail: anchal_d@rediffmail.com

[3]Ex-Scientist (ICAR) and Development Researcher and Consultant, Plot No 786/1720, Sampur, PO: Ghatikia, Bhubaneswar – 751003, Odisha, India, E-mail: prchoudhury@rediffmail.com

CONTENTS

2.1   INTRODUCTION

India receives annual rainfall of 1200 mm (4000 km³), out of which 70–80% occurs during monsoon season. Potential irrigable area is 113.5 Mha with gross and net irrigated area being 75.14 and 54 Mha, respectively, out of which 53% and 40.5% area are fed by groundwater and surface water, respectively. Despite large area under irrigation, stress during crop growth is inevitable as out of 139.7 Mha net cultivated areas, 85.7 Mha still remains rainfed which faces different water management related problems *viz.* extreme spatial and temporal variability in rainfall; higher evaporative demand than rainfall, drought and water scarcity; greater exploitation of already depleted ground water resources; low rainwater use efficiency and low crop productivity; and poor ground water quality, and deterioration of soil health under changing climate. In other way also human population of India will rise to 1500 million by 2025 and 1800 million by 2050. By the same time, livestock population will be 600 million.

The food, green fodder and fuel wood requirements shall be 275 M tons, 1000 M tons, and 235 Mm³, respectively against present status of 259 M tons of food grain, 513 M tons of green fodder and 40 Mm³ of fuel wood. Each year 5 to 6 M tons of food grains shall have to be added to ensure food security in the country. The additional productions are to meet from the declining per capita cultivable land [24]. Table 2.1 shows the water needs

TABLE 2.1   Estimates of Water Needs for India (Mham)

Activity	1990	2000	2025
Domestic	2.5	3.3	5.2
Energy	1.9	2.7	7.1
Industrial	1.5	2.7	12.0
Irrigation	46.0	63.0	77.0
Others	3.3	3.5	3.7
Total	**55.2**	**75.2**	**105.0**

of India by 2025. Entire water potential (i.e., utilizable water) of 1122 km^3 needs to be developed by 2025 through all means of surface and ground water development and efficient management.

Agriculture consumes major quantity of water and is estimated that in 2010, total water withdrawal was 761 km^3 of which 91% (688 km^3), has been for irrigation, about 56 km^3 for municipal and 17 km^3 for industrial use [10]. Scenario analysis shows that approximately 7100 km^3 year^{-1} is consumed globally to produce food, of which 5500 km^3 year^{-1} is used in rainfed agriculture and 1600 km^3 year^{-1} in irrigated agriculture [8]. Limited availability of soil moisture at critical stages of crop growth is major constraint of rainfed agriculture in semi-arid regions [18]. Approximately 40% (600 M ha) of the world cropland area is affected by low and unpredictable rainfall, with 60% of these lands located in developing countries.

This chapter discusses innovations in water conservation for rainfed areas of Eastern India.

2.2   RAINFED AGRICULTURE AND ITS PROBLEMS

Population growth and economic development will lead to an increasing competition for scarce water resources. World population is projected to increase 35% by 2050, which will require a 70–100% rise in food production. Rainfed agriculture plays and will continue to play a dominant role in providing food and livelihoods to ever increasing world population. Rainfed agriculture covers 80% of the world's agricultural area. It produces 62% of staple food. Of the 1.5 billion ha of cropland worldwide, 1.223 billion ha (82%) is rainfed. These regions cover about 40% of the world's land area and host nearly 40% of the world's population. Further, about 70% of the world's staple food continues and will continue to be harvested from rain-fed areas, since the scope for further expansion of irrigation is limited due to growing competition for water and the high investment cost [11]. However, high-untapped productivity and income potential exists in these areas. Growing population numbers and changing consumption patterns in fast-developing economies are increasing global food demand [19].

In developing countries, rainfed grain yields average 1.5 tons ha^{-1}, compared with 3.1 tons ha^{-1} in irrigated agriculture. India ranks first among the countries that practice rainfed agriculture both in terms of extent and value of production. Out of the 329 Mha of total geographical area of the country, about 146 Mha is degraded and 85 Mha is rainfed arable land. Out of 140.3 Mha net cultivated area, 79.44 Mha (57%) is rainfed, contributing 44% of the total food grain production. It is estimated that even after achieving the full irrigation potential; nearly 50% of the net cultivated area will remain dependent on rainfall. Rainfed agriculture supports nearly 40% of India's estimated population of 1.21 billion. Cultivation of coarse cereals (91%), pulses (91%), oilseeds (80%) and cotton (65%) predominates in these rainfed regions [2]. Moreover, two-thirds of livestock and 40% of human population of the country live in rainfed regions.

The "green revolution" era had largely by-passed the rainfed agriculture. Subsequently, several development programs were initiated for improving rainfed farming. In the rainfed area, it is not the amount of rainfall that is the limiting factor of production; rather, it is the extreme variability of rainfall, few rain events with high rainfall intensities and poor spatial and temporal distribution of rainfall. Rainfed crop production, which uses infiltrated rainfall that forms soil moisture in the root zone, accounts for most of the crop water consumption in agriculture. In these regions yield gaps are large, not due to lack of water per se, but due to inefficient management of water, soils and crops. The core of achieving this prospect is resource conservation through watershed management in which rainwater management plays a pivotal role [24].

2.3 LAND DEGRADATION AND SCOPE FOR DEVELOPMENT

Natural resource management (NRM) based on scientific principles plays a crucial role for an inclusive and sustainable growth in India. The Approach Paper to the 12th Five-Year Plan aptly notes that 'Economic development will be sustainable only if it is pursued in a manner which protects the environment. With acceleration of economic growth, these pressures are expected to intensify, and we therefore, need to pay greater attention to the management of natural resources, viz. water, forests and land (*India's looming freshwater crisis. The Financial Express, 18 January 2007*). In other way also land degradation continues to be a threat to the food and nutrition security of the country.

According to the latest estimate by the Indian Council of Agricultural Research (ICAR), New Delhi and National Academy of Agricultural Sciences (NAAS), out of total geographical area of 328.73 Mha, about 120 Mha (37%) is affected by various kinds of land degradation [16]. According to the latest estimate, the primary causal factor of land degradation is water and wind erosion (94.87 Mha), followed by soil acidity (17.93 Mha), sodic soils (3.71 Mha), soil salinity (2.73 Mha), waterlogging (0.91 Mha), and mining and industrial waste (0.26 Mha). The topsoil is most vulnerable to erosion, the loss of which causes depletion in quantity as well as quality of soil. About 1 mm of topsoil is lost every year in the country due to soil (water) erosion at an average rate of 16.4 tons/ha/year against the permissible limit of 4.5–11.2 tons/ha/year. This results in over 5.3 billion tons of soil being lost annually due to soil erosion, causing the loss of around 8 million tons (mt) of plant nutrients every year (http://cssri.nic.in/).

Degradation is particularly severe in regions with sloping and hilly terrains and those affected by unsustainable land management practices such as shifting cultivation. The sloping and hilly regions of eastern India, called eastern Ghats region with a geographical area of 19.8 Mha [21] is such an erosion prone zone, having characteristic link of poor lands with people's poverty. For instance, the share of good quality soil in Odisha is one of the lowest, merely 10.4% of the land area of the state [13]. It also happens to be the most backward state of India with 46.4% of the population below poverty line.

Shifting cultivation is prevalent in the hill slopes of the region. However, reduction in restoration or fallow cycle from 15 to 20 years to the current level of 2–3 years due to population pressure resulted in reduced farm output and increased land denudation [14]. This shifted focus of the people to settled cultivation on the sloping and undulated uplands and medium lands, with average slope varying from 2 to 5% and characterized by coarse textured Alfisols. These lands are located downside the denuded hillocks and are the major alternatives for the predominantly subsistence agriculture practiced in this rainfed region of India. Low soil fertility and erosion due to overland flow from denuded hill slopes do not permit more than one crop per year and a crop yield of more than 1.0 tons ha^{-1} in these lands.

The Eastern Ghats are discontinuous range of mountains, hills and plateaus on the east coast of India, which occupy an area of 19.76 Mha between 77^0 22' and 85^0 20' E longitudes and 11^0 30' to 21^0 0' N latitudes and spread over the states of Odisha, Chhattisgarh, Andhra Pradesh, Tamil Nadu and Karnataka [17]. The region has predominance of tribal people (54 tribal communities) constituting about 30% of the total population of 37.9 M [1]. Koraput and its adjoining districts in Odisha, Andhra Pradesh and Chhattisgarh, situated in midst of this region, are quite vulnerable considering around fifty per cent tribal inhabitants along with a sensitive upland habitat plagued with problems of natural resource erosion. Land use practices like shifting cultivation, uncontrolled grazing, large-scale mining, faulty agriculture and over-exploitation of forests have resulted in severe degradation of the region's natural resource base. High silt production rates, i.e., 2.07 to 8.96 ham/100 sq. km in Koraput region, endangers not only sustainable agriculture but also life of reservoirs in the downstream [23].

Participatory and integrated watershed development approach is viewed as a developmental paradigm to transform the impoverished livelihoods and degraded landscapes in India. With the investment and attention to this developmental approach, it has become imperative to blend outputs of multidisciplinary research into the planning and implementation domains to transfer the intended benefits effectively to the people and their support systems. The shortcomings of available technological options with respect to their nature (outcomes of isolated soil and water conservation research lacking watershed approach) or applicability (legacy of alien environments or recommendations emanating from controlled conditions) or adoptability (formulated which lacks appreciation of the needs of local conditions and people, for which they are intended) points to the need of watershed based interactive research [3]. In ecologically sensitive and economically poor, tribal dominated Eastern Ghats of India, lack of research initiatives in the field of NRM in general and the watershed management in particular poses multiple roadblocks in augmenting productivity for *in-situ* development as well as in rehabilitating these deteriorating upper catchment ecosystems for downstream services. Therefore, innovations in developing different criteria for development, implementation, monitoring, evaluation and adoptability of resources conservation at different topo-sequences were brought through two model watersheds, i.e., Kokriguda and Malipungar in Koraput district of Odisha representing the eastern Ghats shown in Figure 2.1.

2.4   WATER RESOURCE DEVELOPMENT AND CONSERVATION TECHNIQUES

Ever declining per capita land and freshwater availability, soil erosion and land degradation are posing serious threat to food, economic and ecological security of India. Prudent use and management of these basic natural resources and effective conservation practices are the issues of prime importance for the sustainable development of the country. Among various approaches, integrated watershed management could reverse the process of land degradation, conserve water and ensure sustainability of productivity. This approach comprises of appropriate planning of natural resources, especially land, water and vegetation to sub-serve the socio-economic and community needs of the society or the community concerned [9]. Since poverty is strongly correlated with regions prone to drought, deserts, hilly areas and regions of rainfed agriculture, the focus has to be is on augmenting soil conservation and other measures to retain soil moisture in these regions to mitigate the harsh living conditions.

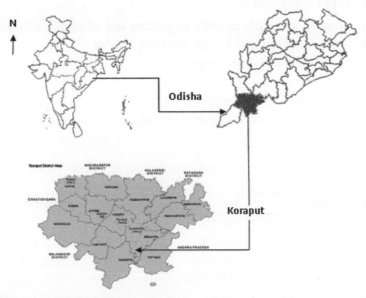

FIGURE 2.1   Location map of the study area.

A brief description of rainwater resources development and management techniques is given below:

- *In-situ* water conservation
- Drainage line treatment
- Water harvesting and its efficient utilization
- Groundwater recharge

These techniques are appropriate for following agro-climatic zones with their priority mentioned below (Table 2.2).

2.4.1 PRE-REQUISITES AND PLANNING FOR SOIL AND WATER CONSERVATION MEASURES

Followings steps can be followed for planning to implement soil and water conservation measures:

1. Prepare the map of the watershed area on a scale of 1:5,000 or 1:10,000 with all topographical features
2. Identify local issues and concern including land and resources inventory
3. Develop goals and objectives
4. Collect information on

    a. Annual and daily rainfall, its pattern and other rainfall analysis
    b. Intensity-Duration-Frequency (IDF) curve
    c. Soil depth and other properties
    d. Soil texture
    e. Land slope of watershed
    f. Vegetative cover and choice of crop

TABLE 2.2 Suitability of Water Conservation Techniques for Different Agro-Climatic Zones

Agro-Climatic Zone	Annual rainfall (mm)	Priority
Arid	100–500	1, 2
Dry Semi-arid	500–750	1, 2, 4
Wet Semi-arid	750–1000	2, 3, 4, 1
Sub-humid	1000–2500	2, 3, 1, 4
Per-humid	>2500	2, 3, 1

g.   Physiography of the area, i.e., size, shape, relief, elevation, slope, drainage density and pattern etc.

h.   Socio-economic factors *viz.* sociological and demographic features, land tenure structures, farm structures, and attitude and behavior of stakeholders

5.   Land capability assessment and erosion potential estimation

6.   Estimate design peak rate of runoff and runoff volume

7.   Investigate and evaluate alternatives

8.   Select best management practices (BMPs) according to its cost-effectiveness, availability, feasibility, durability and compatibility w.r.t. selection of materials with due regard to community acceptance and environmental sensitivity

9.   Design of required structures or measures' various components

10. Mark the structures on the map

11. Mark the layout in the field

12. Adjust for adjustments if any after ground truthing

13. Execute the work and maintain the program with supporting documentation and specifications and construction details

14. Monitoring and quantify impacts on landscape and productivity with following indicators

a.   Increase in yield (crop)

b.   Changes in cropping sequence/rotation/intensity

c.   Increase in number of wells in the vicinity

d.   Increase in groundwater table

e.   Changes in flora and fauna

f.   Changes in micro-organisms

g.   Prolonged life of water harvesting structures at downstream

h.   Watershed health and eco-index

2.4.2 APPROACHES AND METHODOLOGY FOLLOWED FOR IMPLEMENTATION AND ADOPTABILITY OF SOIL AND WATER CONSERVATION MEASURES

The main approach for land and water resources development and conservation in Kokriguda and Malipungar watersheds was inducing and encouraging people's participation. In the beginning, much difficulty was faced in mobilizing the local people to participate because of their past bad experience with other development agencies. Exhaustive Participatory Rural Appraisal (PRA), selection of leader farmers, frequent visits, discussions on problems and government's policy with leader-farmer and the followers, listening to their problems and sharing some of their burdens, helped to build up rapport and induce people's participation.

Most part of the watersheds is sloping and undulated with shallow and poor soil health and having severe erosion problems. The rainfall pattern combined with slope features and presence of denuded hillocks was resulting in enormous soil and nutrient losses and also wastage of water through runoff every year and causing damage to the agricultural fields in the middle and lower reaches. Again this runoff with high velocity breaches the field bunds, compelling the farmers to revamp the earthen bunds every year. Some part of the eroded soil was getting deposited in the *jhola* lands spoiling the standing rice crop. *Jhola* lands (low lands-widened and terraced gully beds) comprising 5% of the landscape of the Eastern Ghat High Land (EGHL) zone of Odisha are rich in water due to flow of perennial streams and springs. Farmers grow rice only in these lands. However, great scope exists for efficient utilization of this perennial water.

Keeping in view the heavy erosion and excess runoff problems, underdevelopment and under-utilization of water resources, unscientific management of natural resources, traditional crops, low yielding *desi* genotypes and absence of use of modern inputs, low cropping intensity and low crop productivity, different soil and water conservation interventions were undertaken based on priorities and technical feasibility. Due to extreme poverty and illiteracy of villagers, emphasis was given on bio-engineering measures instead of conventional masonry structures. The approaches adopted were participatory, eco-friendly and ridge to valley. Structures that are cheap, made up of locally available materials, simple, easily adoptable, feasible, viable and require no/less skill for construction and maintenance were given top priority. Series of such structures were constructed as per the need of the area. The implementation work was started in a participatory decision-making process involving the villagers and the action planning were formulated with involvement of different user groups. These user groups have also been taking up all works including watch & ward and maintenance beyond the project also. They contributed 5% of their earning to the maintenance fund and principal right holders of the usufruct sharing.

Villagers have closed the site to open grazing by observing social fencing. Stopping of stone quarrying was ensured through linkage with revenue department and one year old Mango grafts were procured at subsidized rates through collaboration of the District Horticulture office. Similarly seedlings were obtained at subsidized rate from forest department and grass slips were brought from Department of Animal Resource Development. Awareness created through formal and informal discussions, meetings and exposure visits, considering their traditional knowledge and motivating the farmers to allow such constructions in their lands initially, which in turn created opportunity for wider implementation of these measures through the since long tried and tested principle of "seeing is believing." Following the above criteria, different soil and water conservation measures were implemented in different types of topography.

2.4.2.1   Study Area

The Kokriguda and Malipungar watersheds are located in Semiliguda block of Koraput district in the state of Odisha. More precisely, both the watersheds are situated by the left side of NH-43 at a distance of 5 and 10 km, respectively from Semiliguda towards Visakhapatnam. The total areas of the watersheds are 317 and 275 ha, respectively. These represent topography, socio-cultural set up and agro-ecology of the Eastern Ghats region. The watersheds are sloping and undulating. The altitude ranges from 750–1850 m above mean sea level with an average slope of 12.25–16.43% and drains to river Kolab, a tributary of river Godavari. Hillocks on three sides with presence of *Jhola* land (widened and stabilized terraced gully beds with perennial water) and two more perennial streams running through it make the watersheds, a typical geo-hydrological unit of the region. Basically, both the watersheds are of fern shaped with dendritic type of streams. Groundwater available in plenty has not been used for irrigation purpose. The earthen channels constructed for conveying water of mainstream for irrigation and domestic use suffered seepage loss as high as 70% [28].

Red and lateritic soils are dominant in the watersheds with small patches of fine textured black soil in the low land (*jhola*) patches. Four distinct soil groups could be identified, viz., gravelly and loamy sand in hillocks, sandy loam in the uplands, sandy clay loam in the medium and fine textured silty clay loam to silty loam near low lands. In general, water-holding capacity is low. Soils are acidic (pH 4.5–6.2), low in organic matter, N and P contents but rich in iron and aluminum oxides. Potassium status of the soils is medium to high [18, 22].

Climate of the area is moist sub-tropical. Minimum temperature in winter hovers around 1–2°C. Mean minimum temperature is 10.1°C and mean maximum 34.8°C. Annual rainfall of 1400 mm with 77 rainy days is received, 85% of which occurs during June to October. Relative humidity in monsoon period ranges from 67 to 96% and in post-monsoon season 48 to 95%. Bright sunshine hours in monsoon period range from 1.8 to 6.4 whereas in non-monsoon season from 6.8 to 9.4 hours. Mean evaporation varies between 2.56 and 6.27 mm day^{-1}.

In spite of being located near developed industrial townships, the winds of development seem to have bypassed the Kokriguda and Malipungar villages. Kokroguda and Malipungar villages had 78 and 179 households, respectively with, all belonging to *paraja* tribe in Kokriguda and in Malipungar they are of *paraja* and *Kandh* tribe with Mali community and with 90% people living below poverty line. Literacy rate was only 25%. The villages lacked all the basic amenities, like potable water facilities, electricity facilities, etc. Drinking country liquor was an addiction. Subsistence farming was the main occupation. However, about 40% of families were intermittent wage laborers in nearby towns.

Cultivable area comprises of upland, medium and low lands out of which 80% was rainfed. Non-arable land constituted about 47 to 50.2% of the total geographical area of the watersheds. Forest area was about 90 and 108 ha, respectively out of which 40% was under protected secondary forest of mixed vegetation lying on southern side hillocks and the rest 60% was degraded scrub on north-eastern and eastern side hillocks. Net sown area was 127.25 and 93 ha, respectively in Kokriguda and Malipungar watershed.

2.4.2.2 Topography in the Watersheds of Koraput District

In Koraput district, mainly the topography varies from ridge to valley as shown in Table 2.3.

However, the landforms in the study watersheds are varied from steep hill slope to valley bottomlands as shown in Figure 2.2. Soil and water conservation measures suitable in different topo-sequences are discussed in the following subsections.

A. Steep slope

The main hillocks possess higher slope degree even some places beyond 100%. Being in the proximity of the village habitation, these are totally

TABLE 2.3   Land Situations In the Region as Per Tribal Nomenclature

Name	Position in toposequence	Characteristics
Bada	Back yard	Gently sloping with good soil depth; mostly irrigated put under vegetables and fruit trees, forest trees found in the boundary
Beda	Low land	Wider stabilized streambed
Dangar	Hill top and slopes	May be with or without trees. Not suitable for agriculture
Jhola	Low land	Stabilized streambed used for cultivation. Generally narrower with uplands/mala/dangar and trees on both sides
Mala	Lower hill slopes	Past history of forest/shrub forest. Have been cleared for agriculture. More stones and pebbles
Pada	Upland below Mala	Gently sloping with better soil depth and less of stones and pebbles.
Saria	Medium land surrounded by low land in all sides	Higher soil depth

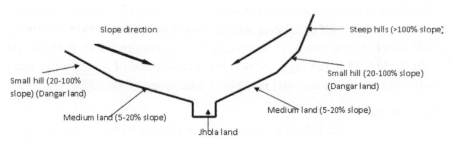

FIGURE 2.2   Schematic cross section of land physiography in study watershed.

deforested. However, in certain pockets regeneration potential in terms of coppicing exists. Soil has washed out and thus at certain places rocky out crops are exposed. But, different perennial streams are emerged from these hills.

For rehabilitation of denuded hillocks in steep slopes, conservation plans in the form of regeneration and protection of secondary forest were adopted. The villagers were motivated not to cut any of the live trees and protect the forest. The forest is being looked after by a watchman appointed from neighboring village who is remunerated collectively in kind by the villagers. The extraction of wood from forest for household purposes was strictly as per need and season specific; only dry wood or bushes are cut. This type of conservation increased 8.3 times the plant population after six years. There was increase in stream flow by 27.8%, whereas perenniality increased by 15.81% due to dense vegetation in steep slope and increase in interception loss.

B. Small hills/*dangar* land

Degraded hills with nil or sparsely covered canopy of denuded/secondary scrub forests and lost top soils are the common features of the eastern Ghat ranges dominated by tribal population. The carrying capacity of this life support system is decreasing while the inhabitants are increasing. "*Podu*" or shifting cultivation was the main livelihood option for the tribal population here, before a few years ago. But now, due to degradation, many hill tops have been left out of frustrations. However, the hill slopes in many patches are under sedentary cultivation now, leading to further loss of soil and aiding to formation of gullies below in flatter lands apart from water scarcity. Greening these barren hills for conservation of precious resources and augmenting production has been the great challenge.

Graded and stone bunds

Graded bunds with top width 0.3 m, bottom width 0.6 m and depth 0.45 m with 0.2 per cent grade were constructed at 1 m vertical interval. A continuous small trench with equalizers was provided in upstream side to capture runoff water. Stone bunds of top width 0.3 m, bottom width 0.45 m and height 0.3 to 0.45 m having provision for safe disposal of runoff water were constructed at 10 m horizontal interval in land slope of 12–15%.

Sunken ponds

At the bottom of the hillocks, sunken ponds of size 30 m × 30 m × 1.5 m were constructed to reduce the runoff velocity as well as increasing the ground water recharge. A temporary storage of 0.03 ha-m resulted from these sunken ponds.

Inward sloping contour bench terrace

On the hillock, inward contour bench terraces of 1 m width were constructed at 5 m horizontal intervals in 12 per cent slope. Guava was planted in the terrace.

Trench-cum-bunds

Continuous contour bunds and trench-cum-bund (trench 0.3 m × 0.3 m) were constructed at 10 m horizontal interval. Vegetative barriers (*sambuta*, hybrid Napier, broom grass) and hedge row species (Assam shade, *Gliricidia* and *Crotolaria*) were planted on bunds, and mango, cashew, teak, etc. were planted in the inter-bund space.

Soil working techniques

Micro catchment, saucer, crescent shaped and tick ditch/v-ditch (1.5 m width and 0.3 m height) *in-situ* water harvesting and conservations were imposed in mango and cashew at different slopes (Table 2.4). A comparison of these showed that the runoff contributing area per plant is highest at 36 sq.m for both micro-catchment and tick ditch (See appendix – I at the end of this chapter), where has the tick ditch has got the maximum water storage area of 0.34 cum, which is marginally higher than that of micro-catchment. But the water receipt by the mango tree under micro-catchment system was highest considering the loss through evaporation and competition for the water resource by the other system component of grasses.

Diversion channel

To divert the excess runoff from the hillock, 200 m long diversion channels having top width 0.65 m, bottom width 0.3 to 0.45 m and depth 0.3 to 0.65 m were made. All these diversion channels were diverted to a large pit resulted from stone quarries for safe disposal.

TABLE 2.4   In-Situ Water Conservation Practices and Additional Components in the Integrated Systems

	In-situ water conservation	Description	Additional components
T 1	Micro catchment	Two diagonal micro ridges diverts the runoff water to a 0.45 m x 0.45 m x 1.5 m half moon concave trench at 0.5 m upstream of the plant.	Grass (*Brachiariahumidicola*) as filter strip before trench in u/s at 0.3 m x 0.3 m in double row staggered. Aonla as filler in quincunnex
T 2	Tick Ditch (slanting trench)	A bund with cross sectional area of 0.1 sq.m formed at a distance of 0.5 m upstream of plant row by scooping soil from further 1.5 m upstream by making a slanting trench for runoff impounding.	Grasses (*Andropogon, B. mutica*) in the Tick ditch in two rows with former tall species near bund at 1 m spacing in single row and later spreading species at the tail of the ditch in single row at 0.3 spacing in single row. Teak as filler in quincunnex
T 3	Saucer and Pitcher (2 liter capacity)	Soil working at 0.5 m diameter around the plant to allow water collection in that region. Pitcher provided only in the first year.	Hedge row of *Indigofera teysmannii* at 1 m contour interval planted at 0.3 m apart

Safe disposal structures

Farmers of the watersheds were not concerned about the disposal of water through a particular outlet. Runoff water was flowing haphazardly and deteriorating the agricultural lands. Therefore, numbers of safe disposal cut outlet structures of stones each with of 15 cm height, 30 cm crest length and slope 3:1 to 4:1 were constructed at different places.

Summer plowing

Summer plowing was promoted by demonstrating its benefits for suppression of thick population of annual and perennial weeds, *in-situ* conservation of pre-monsoon showers, facilitation of early sowing and better establishment of seedlings of rainfed crops like suan, groundnut and upland paddy. Summer plowing was done during last week of April-May and depth of plowing was kept about 20 cm [5] as the farmers use only indigenous plows drawn by small size indigenous bullocks.

Agro-forestry systems/Alternate land use systems

In addition to these mechanical measures, different agro-forestry systems like horti-pasture and horti-silvi-pasture systems were imposed. Plantations of fruit (cashew and mango) and forest trees (*Acacia mangium, Melia azaderach, Teak, aonla* and *Tasmania, Eucalyptus*) were taken up on denuded hill slope with different *in-situ* water harvesting techniques along with other production and conservation components under system approach. Seasonal components like pastures (*Stylosanthushamata*) have been introduced to get returns early. Fuel wood components were added to fruit systems to provide intermittent and quick income. Bio-barriers (*Eulaiopsisbinata, Andropogonspp, Brachiariaspp, hill broom*) are incorporated to add to conservation and production efforts. Bio-fencing was done with sisal on boundaries and minor rill trainings carried out with loose boulder bunds and safe disposals for protection and a group of local youths was formed, trained and involved in all activities to take care of assets in post- projects phase and augment its sustainability dimension. A drastic reduction in runoff and soil loss from second year onwards was observed and this could be due to canopy development of trees coupled with thick grass cover of *stylo* and effects of *in-situ* moisture conservation techniques.

In another system of integrating conservation cum production for rehabilitation of degraded hillocks horti-pastoral models supplemented with *in-situ* moisture conservation techniques were implemented in a participatory mode. Fruit trees like cashew and guava were planted in integration with natural grass and *Stylosanthus hamata* as pasture component with *in-situ* moisture conservation practice like trench (1 m x 0.45 m x 0.45 m) at 0.5 m distance from fruit tree in upstream and no trench in a participatory mode. Three years average runoff and soil loss data showed that Guava+ Stylo + trench was most conservation effective with less runoff and soil loss (7.7 % and 3.6 tons ha^{-1}, respectively) followed by Cashew + Stylo + trench. Lower runoff and soil loss was observed in stylo + trench plots as compared to natural grass + no trench. Further, effectiveness of stylo + no trench in reducing erosion was as good as of grass + trench. However, a considerable difference between stylo + trench and stylo + no trench was observed. This shows that there is the positive effect of trench as well as stylo grass cover on runoff and soil loss [14].

Integration of grain and pulse crop intercropping
Finger millet (*ragi*) + pigeon pea and finger millet + blackgram both in 4:2 row ratio produced 77.7% and 67.4% higher finger millet equivalent yield (FMEY) over broadcast sown finger millet (farmers' practice). The highest land equivalent ratio (1.34), net returns (Rs. 9,665/ha) and benefit cost: ratio (1.00) however, was obtained with finger millet + pigeon pea (6:2 ratio). However, black gram on contours recorded the lowest runoff (10.2%) and soil (4.7 tons/ha) and nutrient (N, P and K) losses through erosion. Ragi + pigeon pea intercropping (6:2) give a ragi equivalent yield of 18.18 q/ha on dangar-1 (lower hill slope), 16.87 q/ha on danger-II (middle hill slope) and 15.71 q/ha on danger-III (upper hill slope: 1 q = 100 kg). It yielded average net returns of Rs. 3459 with benefit–cost ratio of 1.24:1 (1.00 US$ = 60.00 Rs.). The average reduction in runoff and soil loss was observed 5.31% and 41%, respectively over farmer's practice [7].

C. Pediment slope (Medium land)
The foothill slope (up to 10–20%) is under permanent cultivation of rainfed crops. The runoff of water emanating from the upper part of hill accumulates in these varied fields and creates initiation of gully along with rill erosion. The continuous wash out has reduced production potential and arable cropping paddy needs conservation measures for enhancing production potential and water-holding capacity of these utterly degraded soils.

Vegetative barriers with planting methodologies/miniature bunds in the lower reach of sloping medium lands
To reduce the soil erosion in sloping arable lands planting of vegetative barriers of locally available grasses in integration with trench-cum-bund was considered to be an appropriate low cost soil and water conservation

technology. Different grass species viz: *vetiveria zizanoides, Eulaliopsis binata (vern. Sambuta) and lemon grass* were grown under double row staggered planting (30 cm x 30 cm spacing) in association with soil working measures (bund and berm planting). The system showed that lowest average runoff (8.1%) and soil loss (4.0 tons/ha) was observed in *sambuta* + Bund treatment followed by vetiver + Bund (Table 2.5). Between two planting techniques, bund planting of barriers was far better than berm planting [6]. Keeping in view the above results and local availability, sambuta with bund planting is recommended as the most effective barrier in sloping lands. Better performance of sambuta in comparison to vetiver in reducing runoff and soil loss from 11% sloping lands has also been reported by [24] (Table 2.6).

Hedgerows for resource conservation in the upper reach of sloping medium lands

Apart from curbing land and water degradation and augmenting crop yields, factors like soil health improvement as well as increase in fodder and fuel availability are quite desirable for watershed development in this region. Therefore, different hedge row species viz: Assam shade (*Indigofera*),

TABLE 2.5 Runoff and Soil Loss Under Different Vegetative Barriers in 5% Slope

Treatment	Run off (%)	Soil loss (tons ha⁻¹)
Sambuta + Bund	8.1	4.0
Sambuta + Berm	11.0	5.2
Vetiver + Bund	9.8	4.5
Vetiver + Berm	13.0	7.2
Lemon grass + Bund	11.8	5.8
Lemon grass + Berm	14.7	8.3

TABLE 2.6 Conservation Potential of Mechanical and Vegetative Barriers

Treatment	Run off (%)	Soil loss (tons ha⁻¹)
Stone bund	11.9	5.64
Hill broom	13.6	6.7
Vetiver	8.8	4.0
Sambuta	9.5	4.4
Control	25.9	14.0

Gliricidia and Perennial Arhar supplemented with *sambuta* grass in a miniature field bunds of 0.15 m × 0.60 m (height × width) can be use with spacing for shrubs as 0.5 m x 0.5 m and for grass 0.3 m x 0.3 m (double row staggered). *Gliricidia* is a medium size leguminous tree, which yields upto 20 tons ha^{-1} nutrient-rich and easily decomposable biomass. It was observed that hedge row reduced runoff by 33% (10.7% runoff compared to 16.1% in control), soil loss by 35% (6.3 tons ha^{-1} compared to 9.71 tons ha^{-1} in control) (Table 2.7), With higher soil moisture storage by 28–37 mm and 22–43 mm at 12 and 17 days of dry spell, respectively, the grain yield of finger millet increased by 49% from 952 kg ha^{-1} in control to 1413 kg ha^{-1} in *Gliricidia* + *sambuta*. Addition of *sambuta* grass as filter strip significantly reduced the losses of water runoff and soil. The grass filter strip also improved soil moisture storage by 9–12 mm and 6–15 mm at 12 and 17 days of dry spell, respectively [15].

Bunding

Contour bunds of 0.3 m height, 0.3 m top width and 0.6 m bottom width were constructed. To impart stability, these bunds were planted with grasses like *sambuta (*the local grass*)* and vetiver in double row staggered at 0.3 m × 0.3 m spacing; hedge row species like Assam shade and *Gliricdia* (0.5 m plant to plant distance) and fruit species like papaya and drumsticks. At some places where stone were available, stone bunds were also constructed. Continuous trench-cum-bunds with dimension of trench (0.3 m x 0.45 m) were constructed along with equalizers at every 5 m distance.

Land shaping

Most of the land in the watershed was undulated leading to erosion, ill distribution of water and low productivity. Because of shallow soil depth and high cost involved in land leveling, only land shaping was carried out in critical areas.

TABLE 2.7   Runoff and Soil Loss Under Different Hedge Rows System

Treatment	Run off (%)	Soil loss (t ha^{-1})
Indigofera	12.75	7.75
Indigofera + sambuta	8.88	5.04
Gliricidia	11.87	6.41
Gliricidia + sambuta	10.71	6.33
Sole sambuta	13.80	7.48
Control	16.07	9.71

Field bund strengthening
The sloping agricultural lands below the foothills had undergone erosion, and had become undulated with multi directional slopes. Small- shallow gullies and rills were common. Moreover, in the *jhola*, bunds were in dilapidated shape and needed renovation. Thus, field bund strengthening was carried out by enhancing the height and width of bunds and planted with *sambuta* grasses on the bund.

Multi-tier cropping system
Multitier cropping systems with boundary plantation of forest tree and square plantation of two fruit trees Papaya and Drumstick and different intercrop combinations (Ginger + Pigeon pea (8:2), Runner bean + Pigeon pea (8:2), *Ragi* + Pigeon pea (6:2) and Ragi broadcasting (as per farmers' practice) resulted a significant decrease of 50% (on average basis) both in runoff and soil loss in all the treatments in comparison to control. The comparative study shows Ginger cultivation with Pigeon pea had ameliorative effect to the soil by increase in the pH and available nitrogen of the soil. Among different crop combinations maximum percent increase in nutrients availability was found in Ginger + Pigeon pea (8:2) treatment in all the models to the tune of 12–23 for phosphorus and 17–59 for potassium [12].

Drainage line treatment
Interventions for drainage line treatments were made for different orders of gully networks as follows:

i. Loose boulder check dams
The watershed area is ripped to many parts with gullies crisscrossing as a result of inappropriate land use and deforestation. These gullies were expanding and damaging agricultural lands. To curb their further expansion, different gullies were treated. Loose boulder check dam (LBCD) reinforced with *Ipomea* was quite effective in small gullies with head wall extension 0.45 m, height of head wall 0.5 m and bottom width and top width of head wall and head wall extension 0.3 m is suitable. In upstream Ipomoea planted and slope in upstream kept as 1:1 and in downstream 3:1. LBCD reinforced with *Ipomea* having head wall extension 0.6 m, height of head wall 1–1.2 m and bottom width and top width of head wall and head wall extension 0.45 m is very much suitable in medium and deep gullies. In upstream *Ipomea* was planted and slope in upstream kept as 2:1 and in downstream 4:1.

ii. Brushwood check dams

Single and double brushwood check dams consisting of *Gliricidia* and *Simli* posts of 5 cm diameter and *Ipomoea* as binding material, with post-to-post distance 0.15 m, row-to-row distance of 0.5 m, below ground depth of 0.3 m and above ground height 1.5 to 2.0 m, were constructed at suitable locations in small and medium and deep gullies, respectively.

iii. Renovation of earthen channels

Channels passing through the farmers' fields were eroded badly causing damage to the adjacent fields. These channels were renovated, by widening the bed to bottom width of 0.45 m and top width of 0.6 m. For stabilization of the bed of the channel, *sambuta* grass was planted in staggered pattern keeping row to row and plant to plant spacing of 30 cm. Small loose boulder structures were constructed at suitable locations to reduce the velocity of runoff water to a non-erosive level. Due to this, scouring action of flowing water was reduced.

iv. Retards and vegetative measure

Agave sislana, sambuta and bamboo were planted on gully beds, whereas gully banks were protected by planting of bamboo. Retards of sambuta and *Agave* with different rows and alternate arrangements were also constructed in small gully beds. Field bunds were stabilized by planting trees/grasses like Bixa, Simaruba, bamboo, Subabul and Sambuta.

v. Valley bottom lands

These are widened and stabilized terraced gully beds, locally known as *jhola* lands and used for paddy cultivation only as water is available throughout the year. All the excess runoff out of these hills and pediment slopes passes through this *jhola* lands.

Valley bottom lands are *jhola* lands, which are rich in water due to flow of perennial streams and springs. Farmers grow rice only in these lands. Due to heavy flow of runoff water the field bunds were generally damaged, therefore field bund strengthening works with *sambuta* plantation were done. Within the field bunds safe disposal structures of small rectangular weirs were also constructed. The perennial water was also harvested using *jhola kundis* as described in this chapter.

vi. Development of water resources: *jhola* lands

Water resource development is a key element in any watershed development program. Therefore, as per the suitability and need, different water harvesting systems were developed in both the watersheds. In Kokriguda watershed one perennial stream and in Malipungar watershed three perennial streams were cascading down the hillocks. Also there is the existence of *jhola* lands, which are rich of perennial source of water. So all these water along with runoff water were harvested and reutilized as follows:

Masonry water harvesting structure with underground pipeline (UGPL) system: Perennial stream flowing down the hillock had been diverted by the villagers through earthen channel, which had conveyance efficiency of less than 30%. Therefore, one masonry water harvesting structure was constructed at the foothill and then one UGPL system of PVC pipes (internal diameter 150 mm) with a total length of 1248 m was installed to convey the perennial stream for irrigation of remunerative crops and also for domestic use. This system was with eight outlets and potential command area of 35 ha.

Farm pond: The villagers did not have any water storage facility for taking bath and providing water to cattle. So a farm pond was constructed in upper *jhola* land to harvest the water flowing in these lands. After a participatory site selection, villagers took the initiative for constructing the pond. The storage capacity of the pond was 0.06 ham and depth 2 m. The defunct diversion weir at the downstream end of the pond was used as spillway. This pond was used for multipurpose.

Jhola kundi: In Eastern Ghats region, a typical land feature is the existence of *jhola* lands, which remains submerged throughout the year. *Jhola* lands are at 1 to 1.5 m lower elevation than the adjoining medium and uplands. No sound mechanism was available to trap and utilize this water. Even community based structures suffered from lack of maintenance and operational difficulties. In the medium lands, which are located at higher elevations adjacent to *jhola* lands, farming of remunerative crops is not practiced due to unavailability of adequate water resources, except rainfed paddy, finger millet, etc. Taking the advantage of the perennial water resources in the *Jhola* system, a number of *Jhola kundis* were designed and constructed in a participatory mode along the periphery of the *Jhola* system in Malipungar watershed.

Jhola kundi is a low cost water recharging device of circular shape; dug manually with depth varying from 2 m to 4 m and diameter of 3 m approximately for easy lifting of water through traditional water lifting devices at individual farmer's will or need. After observing water availability in *jhola* land (low lands) and the water table fluctuations in *jhola kundi*, it was found that farmers can irrigate 0.5–1 ha in rabi and 0.2–0.4 ha in summer. Only one person is required at a time for operating the water-lifting device and for irrigation on plot-to-plot basis. Traditional water lifting devices (locally called as Tenda) (see appendix at the end of the chapter), whose optimum operation lift is 1.2–4.0 m or paddle operated Krishak Bandhu Pump whose optimum operation lift is 5–7 m are recommended for these structures. The technology is quite affordable as it is cost effective. Cost of excavation of approximately Rs. 2000/unit only is required for construction of capacity of 30 m³ *jhola kundi* (Rs. 65 per m³ of earth work), which can be met out from the farmer's own contribution.

Traditional water lifting device can be fabricated by the farmer himself using locally available materials like wood, bucket, rope and bamboo, thus involving almost no fund. Cost of the water lifting device operated through pedal operated Krishak Bandhu (reciprocating) pump ranges between Rs. 1200 and Rs. 1500 approximately. Being a farmer' friendly water lifting devices, subsidy schemes of the State Govt. are available also. Net income from important cash crops ranged from Rs. 23,027–64,700 per hectare, respectively with increased cropping intensity (Table 2.8). It was attracted the attentions of visitors and farmers of Koraput district. Farmers of other nearby villages and EAS and IWDP watersheds of DRDA, Blocks and KVK are now adopting this technology [26].

TABLE 2.8 Crops, Overall Productivity and Income Before and After Installation of *jhola kundi*

Attributes	Before installation	After installation
Crops	Millets, maize, vegetables (in small patches)	Cash crops (vegetables, ginger, maize etc.) Fishes for consumption in home. Flowers for fulfilling esthetic needs and beautification of hairs of tribal women
Yield (100 kg/ha)	48.5	172
Net returns (Rs./ha/year)	4900	52,178
Cropping intensity (%)	127	270
B:C ratio	-	2.80

Earthen dam-cum–UGPL system (spring pond): There was a perennial stream (perched aquifer) emerged at the bottom of one of the hillocks in Malipungar watershed. Its water was going waste and the downside fields were getting water logged. Looking into the feasibility and *Mali* (OBC) farmers' demand, an earthen dam of 360 m³ capacity was constructed at the foot of the hillock. For efficient conveyance of the harvested/tapped water, 200 m long PVC underground pipeline system was laid. This system could bring 4 ha area under irrigation, which has been put under the cultivation of paddy in *kharif* and vegetables in winter. Farmers earned average net profit of Rs. 10,381/ha (excluding the cost of WHS), whereas before construction of this WHS, farmers were earning only Rs. 974/ha.

Surface pond: Another earthen dam (surface pond) of 36,000 m³ capacity with masonry spillway was constructed to harvest spring and runoff water and 30 m PVC pipeline was laid for conveyance as per the demand of *Paraja* tribe farmers and the stored water was used by the villagers for irrigation to the cultivated land.

Dugout farm pond with lined channel: As perceived during PRA exercise, the capacity of existing farm pond was very small to accumulate water diverted from *jhola* land. Again water from another spring for irrigation of the vegetable crops grown in *bada* (irrigated uplands) lands was not sufficient. Thus, small pond was renovated and its size was increased to accumulate 12,000 m³ water. A lined channel of 100 m length was constructed for increasing the conveyance efficiency and bringing more area under irrigation. Excess water from perennial stream and above earthen dam was also diverted to this pond and stored otherwise it would have gone waste. Total area irrigated by this system is 9 ha belonging to 48 vegetable growing farmers. Farmers are taking up intensive cultivation of a variety of vegetables and earn net income of Rs. 59,713/ha/seasons. Water user's association was formulated for maintenance of these systems [22].

Conversion of stone quarry pits into water harvesting structures: In Kokriguda watershed there were four large and three small but sufficiently deep depressions, which were created by the outsiders through stone quarries in non-arable lands. It was decided to renovate and convert these quarries in to ponds for storage of water to irrigate plantation crops during dry spell periods. A storage capacity of 0.22 ham was created. Application of cow dung and clayey soil mixture helped in water retention in these ponds.

2.4.2.3 Impact of Different Soil and Water Conservation Techniques

2.4.2.3.1 Changes in Surface Water Storage

Due to construction of farm pond, *jhola kundis*, conversion of stone quarries into storage structures, sunken ponds and inlet structure of underground pipeline (UGPL) irrigation system in Kokriguda watershed, the availability of surface water was enhanced. An additional storage capacity of 1.212 ham was created due to these water harvesting structures. The UGPL system was very successful and revolutionized the water supply for domestic needs and irrigating winter vegetables and other crops. It reduced drudgery of women (for fetching potable water) by 55%; saved time could be diverted to income generation activities and family care. It created employment potential of 208 man-days $ha^{-1}year^{-1}$ and reduced male out migration. The overall efficiency of this UGPL system was increased by 122%. Net returns from vegetable cultivation in its command were Rs. 15,625.00 per farmer (Rs. 14,245 ha^{-1}) amounting to a total of Rs. 1.5 x 10^5 $year^{-1}$. The system is being maintained by the Water User's Association (45 members) [22, 28, 29].

Due to construction of *jhola kundis'*, dugout, surface, spring pond with underground pipeline system and lined conveyance channel in Malipungar watershed, the availability of surface water was enhanced and an additional storage capacity of 11.8 ham was created. Due to this, an additional area of 30 ha (mostly medium and uplands) belonging to 62 beneficiaries could be brought under irrigation. This irrigation facility improved crop yield by 56 to 76% and net profit by 281–504% (Table 2.9). Due to underground pipeline and lined channel, conveyance efficiency increased by 89.2%, which increased vegetable area with additional area, also brought under rice [26, 27]. Drudgery of women decreased substantially as water resources developed deliver water for domestic use very near to the village.

2.4.3.2.2 Impact of Soil and Water Conservation Works

Reduction in surface runoff and soil loss
It was observed that there is decrease in water flow in *Jhola* land by 8.52% due to different *in-situ* moisture conservation measures and tapping water from *jhola land* through *jhola kundis*. Runoff from whole watershed during pre project period was 42.5%, which was declined to 24.2% during

TABLE 2.9 Improvement in Crop Yield and Profit Due to Water Resource Development at Malipungar Watershed

Resources	Yield (100 kg/ha)		% increase	Net returns (Rs./ha/season)		Increase in net returns (%)	No. of beneficiaries
	Before	After		Before project	After project		
Dugout pond	115.4	180	55.98	15660	59713	281.3	48
JholaKundi	48.5	172	254.6	4900	52178	964.9	24
Spring pond	12.8	22.5	75.78	974	5881	503.8	10

post project period, might be due to various soil and water conservation interventions taken up in the watershed. Impact of different soil and water conservation measures *viz*: loose boulder check dams, trench cum bund, micro catchments, saucer pits, V-ditch, vegetative barriers and hedgerows in Malipungar watershed on reduction of soil loss was recorded in the water-shed. The soil loss was reduced by 3.5 times due to these measures.

Imposition of different conservation measures including graded bunds, trench-cum bunds, vegetative barriers, hedgerows, etc. in Kokriguda water-shed remarkably improved the yield of different crops; the range of increase varying from 15.4% in little millet to 38.2% in upland paddy. Additional returns due to these measures were Rs. 914, 1867, 330, 3806 and 115 per ha in *ragi*, upland paddy, *Niger*, groundnut and *suan*, respectively, at an expense of about Rs. 3000 per ha (cost of SWC measures). In the first year of land shaping, yield of *ragi*, groundnut and *Niger* decreased by 0.45, 0.2 and 0.22 q ha^{-1}, respectively. However, in the subsequent year, there was a net gain of 1.55, 1.07 and 0.46 q ha^{-1} in the yield of respective crop. Additional monetary returns accruing from this intervention were highest with ground-nut (Rs. 1450 ha^{-1}) followed by *ragi* (Rs. 956 ha^{-1}). Additional returns with *Niger* were very marginal (Rs. 515 ha^{-1}). Benefit: Cost ratio calculated keep-ing in view the 25 years project life and 10% discount rate was 1.37.

Rise in water table depth in open wells

There was 9.42% increase in ground water table in Malipungar watershed, whereas an average rises of 0.325 m in water table in Kokriguda. This rise in water table could be attributed to overall impact of soil and water conser-vation measures. Increased recharge in well ensured year-round sufficient water supply in the village.

Overall impact of conservation measures on watershed water yield

In Kokriguda watershed due to intensive different soil and water conser-vation measures implemented, there was decrease in surface water flow in *jhola* land by 51.16% and increase in water flow in stream 1 and stream 2 by 27.8 and 20.2%, respectively were recorded. Decrease in stream water flow in *jhola* land could be due to more use of water by the villagers as area under cropping increased. There is also decrease in perenniality of stream flow by 37.35% from pre-project period to after-project period near the foothill. Increase in stream flow in two perennial streams might be due to increased secondary forest vegetation in the catchment [18].

Variation in water yield recorded at the outlet of the watershed showed that there was decrease in water yield by 51.16% from pre-project to post-project period. This happened due to use of greater quantity of water by the villagers as crop-sown area increased and brought under remunerative crops. Net area sown, area sown more than once and gross cropped area under different crops before implementation of the project increased from 127, 3.5 and 130.8 ha in pre-project period to 156, 38.6 and 200.4 ha, respectively during post project period [18]. There was also decreased in perenniality of stream flow by 37.35%, because farmers were using more water during the month of May and less water during October; during May, farmers have been growing vegetable crops that required greater volume of water for irrigation and during October month rainy season cereal crops approach maturity requiring no irrigation [4]. This implies that due to different soil conservation treatments in hill slopes and medium lands, there is reduction of runoff at the outlet, but there was increase in water yield within watershed area. The additional water available was used by the farmers by bringing their lands under cultivation, which was not possible before implementation of the project.

2.5 CONTRIBUTION AND AWARENESS AMONG THE STAKEHOLDERS ABOUT SWC MEASURES

Farmer's were convinced to participate in the implementation of SWC measures and their contribution of 4–10% was obtained in the form of labor, which was converted into cash. Some amount from labor wage of farmers as agreed by them was deducted as their contribution towards the developmental activities and deposited in the Watershed Development Fund.

A survey was carried out to find out the awareness among farmers about the soil and water conservation technologies after its implementation. It was found that stakeholder's awareness about the SWC measures was 80%, 100%, 100% and 100% among landless, marginal, small and large farmers, respectively, with an overall figure of 95%. Resource conservation potential actually studied in the watershed was demonstrated to researchers, development officers, students, farmers, state Govt. officials, development workers of NGOs through exposure visits and presenting the performance of these measures in workshops, meetings or trainings. The impact of these efforts was studied by visiting the watersheds developed by NGOs and State Govt.

in the district and discussing with local farmers to know the adoptability and sustainability of these measures.

The soil and water conservation technologies implemented at Kokriguda and Malipungar watersheds hold potential of wider replicability within and outside the eastern Ghats region. Village youths of the watershed were trained to handle such interventions and have capability to replicate in the nearby places. The enthusiastic participation of villagers across the developmental stages of the project led to imparting of sustainability dimension to the project. Most of the assets created in the watershed were fully made use of and adequately maintained by the sensitized villagers. Further the linkages developed with the external agencies resulted in the exposure of villagers to various Govt. and Non-govt. developmental agencies and helped them for maintenance and sustainability of watershed assets.

2.6 SUMMARY

Very high priority has been accorded to the holistic and sustainable development of rainfed areas by the Government of India in last few years following *Watershed Approach*. Indeed, the watershed approach represents the principal vehicle for transfer of rainfed agricultural technology. This approach was adopted for developing two model watersheds Kokriguda and Malipungar in Semiliguda block of Koraput district of Odisha. Different cost effective soil and water conservation measures were implemented from ridge to valley due to which there was a tremendous decrease in runoff and soil, rise in groundwater table and increase in crop yield and cropping intensity during the project of four- to six-years period.

Due to water resource development area under remunerative crops like vegetables increased and conveyance efficiency increased from 23–95%. People's participation in different activities was also very encouraging. From this study, it could be concluded that adoption of participatory bottom-up approach, implementation of easily maintained and low cost bioengineering SWC measures using locally available materials, promoting, refining the ITKs and facilitating its blending with modern technology from ridge to valley and their demonstrations and field visits will surely make the wider adoption of SWC measures in any watershed for reduction of further natural degradation. By adopting the appropriate management practices it is not only possible to attain very high returns from the rainfed areas but also the resources could be put to most optimal use with long-term sustainability.

KEYWORDS

- Agro-forestry
- Alternate land use
- Assam shade
- Awareness
- Bio-engineering
- Brushwood check dam
- Bund
- *Dangar* land
- Diversion structure
- Drainage line treatment
- Dug out farm pond
- Eastern Ghats
- Gliricidia
- Graded bund
- Hedge row
- Intercropping
- Inward sloping contour bench terrace
- *Jhola kundi*
- Jhola land
- KB: Krishak Bandhu
- Land capability classification
- Land degradation
- Land shaping
- Linkages
- Livelihood
- Loose boulder check dam
- Micro-catchment
- Multi-tier cropping
- Participation
- Participatory rural appraisal
- Perenniality

- **Rainfed agriculture**
- **Retard**
- **Safe disposal structure**
- **Soil loss**
- **Soil working technique**
- **Spring pond**
- **Stakeholder**
- **Stone bund**
- **Sunken pond**
- **Surface pond**
- **Surface runoff**
- **Topo-sequence**
- **Trench**
- **Tribal**
- **V-ditch**
- **Vegetative barrier**
- **Water harvesting**
- **Water resource development**
- **Watershed**

REFERENCES

1. Chauhan, K. P. S. (1998). Framework for conservation and sustainable use of biological diversity. Action plan for the Eastern Ghat region. *Proc. National Seminar on Conservation of Eastern Ghats,* Andhra University, Visakhapatnam.

2. CRIDA Vision 2020, (1997). *CRIDA perspective plan.* ICAR-Central Research Institute for Dryland Agriculture, Hyderabad.

3. Das, D. C. (1994). Watershed management in India experience in implementation and challenges ahead. *Proc. 8th ISCO on Soil and Water Conservation Challenges and Opportunities,* Vol 2, IASWC, Dehradun, India, pages 743–774.

4. Dass, A., Patnaik, U. S., Sudhishri, S., and Choudhury P. R. (2002). Vegetable cultivation: A major intervention for tribal watershed development. *Indian Journal of Soil Conservation,* 30(3), 240–247.

5. Dass, Anchal, Sudhishri, S., Lenka, N. K., and Patnaik, U. S. (2009). Effect of agronomic management on watershed productivity, impact indices, crop diversification and soil fertility in Eastern Ghats of Odisha. *Journal of Soil and Water Conservation,* 8(3), 34–42.

6. Dass, Anchal, Sudhishri, S., Lenka, N. K., and Patnaik, U. S. (2011). Runoff capture through vegetative barriers and planting methodologies to reduce erosion, and improve soil moisture, fertility and crop productivity in southern Orissa, India, *Nutrient Cycling in Agro-ecosystems*, 89, 45–57.

7. Dass, Anchal and Sudhishri, S. (2010). Intercropping in finger millet (Eleusinecoracana) with pulses for enhanced productivity, resource conservation and soil fertility in uplands of southern Orissa. *Indian Journal of Agronomy*, 55(2), 89–94.

8. De Fraiture, C., Wichelns, D., Rockström, J., Kemp-Benedict, K., Eriyagama, N., Gordon, L. J., Hanjra, M. A., Hoogeveen, J., Huber-Lee, A., Karlberg, L., (2007). *Looking ahead to 2050: Scenarios of alternative investment approaches Earth scan.* Colombo: International Water Management Institute, London, pages 91–145.

9. Dhruva Narayana, V. V., Sastry, G., and Patnaik, U.S. (1990). *Watershed Management.* Indian Council of Agricultural Research, New Delhi.

10. FAO. (2011). Irrigation in Southern and Eastern Asia in figures – AQUASTAT Survey; http://www.fao.org/nr/water/aquastat/countries_regions/ind/IND-CP_eng.pdf.

11. FAO. (2005). FAOSTAT, Food and Agriculture Organization (FAO), Rome http://faostat.fao.org.

12. Jhakhar, Praveen, Naik, B. S., Berman, D., Hombe, H. C. H., Dass, A., Sudhishri, S., and Gore, K. P. (2010). Evaluation of conservation potential of multitier cropping systems in Eastern Ghats highland zone of Orissa. *Indian Journal of Soil Conservation*, 38(3), 188–193.

13. Kumar, P. (2011). Capacity constraints in operationalization of payment for ecosystem services (PES) in India: evidence from land degradation. *Land Degradation and Development*, 22, 432–443.

14. Lenka, N. K., Choudhury, P. R., Sudhishri, S, Dass, A., and Patnaik, U. S. (2012a). Soil aggregation, carbon build up and root zone soil moisture in degraded sloping lands under selected agro-forestry based rehabilitation systems in eastern India. *Agriculture, Ecosystems and Environment*, 150, 54–62.

15. Lenka, N. K., Dass, Anchal, Sudhishri, S., and Patnaik, U. S. (2012b). Soil carbon sequestration and erosion control potential of hedge rows and grass filter strips in sloping agricultural lands of eastern India. *Agriculture, Ecosystems and Environment*, 158, 31–40.

16. Maji, A. K., Reddy, G. P. O., and Sarkar, D. (2010). *Degraded and wastelands of India: status and spatial distribution.* ICAR, New Delhi, P. 158.

17. Mukherji, S.D. (1998). Forestry situation in Eastern Ghats. *Proc. National Seminar on Conservation of Eastern Ghats*, Andhra University, Visakhapatnam, India, 247–255.

18. Patnaik, U. S., Choudhury, P. R., Sudhishri, S., Dass, A., Paikaraya, N. K. (2004). *Participatory watershed management for Sustainable Development in Kokriguda Watershed, Koraput, Orissa (IWDP, MoRD, GOI).* Technical Bulletin, CSWCRTI, RC, Sunabeda, Koraput, Orissa, India. 123p.

19. Regar, L. P., Rao S. S., Joshi, L. N. (2007). Effect of in-situ moisture-conservation practices on productivity of rainfed Indian mustard (*Brassica juncea*). *Indian Journal of Agronom,* 52(2), 148–150.

20. Rockström, J. and Barron, J. (2007). Water productivity in rainfed systems: overview of challenges and analysis of opportunities in water scarcity prone savannahs. *Irrigation Science,* 25, 299–311.

21. Sikka, A. K., Singh, D. V., Gadekar, H., and Madu, M. (2000). *Diagnostic Survey and Zonation of Eastern Ghats for Natural Resource Conservation and Management.*

Central Soil and Water Conservation Research and Training Institute, Research Center, Ootacamund, India, p. 82.

22. Sudhishri S, Dass, Anchal, and Lenka N. K. (2006). *Final report of the project: Development of regional scale watershed plans and methodologies for identification of critical areas for prioritized land treatment in the watersheds.* NATP: RRPS #17. CSWCRTI, research Center, Koraput, Odisha, 120p.

23. Sudhishri, S., Patnaik, U. S., and Mahapatra, N. (2003). Rainfall – runoff modeling for Upper Kolab catchment of Orissa, *Journal of Applied Hydrology,* XVI(3), 5–9.

24. Sudhishri, S., Dass, Anchal, and Lenka, N. K. (2008). Efficacy of vegetative barriers for rehabilitation of degraded hill slopes in eastern India. *Soil and Tillage Research,* 99, 98–107.

25. Sudhishri, Susama, Dass, Anchal, and Choudhury P. R. (2014). Combating climate change through water resource conserving technologies. *Indian farming,* 64(6), 2–7.

26. Sudhishri, S, Dass, A., Naik, B. S., and Kindal, S. (2010). JholaKundi: Low cost water harvesting device for commercial agriculture in Southern Orissa, *Indian Farmers' Digest.* 43(9), 34–35.

27. Sudhishri, S., and Dass, Anchal, (2012). Study on the impact and adoption of soil and water conservation technologies in Eastern Ghats of India. 2012. *Journal of Agricultural Engineering,* 49(1), 51–59.

28. Sudhishri, S.; Patnaik, U. S., Dass, A., Choudhury, P. R. (2002). Water resource development in Koraput region of Orissa. *Proceedings of IInd International Agronomy Congress (Vol II)* held at New Delhi. Dec. 26–30, 1358–1359.

29. Sudhishri, S., Panaik, U. S., Dass, A., Dash, B. K. (2006). Participatory water resource development for sustainable tribal agriculture: A case study, *Indian Journal of Soil Conservation,* 34(1), 60–64.

APPENDIX I

Examples of Soil Conservation Measures

Micro-catchment in Mango

Tick ditch soil working technique

Tenda for lifting water from *Jhola kundi*

KB pump used in *Jhola kundi*

CHAPTER 3

DESIGN OF WATER HARVESTING POND

S. K. PATTANAIK

Assistant Professor (Senior Scale), Department of Agricultural Engineering, College of Horticulture and Forestry, Central Agricultural University, Pasighat, Arunachal Pradesh–791102, India; Mobile: +91-9436630596; E-mail: saroj_swce@rediffmail.com

CONTENTS

3.1   INTRODUCTION

Water productivity in agriculture can be enhanced by storing water where it is scanty and by disposing water where it is excess. Water harvesting refers to storing rainwater or runoff water in the soil profile or in structures for use mainly by the plants. However, the stored water can also be used for domestic consumption, livestock, poultry and fishery purposes. Contour bunding, trenching, terracing, land leveling and grading, pond construction, etc. are the common structures of water harvesting. They promote groundwater recharge, which becomes available for use. The system of rainfall and runoff water utilization through water harvesting existed in India and other countries long ago before the development of irrigation projects. Due to changes in social structures and as a consequence of availability of water from the reservoirs, barrages, deep tube wells and urbanization with piped potable water supply, gradually the interest on water harvesting waned and they become non-functional. However, due to the ever-increasing population and climate change the decreasing share of freshwater for irrigation, domestic use, and industrial need, the water harvesting is again gaining its importance among the end users and policy makers.

A typical water harvesting system usually consists of a catchment or water collecting area, a water storage facility and auxiliary components like conveyance, sediment control structure and spillway to dispose excess water coming into the storage. In slopping land runoff flows out faster without letting the soil to be soaked to deeper depths. Leveling and grading slows down the runoff flow rate and allows the soil to absorb larger amount of water. Land leveling is done manually in small areas and by machines in large areas. Fast and accurate laser grade controlled leveling equipment are used for precision leveling and grading works. The design grade to slow down the runoff flow rate is less than 0.5%. In shallow soil and steep slope, providing a slope of less than 0.5% will entail large earthwork, which will expose the unproductive soils of the lower depth and hence is not recommended for such conditions.

The design procedure involves surveying of the target land at close interval (usually 30 m) grid points by finding the difference between the plane of the land; calculating the cut and fills at the grid pints by finding the difference between the plane of the desired slope and the existing elevations; identifying these points of cuts and fills and marking the quantum of cut or fill in stakes; followed by manual and machine operation till the desired grade obtained. In this book chapter the details of the design and construction of water harvesting pond especially the low cost technology of in-situ rainwater harvesting is discussed.

Water harvesting ponds are constructed to store rainwater in-situ or runoff water. The stored water is used for gravity irrigation to the lower elevation or by lift irrigation to higher elevation. The catchment treatments to enhance runoff are removing depressions and channelization of runoff through a drain network towards the pond. Water harvesting ponds may be irregular, circular, rectangular or square cross section. Accordingly, a pond may be an inverted frustum of a cone or a pyramid. The side slope is usually not steeper than 1:1. A circular pond has minimum perimeter for given area, however rectangular or square ponds are easy to construct. Pond water is lost due to evaporation and seepage. Seepage may be controlled by compaction and lining.

There is no satisfactory method for evaporation control. For better groundwater recharge, it is desirable to ensure a high infiltration rate from the pond. Percolation tanks are cleaned by scrapping away the deposited silt and the scrapped silt is spread on cropland for soil enrichment. In this way, a good water percolation rate through the pond is maintained. If all the runoff cannot be stored in the pond, the excess water may be diverted or the pond is provided with spillway. Ponds may be sunken type (dug out) or embankment type. In sunken pond, the entire pond depth for water storage lies below the local ground level and in the latter, the pond storage is partially above the ground and partially below.

Ponds are also useful in low-lying plain/foothill areas of high rainfall regions where crops are inundated during the monsoon and face water shortage during dry period. In such areas, a part of the land holding may be excavated to develop a pond and the excavated earth is spread over the remaining part of the holding to raise its elevation above the expected depth of runoff accumulation. Adoption of this option helps farmers to save the *kharif* crop from waterlogging damage and allows giving life- saving irrigation to the crops during the dry season using the pond water.

This chapter presents design of ponds for water harvesting.

3.2   WATER HARVESTING PRINCIPLES

The design principle of water harvesting structures is similar to the other hydraulic structures requiring a wide range of input. In many regions local thumb rules are used for designing the ponds. For hydrological design a more or less universal criterion is followed which is basically "the ratio of the catchment area to the command area." Where this ratio is known or assumed, the possible size of the field to be irrigated, i.e., the command area by the harvested water can easily be determined.

The size of catchment area can be assessed either by conducting field survey to estimate in the field or measured from the topographic map of the catchment, provided that the map is available. In several parts of the world, the value of thumb rule varies from 1:5 to 1:4, depending on rainfall magnitudes and its distribution; watershed characteristics, runoff coefficient and water requirements of the existing crops to be irrigated.

3.3   POND DESIGN CONSIDERATIONS

Assessment of water requirement and gross storage, rainwater availability and estimation of runoff are the three most important hydrologic considerations for the design of water harvesting pond. More often the area available for pond construction may be limited, in which case it is known that for a given depth how much water could be stored in the pond. Then the hydrologic design will comprise estimation of the runoff volume from the catchment area of the pond and deciding the excess water beyond the pond capacity that needs to be diverted or spilled through the pond. Thus, runoff estimation is required whether or not the pond is designed to store the entire runoff. A few simpler methods for ascertaining water requirement and gross storage, rainfall availability and runoff estimation are discussed here.

3.3.1   SIZE

Size of a pond is usually dictated by the availability of adequate land in the vicinity of the village. In rare cases do we have the option to design and build a pond of a desired size to meet the water requirements of the community including irrigation? Where we have such an option, the first step is to work out the water requirement for various needs. The next step is to determine the rainfall availability and the catchment area, above the pond site, from where the monsoon run off would be available to fill the pond. Thereafter the location, alignment and height of the earthen bund are decided, as also the location and size of the spillway to evacuate the surplus monsoon discharge.

3.3.2 WATER REQUIREMENT AND GROSS STORAGE

Unless otherwise prescribed for an area, following general guidelines may be used to determine the water requirements of a village community and the gross storage capacity of the pond.

a. **Irrigation**: Provide about 0.67 hectare meter of capacity for a hectare of irrigation. However, irrigation demand may be calculated from the irrigation requirement of the crops to be grown in the command area and their allotted area.

b. **Animal Needs**: Provide at the following rates:

i. Beef Cattle: 54–68 liters/day
ii. Dairy Cows: 68 liters/day (drinking)
iii. Dairy Cows (drinking + barn needs): 158 liters/day
iv. Pigs: 18 liters/day
v. Sheep: 9 liters/day

c. **Domestic Water Needs**: 40 liters per head per day
d. **Fish Culture**: Ensure about 1.85 m depth to provide proper temperature environments.

The storage capacity should be at least double the total water requirement to take care of evaporation and seepage losses. As a rough guide, 10 per extra storage may be provided for sediment deposition. For example if the total annual water requirement is 3000 cum and pond will have only one filling, its gross capacity should be 6600 cum (2 × 3000 + 10%).

3.3.3 ANALYSIS FOR RAINFALL AVAILABILITY

This is done by using probability concept. After the analysis, one is able to forecast at least or at most what amount of rainfall may be available during a certain time interval (a week, a fortnight, etc.). The probability analysis, for example, would tell us whether a weekly rainfall of 50 mm or more (i.e., at least 50 mm) may be expected to occur every year, every alternate year or every 5th year, and so on. Alternatively, the weekly rainfall of 50 mm is said to have a Recurrence interval (RI) of 1-year, 2-years or 5-years, respectively. The same concept applies to daily, fortnightly, monthly, seasonal and annual rainfall and the concept is used for the design of any work based upon rainfall availability, be it a small water harvesting pond or a large reservoir in an irrigation project.

The simplest probability analysis is done by ranking method. Weekly (or of any chosen duration) rainfall values are culled out from the long-term rainfall record and are arranged in a descending order, taking one value from one year of the same chosen period. Thus, if fortnightly rainfall availability is to be assessed, then the total rainfall for the first fortnight (January 1 to January 15) for all the years of available record are arranged in a descending order table, giving rise to as many entries as there are number of years of record. Actually, however, analysis of data of the whole year is not required, as runoff producing rainfalls occur during the monsoon period only. Thus, effectively one has to analyze the data of four monsoon months, namely June, July, August and September (122 days in a year) of every year of available record.

For a meaningful analysis the length of record should be large, say at least 25 years. The descending order table has the largest value at the top and the lowest value at the bottom. The values are then ranked from 1 (assigned to the highest value) to n (assigned to the lowest value) where n is the number of years of record. The probability (p) is calculated using the simple formula: $p = $ (Rank number)/(n + 1) and the recurrence interval in years is calculated as the reciprocal of p. Thus, if in a descending order table of 35 weekly rainfall totals (n = number of years of record = 35), the sixth rank value of rainfall is 45 mm, then its associated probability is 6/(35 + 1) and the recurrence interval RI = 36/6 = 6 years. The analyst concludes that a weekly rainfall of 45 mm or more may occur at least once on six years (*Alternatively*: A one-week total rainfall of 45 mm may be expected to be equaled or exceeded at least once in six years).

Following the procedure described above, one gets an idea of the maximum rainfall expected in a certain period. A pond however, may store runoff generated from smaller rainfall events also. Such events will not be included in the above analysis. To account for such events, the desired analysis method will be through water budgeting, which may be done for a chosen unit time period of 1-day, 1-week, 1-fortnight, etc. Water budgeting involves calculation of balance extra water that may be diverted towards the pond for it's filling or diverted elsewhere bypassing the pond in the pond is full. In this case, the probability analysis will be done on the available runoff, for the total number of years of record, analysis being done on annual cumulative maximum water available in the chosen time of the monsoon months.

3.3.4   RUNOFF ESTIMATION

Simple and commonly adopted methods for runoff estimation are by using the SCS USDA Curve Number method and Rational method [2–4]. The standard Curve Numbers theoretically vary from 0 to 100 and runoff coefficient varies from 0 to 1.

3.3.5   POND CAPACITY

Pond capacity is determined [1] by taking into consideration the amount of water available for storage, the water requirement, the proposed pond site features, and the economic implication of the pond construction (i.e., its cost and the expected benefits from it). While determining pond capacity, the continuous losses through seepage and evaporation and the expected withdrawals of water from the pond for irrigation or other use are taken into account. It is possible to include the pond design activity within the ambit of the water balance calculation, involving the input and output of water with respect to the pond.

Based on the result of the above analysis to finalize the pond capacity, the pond side slope, ratio of bottom area to top area, inlet and outlet structures, spillways, diversion ditches, lining options, etc. are considered. Finally, to ascertain the economic viability of incurring expenditure on pond construction and maintenance, an economic analysis is done. The following formulae are used in determining the circular, rectangular and square water harvesting ponds.

Circular Water Harvesting Pond

$$\text{Volume} = [h/6][A_1 + A_2 + 4A] \tag{1}$$

$$A_1 = \frac{\pi}{4}d_1^2, \ A_2 = \frac{\pi}{4}d_2^2, \ A = \frac{A_1+A_2}{2}$$

where, A_1 and A_2 are the areas of ends, top and bottom; A= Area of the mid-section parallel to end areas. Sectional view of circular water harvesting pond is shown in Figure 3.1.

Rectangular Water Harvesting Pond

Sectional view of rectangular water harvesting pond is shown in Figure 3.2.

$$\text{Volume} = [h/6][A_1 + A_2 + 4A] \qquad (2)$$

where, A_1 and A_2 are the areas of ends at top and bottom; $A_1 = a_1 \times b_1$; $A_2 = a_2 \times b_2$; A= Area of the mid-section parallel to end areas which can be approximately taken as $= (A_1 + A_2)/2$.

Square Water Harvesting Pond

Sectional view of square water harvesting pond is shown in Figure 3.3.

$$\text{Volume} = [h/3][A_1 + A_2 + (A_1 + A_2)^{0.5}] \qquad (3)$$

$$A_1 = a_1 \times a_1; A_2 = a_2 \times a_2$$

where, A_1 and A_2 are the areas of ends, top and bottom;

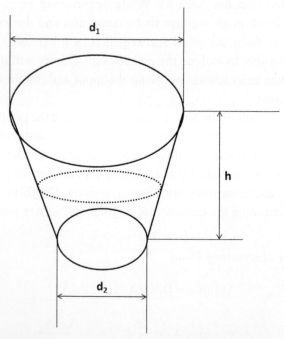

FIGURE 3.1   Sectional view of circular water harvesting pond.

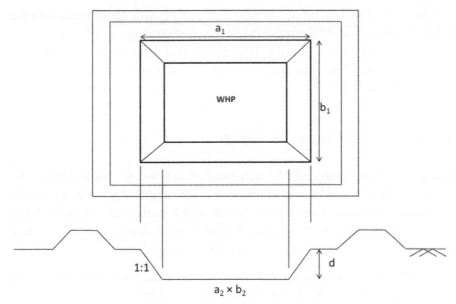

FIGURE 3.2   Sectional view of rectangular water harvesting pond.

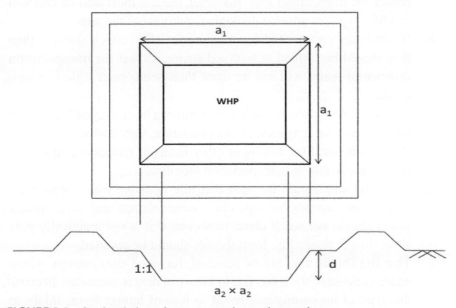

FIGURE 3.3   Sectional view of square water harvesting pond.

However, the capacity of irregular size pond is also determined with the help of contour map of the watershed area, contributing runoff to the site of the pond construction. Usually, Trapezoidal and Simpson's rules are used for this purpose. The Simpson's rule gives more accurate value than the trapezoidal rule [2].

3.3.6 SITE SELECTION OF POND

The selection of suitable site for construction of pond, the preliminary studies on different possible sites are carried out through survey. Each possible site should be studied separately. A site which proves most practical and economical is selected for construction of pond. However, some important features for site selection of a farm pond are given below:

a. From an economic point of view, a pond should be located at that site where the largest storage volume can be obtained with least amount of earthwork. This condition generally occurs at the site, where the valley is narrow and side slopes are relatively steep. Such sites tend to reduce the area of shallow water, which is very conducive to reduce the evaporation loss. However, the site must also be checked carefully, against adverse geologic conditions of the area.

b. If ponds are constructed for the purpose of livestock storage, then they should be formed at such a distance, so that the transportation distance of water will not be more than one-quarter mile in rough areas.

c. The pond site should be such that, water can be conveyed for various uses, such as for irrigation or fire protection, very easily.

d. Ponds to be used for fishing or other forms of recreation; should be readily accessible by transportation facilities.

e. The pond site should be such that, the drainage from farmsteads, feeding lots, corrals, sewage lines, mines dumps and other similar things should not reach there. However, if it is not practically possible, then drainage line from the site should be diverted.

f. That site should also not be selected, for pond construction, where sudden release of the water due to failure of dam is suspected, because this type of happening can result to loss of human lives, injury to the persons or livestock, damaging of residences or industrial buildings, rail roads or highways or interruptions in use of service of public utilities. If there is only one suitable site is available in the area, then one or more of above hazards, should be studied very carefully.

g. The site, where low hanging power lines are present in the area, should be avoided for pond construction because they create the problem for use of farm pond.

h. For selecting the pond site, a check should also be made to ensure that there is no a buried pipe line or cables at the construction site. Otherwise, they might be damaged by the excavating equipments, which can also result injury to the operator of the equipment. Where use of such site is essential, the land-owners must be contacted before starting the construction work.

3.3.7 DESIGN OF MECHANICAL SPILLWAY

Dam design must provide a suitable means for disposing the water from the pond. Most commonly, the spillways are used for disposing the water from water storage bodies. The kind of spillway to be used depends upon the size of watershed contributing the runoff to the pond. For a pond having watershed area less than 4 hectares, the vegetative spillway can be used for the purpose. A combination of vegetative and mechanical spillway should always be preferred if the area of watershed is from 4 to 12 hectares. On the contrast, when watershed area is more than 12 hectares, then the type, design and location of spillway should be performed carefully.

The kind of spillway to be used in the dam also depends upon the watershed characteristics and site conditions. If the drop height is less than 4 meter and there is less possibility of silt deposition, then a drop spillway may be used. The rectangular weir type inlet drop structures are also preferred, as they are less susceptible to make the structure clog as compared to the others. When drop height exceeds 4 meter and there is chance of silt accumulation, the drop inlet type spillways are mostly preferred; it may have a box type inlet or an arched type inlet. Arched inlets are more preferred as they provide additional advantage of arch strength in case of masonry structures.

The outlet of the spillway should be so designed and constructed that the kinetic energy gained by flowing water, as falls from the top of structure to the stilling basin, must be dissipated in maximum amount and or converted into potential energy before the flow is discharged into the stream. The stilling basin or rock riprap is provided for this purpose at the outlet end.

The spillways are generally constructed on a firm foundation and with the help of durable materials, for providing long life. When a drop structure is used in embankment of farm pond, then it is also referred by the name of surplus weir [2, 3]. The design and construction of this structure is the same as discussed earlier in the chapter of permanent gully control structures. The mechanical spillway is shown in Figure 3.4.

3.3.8 DESIGN OF EMERGENCY SPILLWAY

The main function of emergency spillway is to protect the embankment from overtopping action due to unexpected increase of inflow into the pond storage. The emergency spillway should be located at one end of the embankment. It should also be kept in view that the bottom of the emergency spillway should be fixed at the maximum expected flood level for the selected frequency of runoff, used for design of the pond. The dimensions of the emergency spillway are determined on the basis of the runoff rate to be disposed through it. If the peak flow rate for the design of emergency spillway, is known, then the dimension of it can be have, using the weir formula, given as under [2, 4]:

$$Q = CLH^m \tag{4}$$

where, Q = discharge rate, m³/s; C = coefficient of discharge; H = head on the crest, m; L = length of weir's crest, m and m = exponent.

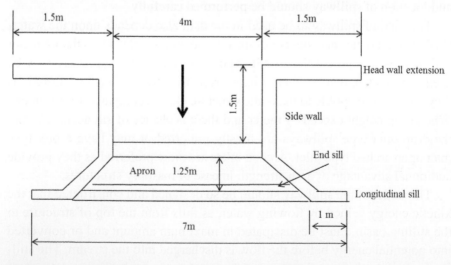

FIGURE 3.4 Sectional view of a typical mechanical spillway.

The values of C and m depend upon the shape of the weir's opening. The value of C varies according to the types of weir. The values of C and m for different types of weir are given in Table 3.1.

Regarding the side slopes of the emergency spillway, 2:1 is recommended. The depth of flow over the spillway should not be more than 30 cm. The spillway and its outlet should be fully protected against scouring caused by water flow, either by establishing the vegetation or by stone pitching. However, the emergency spillway may not be necessary, if a surplus weir has already been installed as mechanical spillway in the farm pond, which is sufficient to discharge the excess runoff.

3.3.9 DESIGN OF PIPE SPILLWAY

The pipe spillway is used as a passage of water under an embankment where storage of water is provided on the upstream side of the embankment. The spillway dissipates the energy of the falling water. A closed conduit is used below the embankment to carry the water under pressure. If low discharge flows and the conduit is only partially full, it behaves like an open channel flow. Figure 3.5 shows the components of a pipe spillway. The main components are inlet, conduit and outlet. An earth dam or embankment constructed across the conduit stores water behind it. The pipe spillway is also used as a culvert under a road or as a passage of surface water through a spoil bank along a drainage ditch. For higher heads, it requires less material for construction than a mechanical spillway.

Where the available storage is high, the capacity of the spillway can be reduced without any difficulty. Due to reduction in discharge, the peak flow in the downstream channel becomes lower and helps in channel grade stabilization and flood prevention. However, drop inlets made of small diameter pipes are subject to clogging by debris. Construction of a satisfactory earthen embankment is a prerequisite for the spillway.

TABLE 3.1 Values of C and m

Type of weir	C	m
Rectangular	0.0184	3/2
Triangular (V-notch)	0.0138	2.48
Trapezoidal	0.0186	3/2

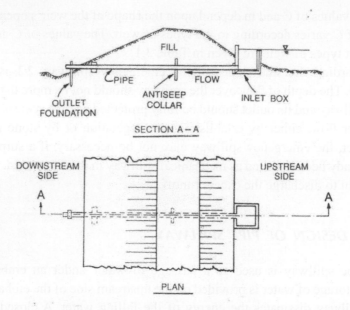

FIGURE 3.5   Plan and elevation sketch of pipe spillway.

3.3.10   DESIGN OF DROP-INLET SPILLWAY

The design of a drop inlet spillway involves the design of the earthen embankment, emergency spillway, inlet, conduit and the outlet of the drop inlet spillway. As in other structures, due consideration should be given to hydrologic, hydraulic and structural designs. In this case hydrologic design should include both peak rate of runoff and also the inflow hydrograph. As the embankment creates storage, inflow and outflow cannot be same at a particular moment. Flood routing procedure is adopted to fix the height of the embankment from the inflow-outflow characteristics. The side slope and width of the embankment are designed by taking into account stability and seepage problems.

Hydraulic design includes the determination of the inlet and the conduit size. Initially the flow is controlled by the inlet when the weir formula is applicable. As the stage (height) of water increases, the pipe flows full and the pipe or orifice formula holds good. The flow through the pipe depends on the neutral slope and slope given to it. Under most situations pipe flow occurs if the slope is less than the neutral slope. The neutral slope, S_n is defined as:

$$S_u = [(K_c V^2)/2g] / \{[1 - ((K_a V^2)/2g)^2]^{0.5}\} \qquad (5)$$

where, K_c = friction loss coefficient; V = velocity of flow, m/s; L = Length of pipe, m.

Frictional head loss, H_f for flow through the pipe is given by

$$H_f = L[(K_c V^2)/2g] \qquad (6)$$

Different flow conditions do exist depending on the position of inlet and outlet of the pipe spillway. The inlet losses may be very large under some situations and pipe flow may not occur even though its slope is less than the neutral slope. Therefore, checking is required to ensure whether the inlet or the conduit controls the flow. For pipe flow condition, the discharge is given by:

$$Q = [a\ (2gH)^{0.5}]/[1 + K_e + K_b + K_c L]^{0.5} \qquad (7)$$

where, Q = discharge, m³/s; a = cross-sectional area of conduit, m²; H = head causing flow, m; K_e = entrance loss coefficient and K_b = loss coefficients for bends.

The values of different loss coefficients are given in Appendix – I in this chapter. In case the slope S is greater than the neutral slope Sn, the outlet is not submerged and the conduit has a short length then orifice flow formula holds good. Discharge is given by"

$$Q = [a\ (C_d)\ (2gH)^{0.5}] \qquad (8)$$

The discharge coefficient, C_d for different inlet conditions can be obtained from standard tables in hydraulics book. For a sharp edged orifice, the value can be taken equal to 0.6. The inlet and the conduit sizes are determined using the above formulae and flood routing procedure. However, in case the runoff becomes higher than the design value, the structure may fail due to overtopping. To provide safety against this failure, an emergency spillway should be provided at suitable location. The spillway should have a good grass cover or stone pitching. Also the upstream side of the embankment and if possible the downstream side beyond the outlet should be stone pitched.

To prevent the failure due to piping and seepage anti-seep collars are provided below the conduit. The collars may be constructed of brick masonry or preferably of concrete. For placement of the collars, the saturated zone should be located as the collars are required to be placed in that zone.

The slope of the seepage line depends upon the type of soil and it may vary from 4:1 to 5:1 for most of the soils. The outlet of the spillway should discharge few meters away from the toe of the embankment to prevent its scouring. For this purpose, the conduit may project out by about 2 meters from the embankment and should be properly supported from the ground.

3.3.11   CONSTRUCTION OF WATER HARVESTING POND

The characteristics of watershed should be studied and analyzed before construction of the water harvesting pond (*WHP*). The surface profile at the construction site should also be surveyed for estimating the earthwork and design of spillways, etc. The contour map of the construction site is also prepared to determine the pond's capacity. Prior to start the construction work, the location of core wall, earth fill for embankment, mechanical and emergency spillways, height of dam etc., are fixed and indicated by stakes. The area to be used as base of the dam is also cleaned from trees, bushes, boulders, root stumps, etc. whatever present on the soil surface and stock piled out of the way which can be used to cover the back side of dam to provide better soil for grass seeding. The entire width of the dam site should be disked or plowed before starting to place the soil materials. While placing the earth materials, a berm towards downstream side, should also be formed to check the Sliding action.

The moisture content of earth fill plays a great role in construction of a stable earthen embankment. Generally, the 'optimum moisture content' is preferred for the purpose. Optimum moisture content may be defined as the moisture content at which soil is sufficient wet for good tilth but not so enough as to cause the moisture out from the soil under compaction. The moisture content of the fill material at the time of compaction should be such that, the maximum density of the soil can be obtained after compaction. A greater density makes the dam section more stable and impervious to seepage.

Construction of earthen embankment should be performed by the construction of core wall first, just by filling the impervious earth material in center and compacting them properly. The compaction should be continued upto the time when 85–100% density is developed there. This is obtained by rolling at optimum moisture content.

For compacting, the earth-fill materials should be placed in layer of 20–25 cm thick at the site. It should have a mild slope, away from the center of the dam. At the same time a test should also be conducted for determining the amount of water needed for spreading over the fill materials and number of passes of roller for compacting the soil, to get the desired density. It is obtained by knowing the value of initial soil moisture content. If the initial moisture content of the fill material is less then the required amount of water is sprinkled and compacting roller is put into operation. In case, when fill material is already sufficient wet, then disking for exposure of fill materials should be done to reduce the moisture content by increasing the evaporation loss.

The nature of the fill material should also be determined as it plays an important role in compaction. From study it has been observed that the soils having high clay content should be compacted at moisture level, slightly lower than the plastic limit to have better compaction. The construction of spillway should be started, when the embankment reaches to the level at which spillway is to be installed. Special care needs to be taken for compacting the soil around the components of spillway. Finally, the embankment should be trimmed, to get the designed slopes.

3.4   CASE STUDY

A low cost in-situ rainwater harvesting technology is developed at the College of Horticulture & Forestry, Central Agricultural University, Pasighat, Arunachal Pradesh is discussed. The climate here at Pasighat is humid subtropical climate, with hot summers and mild winters and receives high average annual rainfall of 4510 mm (Figure 3.6). The soil is porous, gravely and sandy and is characterized by low water-holding capacity and excessive drainage of rain and irrigation water below the root zone, leading to poor water and fertilizer use efficiency by the crops (Figures 3.7a and 3.7b).

Most of the horticultural crops are grown in these soils have low productivity because of excessive drainage of water and leaching of nutrients. The crops are subjected to acute water stress during five months from November to March, which acts as a major cause of low productivity of crops. The pisciculture in these soils is also difficult as storage of water is not possible in such soils. The high rainfall of the state can be managed well through harvesting of rainwater by polyethylene lined water harvesting pond.

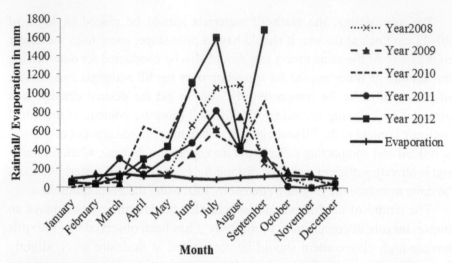

FIGURE 3.6 Monthly variation of rainfall and evaporation showing dry months at Pasighat, Arunachal Pradesh.

FIGURE 3.7A Porous and gravelly soil of East Siang, Arunachal Pradesh (Site-1).

FIGURE 3.7B   Porous and gravelly soil of East Siang, Arunachal Pradesh (Site-2).

The developed location specific low-cost technology, which is adopted by the farmers is found to be a successful one. The construction of reinforced cement concrete (RCC) made water-harvesting pond is costly and needs skill in its design and its construction takes a long time. Utilization of unused naturally depressed areas can be converted into low cost polythene lined water harvesting ponds. The hydrologic, hydraulic and structural design of the water harvesting pond is made as discussed above. Then based on the design the construction of the pond is accomplished within a month.

The site is selected first. Then with the help of earth moving machine the excavation is made based on the design dimensions. The excavated gravely/stony soil is kept as an embankment surrounding the pond by leaving a berm of at least 2 m. A trench of 30 cm × 30 cm is made surrounding the upper surface of the pond as shown in the Figure 3.8.

The surface of the pond is then laid by 5–15 cm soil cushioning with subsequent laying of 250–300 GSM (gram per square meter) finish size silpaulin film as shown in the Figures 3.9 and 3.10. It is found that the pond can be made full of water up to depth of 3 m by in-situ rainwater harvesting as the annual average rainfall is 4510 mm as shown in the Figure 3.11.

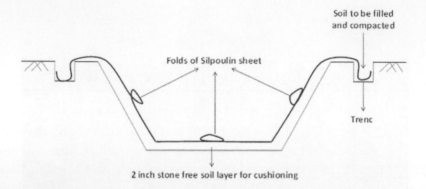

FIGURE 3.8   Sketch of polythene lined water harvesting pond.

FIGURE 3.9   Unfolding of silpaulin film.

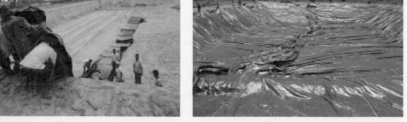

FIGURE 3.10   Spreading of silpaulin film.

FIGURE 3.11   Harvested water in silpaulin lined pond.

No catchment area is needed in this case. This reduces the overall cost of the system. In this case pipe spillway is used to dispose the excess rainfall coming into the water-harvesting pond. A free board of at least 30 cm is found to be useful (Figure 3.12).

The construction work should be completed before the onset of monsoon (prior to March–April). The harvested water can be used for irrigation as well as pisciculture activities successfully. Micro-irrigation system can be effectively used for life saving irrigation. Pisciculture can also be practiced. Oil palm variety Tenera (*Elaesis guineensis*) has shown good growth in the porous and gravelly soils due to the fact that the crops are well-irrigated and fertigated (Figure 3.13). The micro-jet irrigation is successfully provided to the oil palm crop by utilizing the harvested water from the water-harvesting pond. The six year old irrigated oil palm crop has productivity of 10 t/ha in comparison to the without irrigated productivity of 2 t/ha. Mean yield of fishes from the silpaulin-lined pond is 15.1 q/ha by practicing composite fish farming system, which is shown in the Figure 3.14. Fish species Silver carp, Catla and Grass carp shows better growth in the polythene-lined pond in comparison to Common carp, Rohu and Mrigala.

The average life of the silpaulin-lined pond is about six years and the cost per liter of storage of the water is less than 50 paisa even if depreciation cost of the silpaulin film is taken into account. One of the major causes for its low cost is that the water harvesting system does not need any extra catchment area as the high rainfall is able to fill a water-harvesting pond up to 3 m depth. As a micro-mode adoption of the technology by the small farmers are carried out. Series of small water harvesting ponds of sizes 10 m ×10 m or less are constructed in their hilly terraces. The excavation work is done manually and the harvested water is utilized to irrigate the fruit trees planted

FIGURE 3.12   Pipe Spillway disposing of excess water.

FIGURE 3.13   Micro-jet irrigation in the Oil palm.

FIGURE 3.14   Pisciculture in the Silpaulin lined water harvesting pond.

in lower terraces through gravity operated micro-irrigation system during lean period only.

3.5   SUMMARY

Water harvesting refers to storing rainwater or runoff water in the soil profile or in structures for use mainly by the plants. Water harvesting in the porous, gravely and sandy soils is a unique experience in the humid sub-tropical climate, with hot summers and mild winters climatic conditions, where the average annual rainfall is 4510 mm. The crops in these soils suffer from severe water stress during the water scarce months from October to March. The rainwater in-situ can be harvested in low-cost silpaulin lined water harvesting ponds for utilizing water in lean periods mainly for irrigation. One of the major factors in making its cost low is that it does not need any extra catchment area as the high rainfall is sufficient to fill the water harvesting pond upto a height of 2.5 m. The harvested water can be used economically through micro-irrigation system as well as pisciculture. The technology is proven to be successful one and is adopted by the citrus growers of the state.

KEYWORDS

- Capacity
- Catchment
- Command area
- Crops
- Discharge
- Dry season
- Field survey
- Fruit trees
- Infiltration rate
- Irrigation
- Kharif
- Life saving irrigation
- Micro-irrigation
- Oil palm
- Pipe inlet structure
- Pisciculture
- Polythene
- Rainfall
- Reinforced cement concrete
- Runoff
- Sharp edged orifice
- Silpaulin film
- Spillway
- Sub-tropical climate
- Sunken pond
- Topographic map
- Water harvesting
- Water harvesting pond
- Water stress

REFERENCES

1. Khanna, P. N. (1997). *Indian Practical Civil Engineers' Hand Book*. Swantantra Offset Printers, Delhi-6.
2. Murty, V. V. N. (2008). *Land and Water Management Engineering*. Kalyani Publishers, Ludhiana, India.
3. Schwab, G. O., Richard, K. F., Talkoff, W. E., and Knneth, K. B. (1966). *Soil and Water Conservation Engineering*. John Willy & Sons, New York.
4. Suresh, R. (2006). *Soil and Water Conservation Engineering*. Standard Publishers Distributors, Delhi.
5. USDA (1957). *Hydraulic Section 5, Engineering Hand Book*, Soil Conservation Service, Washington D.C.

APPENDIX I

TABLE 1A   Friction Loss Coefficient for Circular or Square Pipe at Bends

[R/D]	Bend coefficient, K_b	
	45° bend	90° bend
0.5	0.7	1.0
1.0	0.4	0.6
2.0	0.3	0.4
5.0	0.2	0.3

[R/D] = [Bend radius to pipe center line/pipe diameter].

TABLE 1B   Head Coefficients for Circular Pipe Flowing Full: $K_c = \{[57765\ n^2]/[d]^{4/3}\}$ where, d = pipe diameter, cm

Pipe diameter (cm)	Manning's coefficient of roughness, n				
	0.010	0.013	0.016	0.020	0.025
	Head coefficient				
1.25	4.2899	7.2500	10.9822	17.1598	26.8121
2.50	1.7024	2.8772	4.3583	6.8099	10.6404
5.00	0.6756	1.1418	1.7296	2.7025	4.2226
7.50	0.3935	0.6650	1.0073	1.5739	2.4592
10.00	0.2681	0.4531	0.6864	1.0725	1.6758
12.50	0.1991	0.3365	0.5098	0.7965	1.2445

TABLE 1B   Continued

Pipe diameter (cm)	Manning's coefficient of roughness, n				
	0.010	0.013	0.016	0.020	0.025
	Head coefficient				
15.00	0.1562	0.2639	0.3997	0.6246	0.9757
20.00	0.1064	0.1798	0.2724	0.4256	0.6650
25.00	0.0790	0.1335	0.2023	0.3161	0.4939
30.00	0.0619	0.1047	0.1586	0.2479	0.3873
37.50	0.0460	0.0778	0.1178	0.1841	0.2876
45.00	0.0361	0.0610	0.0924	0.1444	0.2256
50.00	0.0314	0.0530	0.0803	0.1254	0.1960
60.00	0.0246	0.0416	0.0630	0.0984	0.1537
75.00	0.0183	0.0309	0.0468	0.0731	0.1141
90.00	0.0143	0.0242	0.0367	0.0573	0.0895
120.00	0.0098	0.0165	0.0250	0.0390	0.0610
150.00	0.0072	0.0122	0.0185	0.0290	0.0453

TABLE 1C   Head Loss Coefficients for Square Conduits Flowing Full: $K_c = [19.60\ n^2]/[R^{4/3}]$, where, R = Hydraulic radius (m) = Wetted area/Wetted perimeter

Conduit size (m×m)	Manning's roughness coefficient, n			
	0.012	0.014	0.016	0.020
	Head loss coefficient for square conduit			
0.61	0.0347	0.0472	0.0616	0.0963
0.91	0.0203	0.0277	0.0361	0.0564
1.22	0.0138	0.0187	0.0245	0.0382
1.52	0.0103	0.0140	0.0182	0.0285
1.83	0.0080	0.0109	0.0142	0.0222
2.13	0.0065	0.0089	0.0116	0.0181
2.44	0.0055	0.0074	0.0097	0.0152
2.74	0.0047	0.0064	0.0083	0.0130
3.05	0.0041	0.0055	0.0072	0.0113

WATERSHED PLANNING USING INTERACTIVE MULTI OBJECTIVE LINEAR PROGRAMMING APPROACH: A CASE STUDY

RAKESH K. SAHU,[1] J. C. PAUL,[2] and BALRAM PANIGRAHI[3]

Former MTech student, Department of Soil and Water Conservation Engineering, CAET, OUAT, Bhubaneswar–751003, Odisha, India. Mobile: +91-9776356362, E-mail: rakeshkumarsahu57@gmail.com

Associate Professor, Department of Soil and Water Conservation Engineering, CAET, OUAT, Bhubaneswar–751003, Odisha, India. Mobile: +91-9437762584, E-mail: jcpaul66@gmail.com

Professor and Head, Department of Soil and Water Conservation Engineering, College of Agricultural Engineering and Technology, OUAT, Bhubaneswar, Odisha, India. Mobile: +91-9437882699; E-mail: kajal_bp@yahoo.co.in

In this chapter: One US$ = 60.00 Rs. (Indian currency); 1.00 Quintal (qtl.) = 100 kg.

Edited Version of, "*Rakesh K. Sahu, (2015). Watershed Planning Using Interactive Multi Objective Linear Programming Approach: A Case Study. Unpublished M. Tech Thesis, Department of Soil and Water Conservation Engineering, College of Agricultural Engineering and Technology, Odisha University of Agric. & Technology (OUAT), Bhubaneswar, 751003, Odisha, India.*"

CONTENTS

4.1   INTRODUCTION

Natural resource bases of agriculture, which provides for sustainable production, is shrinking and degrading, and is adversely affecting production capacity of the ecosystem. However, demand for agriculture is rising rapidly with increase in population and per caput income and growing demand from industry sector. India's economic security continues to be predicated upon the agriculture sector, and the situation is not likely to change in the foreseeable future. Even now, agriculture supports 58% of the population, compared to 75% in 1947. In the same period, the contribution of agriculture and allied sector to the *Gross Domestic Product* (GDP) has fallen from 61 to 19%. As of today, India supports 16.8% of world's population on 4.2% of world's water resources and 2.3% of global land and per capita availability of resources is about 4 to 6 times less as compared to world average. This will decrease further due to increasing demographic pressure and consequent diversion of the land for non-agricultural uses. Around 51% of India's geographical area is already under cultivation as compared to 11% of the world average. There is also an unprecedented degradation of land (107 M-ha) and groundwater resource, and also fall in the rate of growth of total factor productivity. This deceleration needs to be arrested and agricultural productivity has to be doubled to meet growing demands of the population by 2050. Efficiency-mediated improvement in productivity is the most viable option to raise production.

There is, thus, an urgent need to identify severity of problem confronting agriculture sector to restore its vitality and put it back on higher growth trajectory. Optimal utilization of available resources is very much necessary in order to meet the demand of the people. There may be several objectives as per the choice of decision maker (DM), which is to be selected as objective function. As there are many objective functions, therefore it comes under multi-objective programming. The solution of these multi-objective programming equations will give rise to the optimum planning for that area.

4.1.1 WATERSHED MANAGEMENT

Watershed management is the study of the relevant characteristics of a watershed aimed at the sustainable distribution of its resources and the process of creating and implementing plans, programs, and projects to sustain and enhance watershed functions that affect the plant, animal, and human communities within a watershed boundary. The different objectives of watershed management program are: proper land use, soil conservation, control damaging runoff, flood protection, enhancement of the ground water storage, increase of food production, appropriate use of the land resource of the watershed, thus developing forest and fodder resources, improving socio-economic condition of the inhabitants.

These management activities are within the boundaries of a drainage basin, which includes agricultural lands, forest lands, grass lands and land deteriorated by erosion. Integrated planning of watershed helps in establishment and maintenance of the ecological balance and equilibrium between man and environment. Before preparing the comprehensive watershed management plan, it is necessary to collect and analyze the rainfall data. The analysis of rainfall data helps in knowing the characteristics of rainfall, its occurrence and peak rate of runoff at different probability levels. The analysis of rainfall data is also used to assess the maximum water available from rainfall in *kharif* season. The land use and land capability are to be analyzed for planning of the cultivation practices and land treatments of the watershed.

The existing water resources of the watershed should be assessed and the total need for life saving irrigation for crops in *rabi* season must be determined.

4.1.2 MULTI-OBJECTIVE PROGRAMMING APPROACH

Managing watershed for satisfying the inhabitant's demand is a difficult task if one has to maintain a reasonable balance between usually conflicting environmental flows and demands. The solution to these complex issues requires the use of mathematical techniques to take into account conflicting objectives. Many optimization models exist for general management systems but there is a knowledge gap in linking practical problems with the optimum use of all land resources under conflicting demands in a watershed. There are a number of interactive techniques used for land and water management planning problems. The step method is one of them. It is commonly used for its efficiency, simplicity, and capacity to handle and accommodate problems of the size encountered. Further it uses the efficient simplex method, which is familiar to the watershed planners. This interactive multi-objective model seeks to identify the best compromise solution to the decision maker with each solution reflecting the decision maker's preferences.

Today, the application of all of these methods that are termed a systems approach remains critical. Perhaps now more than ever before, system approach is needed to solve watershed management problems due to the emergence of numerous new concerns relating to stakeholder participation, environmental ethics, life-cycle analysis, sustainability, industrial ecology, and design for ecological (as opposed to engineering) resilience [3].

4.1.3 OBJECTIVE OF THE STUDY

The watershed selected for the present study is Badabandha Nala Watershed of Banapur block of Khurda district, Odisha, India. It is located at latitude: 85° 10′ 30″ N and longitude: 19° 49′ 0″ E. The distance of the watershed from blockhead quarter is 10 kms. The major problems of the watershed are: though the annual rainfall of the area is very high, water is available only in monsoon; a huge water crisis is there in other seasons; present grain production is insufficient to meet the needs of the people; present fodder production is insufficient for the live-stocks of the area; most of the area is barren, which can be bought under either grass land or forestry which may solve the fodder and fuel wood problem of the area and the economic condition of the farmers is not good, etc. Keeping all these problems of the watershed in view, a research study was under taken with the following objectives:

1. to analyze the rainfall data of the watershed.
2. to analyze the land use and land capability maps of the watershed.
3. to develop an efficient compromise land allocation plan for different crops and plantation activities using multi-objective programming approach.

4.2 REVIEW OF LITERATURE

4.2.1 WATERSHED MANAGEMENT

Research studies on resource management have been carried out in Padalsinghi watershed in Beed district of Maharashtra, India in rainfed ecosystem [21] and productivity of tall traditionally grown crops like bajra, red gram, etc. have been found to increase due to the resource management. The area under irrigation as well as the water table was increased due to construction of cement plugs, nala bunds and percolation tanks. The per capita income of farmers was increased by large extend. Watershed management program, in a typically hilly watershed in Hoshiarpur, Shiwaliks, Punjab [1], demonstrated the minimization of soil erosion from the hills and flood problems, in the plain while boosting hill economy through development of hill resources with community participation.

Impact of operational research project on agricultural production through integrated watershed management in Rabni watershed situated in Panchamalin Gujarat was assessed [11]. Under the integrated watershed management program, the area was treated with different soil and water conservation measures like land leveling, contour bunding, gully plug, check dam, etc. The study revealed that the integrated watershed management program increased the ground after recharge due to different soil conservation measures. The management program gave a positive impact in increasing the cropping intensity in food grain production, fodder availability and animal population, etc. in the watershed. Efficient management of soil and water resource can act as a tool for rehabilitation of degraded watershed. Research works [18] in Bagar-Ganiyar watershed in Mohindergarh district of Haryana revealed that demand of fertilizers, seeds, plant protection measures and agricultural implements increased with time, which resulted in an increase in crop production by 20 to 25% in the given watershed. The watershed management program provided about 22% higher employment opportunity in various sectors, income of village Panchayat increased three fold during the period.

Impact of watershed management on runoff, water resource development and productivity of arable lands in Chhajawa watershed of Bara district in eastern Rajasthan [22] indicated that integrated watershed development treatments such as provision of graded bunds, gully control structures, etc. halted the process of land degradation and improved the ground water recharge, which could be successfully exploited and utilized for increasing the productivity of arable lands. The investment made was recovered in 4 years indicating that the program is economically viable. The sustainability and equity issues in the watershed management program in semi-arid tropics of Gujarat [19] revealed that the net returns were not only by 2.3 to 2.4 times but also had fair distribution across the community. Watershed management further resulted in higher investment on farm assets ranging from two to five times with better distribution in post project period as compared to project period.

A study to assess the impact of participation and integration in watershed management program in eastern African region [12] reported that 'participation' in problem diagnosis and program implementation must move beyond community-level for a socially-disaggregated processes and explicit management of trade-offs to diverse groups. Secondly, 'integration' does not come about through implementation of parallel interventions, but rather through an explicit analysis of potential trade-offs and synergies of interventions to diverse system components, and strategies to define and reach systems-level goals.

Alemayehu et al. [2] conducted a study to assess the impact of integrated watershed management system (IWSM) and to determine the land use and cover dynamics that is induced in upper Agula watershed, in semi-arid Eastern Tigray (Ethiopia). The results revealed significant modification and conversion of land use and cover of the watershed over the last four decades (1965–2005). The study further showed that IWSM decreased soil erosion, increased soil moisture, reduced sedimentation and run off, set the scene for a number of positive knock-on effects such as stabilization of gullies and river banks, rehabilitation of degraded lands. IWSM also resulted in increased recharge in the subsurface water.

The core of the proposed methodology the impact of watershed management on coastal morphology using numerical modeling in North Greece [23] refers to a coupled-calibration approach of the watershed and the coastal models, incorporating three scenarios of data availability regarding the parameters of interest. To support the applicability of this approach, a brief presentation of its successful application for an area in North Greece is also presented. The study retains the viewpoint of Integrated Coastal Zone Management and is deemed to provide an operational tool for future researchers and policy planners. The impact of rainwater harvesting (RWH) has been assessed inside the watersheds within the Albemarle-Pamlico river basins in the South-eastern USA [8]. The design strategy of runoff water harvesting (RWH) and use of the Soil and Water Assessment Tool (SWAT) model to simulate baseline and RWH scenarios for urban and agricultural land uses was proposed by them. A high adoption rate (75–100%) of RWH throughout the watersheds reduced the downstream average monthly water yield up to 16%. A lower adoption rate (25%) reduced water yield approximately 6% for the Back Creek watershed.

4.2.2   ANALYSIS OF RAINFALL

The two main components of watershed are land and water and thus for watershed management the study of both the land and water available is necessary. The water available can be studied by analyzing the rainfall of the area. The study of rainfall is stochastic in nature and is essential for forecasting of water availability of the area, crop planning, design of water harvesting structures and soil conservation structures.

Annual maximum rainfall data for Amaravati district of Maharashtra has been analyzed [17]. In this analysis authors considered annual daily rainfall data for 30 years (1966 to 1995) and fitted these data to four different probability distribution functions, i.e., Normal, Log normal, Extreme value type-I, Log Pearson type-III distribution; and probable rainfall values for different return periods were estimated and compared with the values obtained by Weibull's Method. The analysis indicated that the Log Pearson type-III distributions gave the closest fit to the observed data, hence it may be used to predict maximum rainfall. Sheng [28] employed bivariate extreme value distribution, namely the Gumbel mixed model constructed from Gumbel marginal distributions to analyze the joint distribution of correlated storm peak (maximum rainfall intensity) and amount. Based on its marginal distributions, the joint distribution, the conditional probability distribution and the associated return periods can be deduced. Parameters of the bivariate distribution model were estimated based on its marginal distributions by the method of moments (MM). The usefulness of the model is demonstrated by using it to represent multivariate storm events at the Niigata meteorological station in Japan.

Effect of global warming in the daily rainfall distribution used a mixed gamma distribution [4]. A mixed distribution was used to overcome the limitation of conventional frequency analysis, which uses a continuous distribution, as this is not applicable for the assessment of the effects of global warming. It is summarized that even though the variation of daily rainfall distribution is high due to the variation of monthly rainfall amounts, the scale parameter and the wet probability of a mixed Gamma distribution are found to be closely related to the monthly rainfall amounts. On the other hand, the shape factor remains almost the same regardless of the monthly rainfall amount. The rainfall quantities are estimated using the daily rainfall data from June to September were found to be the most similar to those using the annual maximum data. Regardless of the increasing uncertainty as the return period becomes longer, flood risk is found to be increasing as a result of global warming.

Intensity-duration-frequency curves of rainfall at Najran and HafrAlbatin regions in the Kingdom of Saudi Arabia (KSA) were derived [6] by conducting frequency analysis with 34 years rainfall (1967–2001) and using Gumbel and Log Pearson type-III distribution. The chi-square goodness-of-fit test was used to determine the best-fit probability distribution. The results showed that Gumbell distribution was best fitted than the Log Pearson type-III.

4.2.3 LAND USE AND LAND CAPABILITY

The two principal ways of increasing crop production are to increase new lands or to improve the productivity of present cropland. The development of new land is brought about primarily by drainage, irrigation and removal of shrubs, trees and rocks. A major challenge is to develop systems for greater precision in water and plant control so as to increase use efficiency of soil, water, energy resources and to improve the environment for humans.

Land capability of land areas for sustaining crops differ depending on the purpose for which the land is to be used. The value of land capability assessment lies in identifying the risks attached to cultivating the land and in indicting the soil conservation measures, which are required. Improvements to the classification rest on making the conservation recommendations more specific as in the case with treatment oriented scheme developed in Taiwan and tested in hilly islands in Jamaica [27]. Over population, decreased crop production, energy crisis and pollution (agricultural and industrial) problems in many countries are becoming much more serious. Compared to developed countries, under developed countries of the world have a higher population, much lower economic growth rate per capita and greater need for an increase in food production. The availability of tillable and pasture lands, which must produce most of our foods will require ever increasing soil and water conservation measures and more intensive land use to meet the future food demand of the people [23].

Land use varies from region to region, state to state and so also from country to country, which is dependent upon the geological structure, climate, hydrology, soils, and vegetation, human and animal life. In passing from the macro to the micro scale, gradual changes occur in the dominant variable. As far as soil erosion is concerned, climate is dominant at the macro-scale but at the smaller scales, is fairly uniform over the size of the areas being considered, and soils and vegetation becomes important (Morgan, 1986). A field study was conducted [29] to determine the soil erosion problems and the factors that affect the adoption of soil and water conservation measures in Fincha watershed, Western Ethiopia. The study showed that the annual soil loss ranges between 24 and 160 Mg ha^{-1}. The soil erosion has a significant effect on the land capability. Due to this reason, they proposed integrated soil and water conservation planning at the watershed scale.

4.2.4 MULTI-OBJECTIVE PROGRAMMING

Multiple objective programming (MOP) is related with planning models where several conflicting goals and objectives are to be optimized simultaneously. Most of the research studies for multi objective planning are based on linear programming and Goal programming approaches. Besides these, there are some interactive approaches are also developed for analysis of MOP problems formulated. A linear programming model was formulated [25] to find optimum cropping patterns subjected to land, water and labor constraints with the objective to maximize net return and production under various levels of canal release for assisting management. The model was applied to command area of distributary No. 6 of Kendrapara canal, Odisha, India. Incorporating the price fluctuations of yield, labor and fertilizer, an LP model was developed to allocate land under different crops for Kharagpur Block-I area for the year 2001 [9]. In this model, it was found that net benefits obtained by the linear programming model without considering the price fluctuations seems to be exaggerated whereas that given by the model considering price fluctuations results in reasonable and reliable estimates.

A suitable cropping pattern based on the availability of surface and ground water for the command has been developed using multi objective approach for the planning of Mahanadi delta command [3]. The objective functions were maximization of production, maximization of benefit, minimization of labor under consideration of constraints like area, water, labor, fertilizer and capital. An interactive multi-objective linear programming approach to watershed planning for the Bishunpur watershed of Gumla district of Bihar [24] took different objective functions such as maximization of food production, fodder production, fuel wood production, labor employment generation, net income generation from field crops and runoff water augmentation. The above objectives were to be maximized under a set of resource constraints like land, water, labor and nutrients, etc. The multi-objective planning for the watershed was analyzed with an interactive technique (STEP method) and an efficient and compromise solution was generated.

To derive optimal crop plans for an irrigation system with conjunctive utilization of water from surface reservoir and ground water aquifer, a fuzzy linear programming (FLP) model was [16] compared with classical linear programming (LP) model. The LP model maximizes the net benefits from irrigation activities subjected to various physical economical and water availability constraints. The FLP model maximizes the degree of satisfaction subjected to physical and economic constraints. The increase in the degree of satisfaction or truthiness with increase in number of fuzzy variables was studied and the results were reported. It was found that the fuzziness in the ground water pumping plays a prominent role in deriving the optimal operational strategies. From the optimal results, it was found that the FLP model resulted an optimal crop plan with a degree of truthiness of 0.78 taking into account the fuzziness in different variables.

A study was initiated to estimate the sediment yields under prevailing resource management systems and to design a LP based optimized land use plan for soil loss reduction for the Nagwan watershed situated in the Damodar-Barakar catchment in India [10]. The proposed spatial decision support system (DSS) was validated on 9 years (1981–1983, 1985–1989 and 1991) of sediment data yield for the watershed. The results showed that not only decrease in sediment yield was about 14.61% but also an increase in its paddy and corn crop productivities by 2.80 and 68.14%, respectively. Paul et al. [20] conducted a study, for the optimal crop planning in the Barapitanallah mini-watershed, Odisha using multi objective programming approach. In this study, the steps were taken for optimal utilization of land, water and human resources. The result indicates that benefit-cost ratio for the proposed plan was 1.3:1 and the cropping intensity was found to be 142% (kharif 86% and rabi 56%).

A model that aims at the simultaneous maximization of farmer's welfare and the minimization of environmental burden using goal programming has been proposed [12]. These techniques are implemented in Luodias River Basin in Greece to seek compromise solution in terms of area and water allocation (under different crops) resulting an figure that will come as close as possible to the DM's economic, social and environmental goals. A timber allocation model using data envelopment analysis (DEA) [13] was able to allocate forest stands, referred to as stewardship units, to different forest products companies without the need for weighting or prioritizing the allocation criteria. The allocation procedure was demonstrated in a case considering two allocation criteria: profit and employment. The allocation generated by the model was compared with random, profit based and employment-based allocations. The results showed that the model was capable of producing practical solutions and balancing the two allocation criteria. However, adding other allocation criteria was complicated by procedural concerns. Despite its current limitations, the model opens the door to future applications of DEA in forest resource allocation problems.

A technique to generate optimal results during resource allocation has been presented [26], discussed and solved. A new integer programming (IP) model was presented that supports the optimal allocation or useful resources in a university environment during the process of daily timetable generation keeping in view the priorities of both the teachers and administration. New models of planning problems based on the framework of Markov decision processes (MDPs) was developed [5]. Given these models, authors designed algorithms based on linear and integer programming that simultaneously solve for optimal allocations of resources and strategies for acting in the stochastic environments. These algorithms then formed the core of the mechanisms for allocating resources in cooperative as well as competitive multi agent settings. They showed analytically and empirically that the integrated approach leads to drastic (in many cases, exponential) improvements in computational efficiency over methods that consider the problems separately.

Massimo and Maurizio [14] presented a lexicographic goal programming (LGP) approach to define the best strategies for the maintenance of critical centrifugal pumps in an oil refinery. For each pump failure mode, the model allows to take into account the maintenance policy burden in terms of inspection or repair and in terms of the manpower involved, linking them to efficiency-risk aspects quantified as in FMECA methodology through the use of the classic parameters occurrence (O), severity (S) and detect ability (D), evaluated through an adequate application of the Analytic Hierarchy Process (AHP) technique. An extended presentation of the data and results of the case analyzed was proposed in order to show the characteristics and performance of this approach.

A fuzzy goal-programming model having multiple conflicting objectives and constraints pertaining to the machine-tool selection and operation allocation problem, and a new random search optimization methodology termed Quick Converging Simulated Annealing (QCSA) has been used [15]. The main feature of the proposed QCSA algorithm is that it outperforms genetic algorithm and simulated annealing approaches as far as convergence to the near optimal solution was concerned. Moreover, it is also capable of eluding local optima. Extensive experiments were performed on a problem involving real-life complexities, and some of the computational results were reported to validate the efficacy of the proposed algorithm.

The components of the study to find out a preferable solution for the local agencies in Lake Qionghai watershed in China [30] were agriculture, tourism, macroeconomics, cropland use, water supply, forest coverage, soil erosion and water pollution using an interval fuzzy multi objective programming (IFMOP). This study showed that the interval fuzzy multi objective programming for Lake Watershed system (IFMOPLWS) is a powerful tool for integrated watershed management planning and can provide a solid base for sustainable watershed management. William et al. [31] applied an integrated multiple criteria decision making approach to the resource allocation problem. In the approach, the Analytic Hierarchy Process (AHP) was first used to determine the priority or relative importance of proposed projects with respect to the goals of the universities. Then the Goal Programming (GP) model incorporating the constraints of AHP priority, system and resource was formulated for selecting the best set of projects without exceeding the limited available resources. The projects include 'hardware' (tangible university's infrastructures), and 'software' (intangible effects that can be beneficial to the university, its members, and its students). In this paper, two commercial packages were used: Expert choice for determining the AHP priority ranking of the projects and LINDO for solving the GP model.

The study in this chapter was undertaken for the management of such a watershed identified by soil conservation department of Govt. of Odisha. The prime aim is to maximize the different basic objectives like food, fodder, fuel wood, net income generation, labor employment generation and runoff water augmentation of the watershed using multi-objective programming approach.

4.3   THEORETICAL CONSIDERATIONS

Multi-objective programming (MOP) involves optimization of two or more objective functions. The MOP differs from the single objective optimization problem only in the expression of respective objective functions. Watershed management involves a number of conflicting goals and objectives. These goals are to be treated with a set of operational constraints, in order to find the best compromise solution. In this present study an interactive multi-objective mathematical model for the watershed was developed and analyzed by an interactive step method.

4.3.1   ASSUMPTIONS IN THE MODEL

1. The relationship between the variables in objective functions and the constraints are linear.
2. The soil and climate characteristics of the watershed are uniform throughout.
3. Planning is done for two seasons in a year, i.e., *kharif* and *rabi* season.
4. The water requirement in the *kharif* season is met from rainfall at different probability of occurrence and in *rabi* season assured water from ponds is provided.
5. The management practices of the land and cropping pattern is similar. Hence the yield and benefit under a particular crop is constant.

4.3.2 MATHEMATICAL PROGRAMMING MODEL

4.3.2.1 Single Objective Models

The most commonly used optimization technique in any sphere of management is linear programming model that consists of a linear algebraic objective function and linear algebraic constraints. The general linear programming model can be defined as follows:

$$\text{Maximize (minimize)}, Z = \sum C_j x_j \tag{1}$$

$$\text{Subjected to } \sum\sum a_{ij} x_j \ (\leq = \geq) \ b_j \tag{2}$$

$$x_j \geq 0 \tag{3}$$

For i = 1, 2, 3, x and j = 1, 2, 3, n

Here Eq. (1) is the objective function, which is to be maximized or minimized under a set of constraints given by Eqs. (2) and (3). The Eq. (3) is known as non-negative constraints. The most common method used to solve linear programming is simplex method. There are a number of standard computer programs now a day available to solve the problems. A figure showing ingredients of linear programming model is shown in Figure 4.1.

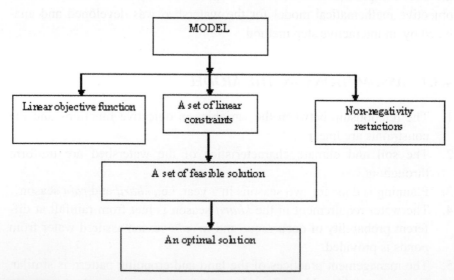

FIGURE 4.1 Components of a linear programming model.

4.3.2.2   Multi-Objective Models

Multiple objective programming (MOP) model is a mathematical technique concerned with a problem, in which several functions are to be optimized simultaneously under a set of management and operational constraints. The general description of a MOP problem involving, 'p' objectives, 'n' decision variables and m constraints can be expressed as:

$$\text{Max } Z(x) = [Z_k(x)], \text{ for } k=1, 2, ..., p \tag{4}$$

$$\text{Subjected to: } g_i(x) \ (\leq=\geq) \ b, \text{ for } i=1,2, ..., m \tag{5}$$

$$\text{and } x \geq 0 \tag{6}$$

where, 'Z' is a vector valued function consisting of the objective functions $Z_k(x)$, for $k = 1,2,...,$ p and 'x' is a vector consisting of decision variables, they are $x_1, x_2, x_3 ..., x_n$. Equation (5) is a set of constraints, defining the feasible regions of the decision variables.

If $Z_k(x)$ and $g_i(x)$ for i=1, 2, ..., x and k=1, 2, ..., p are linear, then the MOP formulation is referred as multiple objective linear programming (MOLP).

The concept of optimal solution in single objective optimization has a different interpretation in MOP. In MOP, the compromise solution concept is more important than the optimal solution, because the solution, which maximizes one objective will not in general maximize the other objectives. Therefore, a non-inferior solution is obtained by making trade-offs between the different objectives to improve and attain a satisfactory level for each objectives.

4.3.2.3   Classification of Multi Objective Programming Techniques

Multiple objective programming techniques have been classified into four major categories as follows.

I. Generating techniques:
   i. Weighting method.
   ii. Constraint method.
   iii. Derivation of a functional relationship for the non-dominated set.

II. Non interactive technique:

(Techniques which rely on prior articulation of preferences)
   i. Goal programming.
  ii. Assessing utility functions.
 iii. Estimation of optimal weights.
  iv. Electro method.
   v. Surrogate worth trade-off method (SWT).

III. Interactive technique:

(Techniques which rely on progressive articulation of preferences)
   i. Step method.
  ii. Iterative weighting method.
 iii. Sequential multi objective problem solving (Sempos).

IV. Techniques that generate alternatives:

i.Hop skip and Jump (HSJ) method.

In the present study, the step method (one of the interactive technique) was used to analyze the multi-objective approach for watershed management, because of its simplicity, easy to understand and also its capacity to accommodate the size of the problem.

4.3.2.4 The Step Method

The step method is an interactive technique that converges to the best compromise solution, in no more than 'p' iterations, where 'p' is the number of objectives. The method is based on a geometric notation of the best fit: The minimum distance from an ideal solution with modifications to a generated solution. The step method involves following algorithm:

Step 1: Construction of the payoff table
   i. All the 'p' individual maximization problems are solved and optimal solutions for each of the 'p' objectives are found out.
  ii. Let the solution that maximizes objective 'K' where K = 1,2, ..., p be $X^k = (X_1^k, X_2^k, X_3^k, ..., X_n^k)$.
 iii. If there are alternative optima for any of these problems, then those solutions are chosen among the alternative optima that are non-inferior.
  iv. The values of each objective, at each of the 'p' optimal solutions are computed. Let they are: $Z_1 (X^k), Z_2 (X^k), Z_3 (X^k), ..., Z_n (X^k)$, for K = 1,2,3, ..., p. which gives us p values of each of the 'p' objectives.

v. The 'p' values of each of the 'p' objectives are arranged in a table, in which the rows correspond to X^1, X^2, X^3, ..., X^p and the columns are labeled by the objectives $Z_1(X^k)$, Z_2 (X^k), Z_3 (X^k), ..., Z_n (X^k).

Payoff Table (Matrix):

Solution Objectives

$$Z_1(x^k)\ Z_2(x^k)\ ...,\ Z_p(x^k)$$

$$X^1\ Z_1(x^1)\ Z_2(x^1)\ ...,\ Z_p(x^1)$$

$$X^2\ Z_1(x^2)\ Z_2(x^2)\ ...,\ Z_p(x^2)$$

$$...$$

$$X^p\ Z_1(x^p)\ Z_2(x^p)\ ...,\ Z_p(x^p)$$

Step 2: From the payoff table, the maximum and the minimum values for each individual optimization of the kth objective is found out. Let M_k be the maximum and n_k be the minimum value.

Step 3: From the objective function, the value of normalizing tem, $\left[\Sigma(Cj^k)^2\right]^{\frac{-1}{2}}$ i.e., is computed, where Cj^{k^2} are the coefficients of the objective functions assuming that they are linear:

$$Z_k(x) = C_1{}^kX + C_2{}^kX + \text{-------} + C_n{}^kX$$

$$C_n{}^{k^2} = C_1{}^2 + C_2{}^2 + C_3{}^2 + \text{-----} + C_n{}^2,\ \text{where:}$$

$$K = 1,2,\ \text{-------},\ p\ \text{and}\ j = 1,2,\ \text{------},\ n$$

Step 4: Calculation of 'α' value

$$\alpha_k\ (\text{alpha}) = \frac{M_k - n_k}{M_k}\left[\sum(Cj^k)^2\right]^{\frac{-1}{2}}, \tag{7}$$

$$\frac{M_k - n_k}{M_k} = \text{Scaling term and}\ \left[\sum(Cj^k)^2\right]^{\frac{-1}{2}} = \text{Normalizing term}$$

Step 5: Computation of the weights W_k

$$W_k = \frac{\alpha_k}{\sum \alpha_k} \tag{8}$$

Step 6: From the values of W_k, the values, which are zero or nearer to zero such objectives have already attained maximum value. For other objectives the compromise solution is to be obtained.

Step 7: Solve the linear programming

$$\text{Minimize 'd'} \tag{9}$$

$$\text{Subjected to } W_k \, [M_k - Z_k(x)] - d \leq 0 \tag{10}$$

For k= 1, 2, ...p

$$d \geq 0 \tag{11}$$

Step 8: Let the solution obtained be *x(i)*. Then the value of each objective function is calculated.

Step 9: The above solution, $Z_k[x(i)]$, is showed to the decision maker (DM).

    a.   If the DM is satisfied then the process will be stopped and the current solution is the best compromise solution.

    b.   If the DM is not satisfied, then some trade-offs are done and a new set of compromise solution may be obtained.

    c.   If the DM is not satisfied, then some other interactive techniques may be followed.

4.3.3   MULTI OBJECTIVE WATERSHED MANAGEMENT

4.3.3.1   Objective Functions

The present study considered the following objectives for proper management of land and water resources of Badabandha Nala Watershed (0407010801100104) of Banapur block of Khurda district (Odisha):

1. Food production.
2. Fodder production.
3. Fuel wood production.
4. Net income generation.
5. Labor employment generation.
6. Runoff water augmentation.

Food production

According to Maslaw's theory, food ranks first among other objectives considered for land and water management problems. Therefore, to meet the food demands of the people of the watershed, and to meet their nutritional requirement, different crops are proposed. Among the crops two cereals, two pulses and one oilseed crop is the preferred choice, basing on the requirement of the people. It is presumed that the existing water potential and created water potential will meet the future water demands of the crops that would be eventually taken in that area. The objective function for maximization of food production is given by

$$\text{Max } Z_1(X) = \sum\sum y_{ij} x_{ij} \tag{12}$$

where, y_{ij} is the yield of jth crop in i^{th} season in tons/ha, and x_{ij} is the area under j^{th} crop in i^{th} season in ha; j= 1, 2, 3, 4, 5 represents upland paddy (paddy-I), medium land paddy (paddy-II), low land paddy (paddy-III), maize, arhar in kharif, and j=1, 2, 3 represents paddy-III, mustard and mung in rabi season.

Fodder production

The fodder production is to be maximized in order to meet the demand of the livestock in the watershed. Livestock management is considered as an important parameter in the process of integrated watershed management planning. The green fodder produced from the developed grass and the dry fodder produced from the cereal crops would meet the requirement of the bovine population of the watershed.

The objective function for maximization of fodder production is given by

$$\text{Max } Z_2(X) = \sum\sum f_{ij} x_{ij} \tag{13}$$

where, f_{ij} is the fodder yield from j^{th} crop or plantation in i^{th} season, tons/ha, x_{ij} is the area under j^{th} crop/ plantation in i^{th} season in ha; j=1, 2, 3, 4 represents paddy-I, paddy-II, paddy-III and maize in kharif, and j = 1 for paddy-III in rabi season, j= 6, 7 represents hybrid Napierbajra grass and subabool in kharif and 4, 5 represents the above in rabi season.

Fuel wood production

To meet the demands of fuel wood and other domestic uses of the people of the watershed, the plantation of fuel wood and timber trees are mostly necessary. For this purpose, the plantation crop Subabool is preferred as it can serve the purpose of fuel wood and fodder for the livestock. It also utilizes the wasteland and contributes to the economic upliftment of the individual farmers as well as the people in the watershed. The objective function for the maximization of fuel wood production is given by

$$\text{Max } Z_3(X) = \sum\sum w_{ij} x_{ij} \tag{14}$$

where, w_{ij} is the fuel wood production from j^{th} plantation in i^{th} season in t/ha and x_{ij} represents area under fuel wood plantation in ha; $j = 7$ represents subabool plantation in kharif and 5 in rabi season.

Net income generation from field crops and plantation

Maximization of net income from cereals, pulses and oilseed crops and plantation helps the farmers to boost their economic status. The farmers are provided with 12% interest loans for their agricultural inputs. The objective function for the maximization of net income generation is

$$\text{Max } Z_4(X) = \sum\sum N_{ij} x_{ij} \tag{15}$$

where, N_{ij} is the net income from j^{th} crop in i^{th} season in Rs/ha and x_{ij} is area under jth crop/plantation in i^{th} season in ha; j=1, 2, 3, 4, 5, 6, 7 represents paddy-I, paddy-II, paddy-III, maize, arhar, hybrid Napierbajra grass and subabool in kharif, and j=1, 2, 3, 4, 5 represents paddy-III, mustard, mung, hybrid Napierbajra grass and subabool in rabi season.

Labor employment generation

In order to boost the job prospects of the rural population like small, marginal farmers and landless laborers, generation of employment and enhancement of present employment facility is considered in this planning. The objective function for the maximization of labor employment generation is given by

$$\text{Max } Z_5(X) = \sum\sum L_{ij} x_{ij} \tag{16}$$

where, L_{ij} is the labor need of the jth crop in i^{th} season in man-days/ha and x_{ij} is area under j^{th} crop in i^{th} season in ha; j=1, 2, 3, 4, 5, 6, 7 represents crops like paddy-I, paddy-II, paddy-III, maize, arhar, hybrid Napier bajra grass and subabool in kharif, and j=1, 2, 3, 4, 5 represents the crop like paddy-III, mustard, mung, hybrid Napier bajra grass and subabool in rabi.

Runoff water augmentation

It is assumed that water requirement in kharif season is to be met from the rainfall at desired probability levels, whereas the water requirement in rabi season is met from the storage structures. As the existing source of water in the watershed is insufficient to meet the demand of the crops, so an effort is made to augment (store) the runoff water in ponds for use by the crops in rabi season. The objective function for maximization of runoff water augmentation is given by

$$\text{Max } Z_6(X) = \sum C_j x_{ij} \qquad (17)$$

where, C_j=capacity per unit ha, i.e., ha-m/ha and x_j=Area under ponds in ha.

4.3.3.2 Constraints

1. **Land**
   a. **Total treatable watershed area**
   The area allocated for different land practices and treatments should be less than or equal to the total treatable area of the watershed.

$$\sum\sum x_{ij} + 2\sum x_j \leq 2TA \qquad (18)$$

where, TA is the total treatable area of watershed, in ha. The coefficient 2 is coming because the area is used in both the seasons: Kharif and Rabi.

   b. **Total agricultural land**
   The area allocated for different field crops should be equal to or less than the total cultivable area of the watershed.

$$\sum\sum x_{ij} \leq 2CA \qquad (19)$$

where, CA is the total cultivable area available for field crops in the watershed in ha.

   c. **Total forest and Gochar land**
   The land allocated for the grass cultivation and forestry plantation should be less than or equal to the total land available in the watershed for forest and grass cultivation.

$$\sum\sum x_{ij} + 2\sum x_j \leq 2FA \qquad (20)$$

where, FA is the total forest land, land to be used for grass and forestry cultivation which also include area required for ponds.

d. Total cultivable upland

Here in the present study crops proposed for upland are upland paddy (paddy-I), maize and arhar in kharif. So the total area allocated to these crops should be equal to or less than total cultivable upland.

$$\sum x_j \leq UL \tag{21}$$

where, j=1, 4, 5 represents to the crops paddy-I, maize, arhar; UL= total cultivable up land of the watershed.

e. Total cultivable medium land and low land of the watershed

The crops proposed for medium land and low land are medium land paddy (paddy-II), arhar and low land paddy (paddy-III) respectively. The total area allocated to these crops should be less than or equal to the total cultivable medium land and low land respectively.

$$\sum x_{ij} \leq ML \tag{22}$$

where, j=2, 5 represents paddy-II and arhar, I = 1, i.e., kharif season; j= 2,3 represents mustard and mung, I = 2, i.e., rabi season; ML=Total cultivable medium land.

$$\sum x_j \leq LL \tag{23}$$

where, j= 3 represents paddy-III, LL=Total cultivable low land.

2. Water requirement

The quantity of water needed for each crop and plantation should be less than or equal to the water resource available from rainfall and different storage structures. It was assumed that the water requirement in kharif will be met from the rainfall and during rabi it will be met from the water storage structures.

a. For *kharif* season

$$\sum WR_{1j} x_{1j} \leq WR_e \tag{24}$$

where, WR_{1j} is the water requirement of jth crop in kharif, m; x_{1j} is the area under j^{th} crop in kharif, ha; WR_e is the total water resource available at desired probability level (%P), ha-m.

b. For *rabi* season

$$\sum WR_{2j}x_{2j} \leq WR_e \text{(tar)} \tag{25}$$

where, WR_{2j} is the water requirement of jth crop in rabi, m; x_{2j} is the area under j^{th} crop in rabi, ha; WR_e(tar) is the total water made available from different storage structure, i.e., ponds.

3. Nutritional requirements

For good health the basic source of nutrition is protein and calorie. The total nutritional component available from individual crops must be equal to or greater than the total requirement of people.

a. Protein

$$\sum\sum P_{ij}x_{ij} \geq PR \tag{26}$$

where, P_{ij} is the protein available from j^{th} crop in i^{th} season in kg/ha; x_{ij} is the area under j^{th} crop in i^{th} season in ha; and PR is the total protein requirement of the people in kg.

b. Calorie

$$\sum\sum C_{ij}x_{ij} \geq CR \tag{27}$$

where, C_{ij} is the total calorie available from j^{th} crop in i^{th} season in Kcal/ha; x_{ij} is the area under j^{th} crop in i^{th} season in ha; CR is the total calorie requirement of the people in Kcal.

4. Labor

The manpower required for cultivation of different crops should be less than or equal to labor force available in the watershed. The labor need by different crops monthly and total manpower available in a month is considered.

$$\sum\sum L_{ij}x_{ij} \leq LA \tag{28}$$

where, L_{ij} is the labor need for j^{th} crop in every month in i^{th} season in man-days/ha; x_{ij} is the area under j^{th} crop in i^{th} month in ha; LA is the total labor force available in man-days.

5. **Fodder requirement**

The fodder available from the field crops (i.e. dry fodder) and plantation (i.e. green fodder) should be more than or equal to the requirement of the livestock population of the area.

a. **Field crops**

$$\sum\sum f_{ij}x_{ij} \geq FDR \qquad (29)$$

b. **Plantation crops**

$$\sum\sum f_{ij}x_{ij} \geq FDR \qquad (30)$$

where, f_{ij} is the fodder available from j^{th} crop/plantation in i^{th} season in t/ha and x_{ij} is the area under j^{th} crop in ith season in ha; FDR is the total fodder requirement of the livestock in the area, ton

6. **Fuel wood requirement**

The fuel wood produced from the plantation crops must be more than or equal to the fuel wood requirement of the people of the watershed.

$$\sum\sum W_{ij}x_{ij} \geq W_r \qquad (31)$$

where, W_{ij} is the wood available from j^{th} plantation in i^{th} season in t/ha, x_{ij} is the area under j^{th} crop in i^{th} season in ha and W_r is bulk fuel wood need of the people, ton.

7. **Food requirement**

Total production of paddy, maize, arhar, mustard, mung should be greater than or equal to the actual demand of the total population in the watershed area.

a. **Paddy requirement**

$$y_{11}x_{11} + y_{12}x_{12} + y_{13}x_{13} + y_{21}x_{21} \geq Pd \qquad (32)$$

where, y_{11}, y_{12}, y_{13}, yield of paddy –I, paddy-II, paddy-III in kharif respectively, tons/ha; y_{21} represents yield of paddy-III in rabi season, tons/ha; $x_{11}, x_{12}, x_{13}, x_{21}$ area under different paddy varieties in, ha; Pd bulk requirement of paddy in tons

b. Maize requirement

$$y_{14}x_{14} \geq M_z \tag{33}$$

where, M_d is the bulk need of maize, tons.

c. Arhar requirement

$$y_{15}x_{15} \geq A_r \tag{34}$$

where, A_ds the bulk need of arhar, tons.

d. Mustard requirement

$$y_{22}x_{22} \geq M_{sd} \tag{35}$$

where, M_{sd} is the bulk requirement of mustard, tons.

e. Mung requirement

$$y_{23}x_{23} \geq M_{ng} \tag{36}$$

where, M_{ng} is the bulk requirement of mung, tons.

8. Area under ponds

$$x_3 \geq A_{tar} \tag{37}$$

where, A_{tar} is the area required for proposed ponds in ha.

9. Non-negative constraints
Area under different crop/plantation, and reservoir should be either positive or zero, it should not be negative.

$$\sum\sum x_{ij} \geq 0 \tag{38}$$

$$\sum x_j \geq 0 \tag{39}$$

The above multi-objective watershed management model consists of six objectives, thirteen decision variables subjected to a set of constraints. Steps are taken to solve these multi-objective problems using computer software and analyze the same by an interactive technique (step method).

4.4   DATA AND METHODOLOGY

4.4.1   WATERSHED

4.4.1.1   Location

The watershed selected for this study is Badabandha Nala watershed a micro watershed (code No.-0407010801100104) which lies in Banapur block of Khurda district (Odisha). The watershed is located at a distance of 10 Kms. From block headquarter. The watershed has latitude: 85^0 $10'$ $30''$ N and longitude: $19°$ $49'$ $0''$ E.

4.4.1.2   General Information

The selected watershed, i.e., Badabandha Nala micro watershed comprises of five villages namely Jadupur, Ghasediha, Panchugaon, Barapatana and Nachhipur with total geographical area of 240 ha. Out of this total area, the treatable area is 234 ha. The project area comes under the east and south-eastern coastal plain agro climatic zone of the state. The index map of the watershed is given in Figure 4.2. All raw data are presented in appendices I to XIV at the end of this chapter.

4.4.1.3   Land use Pattern

The total geographical area of the watershed is 240 ha, out of which 6 ha comes under homestead, nallahs, roads, etc. So the total treatable area of the watershed is 234 ha. It is around 97.5 % of the total area. Under agricultural land upland consists of 22.19 ha which is the main foci of soil erosion as these are present in upper reaches of the watershed. The medium and low lands of the watershed are 130.19 and 18.28 ha respectively. The details are shown in Table 4.1. The forestland which constitutes 14.80 ha is now completely depleted due to over exploitation. So these areas need fresh plantation and conservation practices. The gochar land constitutes 10.23 ha and this is almost barren. It needs treatment and proposed to be under fodder cultivation. The culturable wasteland constitutes 21.56 ha.

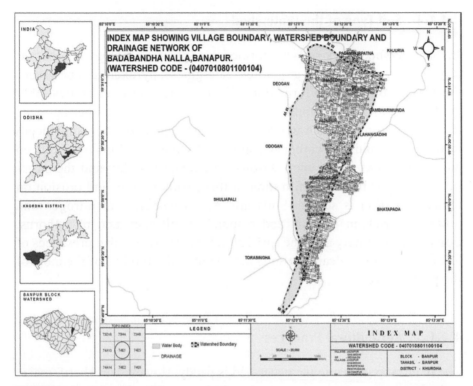

FIGURE 4.2 Index map of the watershed.

TABLE 4.1 Land Use Pattern of the Watershed

S. No.	Land use type	Area (ha)	% of total Area
1.	i) Up land	22.19	9.25
	ii) Medium land	130.19	54.25
	iii) Low land	18.28	7.61
	Total cultivable land	170.66	71.11
2.	Forest land	14.80	6.17
3.	Gochar land	10.23	4.26
4.	Culturable wasteland	21.56	8.98
5.	Unculturable wasteland	16.75	6.98
6.	Home stead, Nallah ground, roads, etc.	6.00	2.5
Grand Total		**240.00**	**100**

4.4.1.4   Soil Type of the Watershed

In the watershed black soils are formed due specific lithology or topography. Parent rocks of basic granulites, calc-gniess, pyroxenites and granodorites are conducive to formation of black soils. These rocks invariably contain plagioclases in appreciable amounts, which on weathering make the soil environment rich in calcium. A lime kankar zone at some depth in the profile and free carbonates are usually present. Soils exhibit deep and wide cracks in summer seasons. The texture is clay and the structure is angular blocky. Infiltration in these soils is slow and erosion on upland situation is severe. Soils are low to moderate in nitrogen and potassium, rich in calcium and respond 10 nitrogen and phosphorus. Soils are moderately alkaline (pH 7.9 to 8.5). There are also presences of hillock the white calcareous soils are found at the foothill of the most of the project area.

4.4.1.5   Agricultural Practices

The main crop grown in the watershed is paddy. However, in kharif, maize and arhar are grown in some areas. In rabi season, crops like mustard, mung and low land paddy are grown in some area. Due to lack of irrigation facility, crops grown in limited areas in rabi season. Most of the rainfall in monsoon flows as runoff and ultimately lossed due to limited storage structures. This causes moisture deficit in upland and medium land areas. The area under crop cover is more in kharif than rabi season.

The cropping intensity is highest in Kharif season (65%) covering most of the lands. But it is less than one third (28%) in Rabi season and in summer season (0%). The cropping intensity in Rabi season is contributed mostly from the relay and mixed cropping in uplands during rainy season and due to fresh crops in medium lands. Seasonal cropping pattern is rich in Kharif and lean in Rabi season. It is poor in summer season. The entire seasonal cropping pattern is dependent of the receipt and harvest of rainwater. So the measures to improve the moisture storage, recharge of soil pores, and storage of runoff and considerate use of moisture by cropping patterns as per the water availability would miraculously prosper the cropping pattern of the area.

4.4.1.6   Socio-Economic and Population

The socio-economic study reveals that most of the people in the watershed have poor economic status. They are annual income is very low. Most of the farmers have very less land holding, i.e., within one hectare. They come under small and marginal farmer's category. Moreover the land is very much fragmented and scattered. Most of the people depend upon the agriculture for their income, though some of them are engaged in other activities like industrial worker, smithy and pottery, etc. Productivity is less in the area due to lack of irrigation facility, agricultural inputs, traditional farming system, and restricted use of improved technology. Human resource is the important on as he is the user as well as the abuser of all the above resources. It is his requirement and decision that can help to keep the resources sustainably productive. The population base of the watershed area is given in the Table 4.2.

4.4.1.7   Livestock Population of the Watershed

For the economic upliftment of the people in watershed livestock plays an important role. To estimate the total fodder requirement of the watershed, knowledge of livestock population is utmost necessary. The village wise livestock population of the watershed is given in the Table 4.3.

4.4.1.8   Climate of the Watershed

The climate of the watershed area is hot and dry sub humid. The mean maximum temperature is 39°C during the month April and may and the mean minimum temperature is 15°C during the month of December and January.

TABLE 4.2   Population Details of the Watershed

Name of the village	Male	Female	Children	Total
Ghasedehi	370	328	74	772
Barapatana	175	147	21	343
Panchugaon	459	430	73	962
Nachhipur	120	90	17	227
Jadupur	0	0	0	0
Total	**1124**	**995**	**185**	**2304**

TABLE 4.3 Livestock population of the watershed

Name of the village	Livestock population(Nos.)				
	Cow	Bullock	Buffalo	Sheep	Goat
Ghasedehi	68	42	8	17	33
Barapatana	25	12	2	7	17
Panchugaon	81	20	7	25	38
Nachhipur	27	8	3	9	24
Jadupur	0	0	0	0	0
Total	201	82	20	58	112

The average annual rainfall is 1250.43 mm and the portion of rainfall received during the monsoon (June September) constitutes about 70–75% of the total annual rainfall. As such irrigation becomes extremely momentous not only for overcoming the enigma of moisture stress for the rest part of the year, but also at the time of failure of monsoon.

4.4.1.9 Types of Existing Water Source of the Watershed

1. Dugout pond constructed by soil conservation department.
2. Dugout pond owned by village panchayat

From these two water sources, small patches of land have been covered in rabi season. Therefore, it is necessary to quantify the water available for its reuse to rabi season crops.

4.4.1.9.1 Pond No.1

The inside dimensions, side slope and the depth from bottom up to the crest level of the spillway were measured for quantifying the storage volume. The storage capacity of the water harvesting pond no.1 up to crest level was estimated following trapezoidal rule. Assuming 20% as losses occurred due to evaporation, seepage and percolation, the rest 80% is considered as utilizable water for irrigation purpose.

4.4.1.9.2 Pond No.2

The dimensions of the second pond are collected to estimate the capacity of the pond. The capacity of the pond was quantified at different depths taking contour interval of 0.5 m through trapezoidal rule. A water depth of 1.5 m is kept as dead storage for bathing and other purposes of the people and the domestic animals. The storage volume between the crest level of the spillway and the dead storage level was considered to be available for crop use. Net quantity of water available for its utilization is obtained after satisfying the unavoidable losses (assumed 20%).

4.4.1.10   Collection and Analysis of Rainfall Data

The monthly rainfall data for the watershed was collected from OUAT metrological laboratory, Bhubaneswar for 34 years, i.e., from 1978 to 2012. These rainfall data has been fitted into different probability distribution functions, i.e., Normal, Log Normal, Gumbel extreme value maximum, Log Pearson type-III and Weibull and probable rainfall values are obtained according to their best fit distribution. The probability analysis of the rainfall is carried out with the help of "FLOOD" software.

4.4.1.10.1   Types of Probability Distribution Functions

All the probability functions are taken while checking for the best fit distribution. Then depending upon the RMSE value and Mean error value, the best distribution for a particular month of 34 years is tested, using the "FLOOD" software. The results obtained are shown in the results and discussions section.

1. Normal distribution

$$f(x,\mu,\sigma)=\frac{1}{\sigma\sqrt{2\pi}}e^{-\frac{(x-\mu)^2}{2\sigma^2}} \tag{40}$$

where, μ and σ are mean and standard deviation of the variate 'x' respectively.

2. Log normal distribution

$$f(x,\mu,\sigma)=\frac{1}{x\sigma_y\sqrt{2\pi}}e^{\frac{-(\ln x-\mu_y)^2}{2\sigma_y^2}}, x>0 \tag{41}$$

where, y = logx, μ_y and σ_y are the mean and standard deviation of variate, respectively.

3. Gumbel extreme value maximum

$$f(x) = \frac{1}{\alpha} e^{\left[\frac{-(x-\mu)}{\alpha} - e^{\frac{-(x-\mu)}{\alpha}}\right]}$$

(42)

where, $\alpha = \frac{\sqrt{\sigma S_x}}{\pi}$, $U = x' - 0.57772\alpha$; x' and S_x are mean and standard deviation of variate x, respectively.

4. Log Pearson type-III distribution

$$f(x) = \frac{\lambda^\beta (y-\varepsilon)^{\beta-1}}{x\{\beta\}} e^{-\lambda(\tau-t)}$$

(43)

where, $\lambda = \frac{Sy}{\sqrt{\beta}}, \beta = \left[\frac{2}{Cs(y)}\right]^2$; $\varepsilon = y' - s_y \sqrt{\beta}$; and $y = \log x$; y', s_y and $c_s(y)$ are the mean, standard deviation and skewness coefficient of variate, respectively.

5. Weibull's distribution

$$f(x) = \left\{ \frac{k}{\lambda} \left(\frac{x}{\lambda}\right)^{k-1} \times e^{\left(-\frac{x}{\lambda}\right)^k} \right\}$$

(44)

where, k>0 is the shape parameter and λ >0 is the scale parameter.

4.4.2   LAND AND WATER MANAGEMENT PLAN FOR THE WATERSHED

Successes of any project depend on its integrated work plan. The planning should be such that it includes the details of the land treatment measures, their design, operation and maintenance of the measures. The different categories of land and their main treatment program are briefed below:

I.   Arable Land   **Suggested Treatment Programs**

    a. Eroded up land   1. Crop with specified rotation

    b. Weakly bunded medium l and   2. Contour bunding and graded

    c. Some portion of low land that   bunding

       are subjected to water logging.   3. Strip cropping

4. Conservation farming

5. Provision of disposal system from one field to another field.

6. Crop demonstrations.

7. Reclamation of waterlogged area.

II. Non Arable Land

a. Area under forest

b. Culturable wasteland

c. Hills

d. Hill slopes

e. Pediments

f. Gochar

Suggested Treatment Programs

1. Forestry promotion

2. Horticultural development

3. Pasture development

4. Breaking of slopes by terracing

5. Reclamation of wastelands

6. Reclamation of wastelands

III. Natural Drainage Line

It includes the natural drains of the watershed which carries run-off to the outlet point.

Suggested Treatment Programs

1. Gully control structures.

2. Construction of runoff harvesting structures

3. Renovation of tanks.

4.4.2.1 Suggested Measures for the Slope Groups of the Watershed

By analyzing the data available for the watershed, the lands covered under different slope groups are distinguished and appropriate measures for each slope group are recommended below in Table 4.4.

TABLE 4.4 Suggested Measures for Different Slope Groups of the Watershed

Slope groups (%)	Suggested land measures
0–1%	Any crop with crop rotation, contour farming
1–3	Some specified low duty crops with intensive agronomical measures such as strip cropping, contour bunding with some areas under pasture grasses
3–5	Pasture with control grazing, forestry with restricted cutting, contour trenching and terracing
5–7	Forest plantation with restricted cutting of trees

4.4.3 DEVELOPMENT OF MULTI-OBJECTIVE MODEL

A Multiple objective programming model has been developed for optimum allocation of land and water to different activities of agriculture and forestry in the watershed. The proposed crops in the planning are paddy, maize, arhar in kharif and paddy, mustard and mung in rabi season. Besides this there was a proposal for hybrid Napierbajra grass as a fodder and subabool as a forestry crop for the watershed throughout the year.

The decision variables are x_{ij}, that is the area under j^{th} crop in i^{th} season:

- $j = 1,2,3,4,5,6,7$ represents upland paddy (paddy-I), medium land paddy (paddy-II), lowland paddy (paddy-III), maize, arhar, hybrid Napierbajra grass and subabool in kharif season ($i = 1$)
- $j = 1, 2, 3, 4, 5$ represents paddy-III (lowland), mustard, mung, hybrid Napierbajra grass and subabool in rabi season ($j = 2$).

4.4.3.1 Multiple Objective Functions

For the watershed management program six basic objectives have been taken in the model. They are maximization of food production, maximization of fodder production, maximization of fuel wood and timber production, maximization of net income generation from field crops, maximization of labor employment generation and maximization of runoff water augmentation. These objectives aim at improving the status of the farmers and ensure judicious utilization of land, water and human resources of the watershed. There are 13 variables taken in the MOP model they are listed in Table 4.5.

TABLE 4.5   Variables With Their Notations

S. No.	Variables	Used for the crop in ha
A. *Kharif*		
1.	x_{11}	Area under up land paddy(paddy-I)
2.	x_{12}	Area under medium land paddy(paddy-II)
3.	x_{13}	Area under low land paddy(paddy-III)
4.	x_{14}	Area under maize
5.	x_{15}	Area under arhar
6.	x_{16}	Area under hybrid Napierbajra grass
7.	x_{17}	Area under subabool plantation

TABLE 4.5 Continued

S. No.	Variables	Used for the crop in ha
B. *Rabi*		
1.	x_{21}	Area under low land paddy(paddy-III)
2.	x_{22}	Area under mustard
3	x_{23}	Area under mung
4.	x_{24}	Area under hybrid Napierbajra grass
5.	x_{25}	Area under subabool plantation
6.	x_3	Are under ponds

I. Food production

Food is one of the basic needs of human being. So maximization of food production is considered as one of the prime objectives, which will help in achieving the self-sufficiency of the food grains for the people of the watershed. The yield of the different field crops in tons/ha is given in Table 4.6. These values are used for fixing the objective function. The objective function for maximization of food production is expressed as:

$$Z_1(x) = 2.4x_{11}+3.4x_{12}+3.9x_{13}+3.6x_{14}+1.6x_{15}+4.2x_{21}+1.3x_{22}+x_{23} \quad (45)$$

II. Fodder production

Fodder is inevitable for the livestock population of the watershed. Keeping in view, the fodder requirement of the livestock's of the watershed, which is an important component of farming is to be maximized. The fodder yield from different crops and plantation crops are given in Table 4.6. The objective function for maximization of fodder production is given by:

$$Z_2(x) = 6.5x_{11}+7.1x_{12}+8x_{13}+2x_{14}+30x_{16}+7x_{17}+8.5x_{21}+20x_{24}+3x_{25} \quad (46)$$

III. Fuel wood production

Fuel wood and timber is essential for the livelihood of the rural people. To satisfy the requirement of the people and also to utilize the culturable wasteland for the purpose of plantation activities, the objective function for the maximization of the fuel wood production is considered in the planning. The fuel wood yield from the plantation trees is given in Table 4.6. The objective function for maximization of fuel wood production is given by:

$$Z_3(x) = 15x_{17} + 15x_{25} \quad (47)$$

TABLE 4.6   Grain, Fodder, Fuel Wood Yield of Different Crops and Plantation

S. No.	Crop/plantation	Variable name	Grain yield (tons/ha)	Fodder yield (tons/ha)	Fuel wood yield (tons/ha)
A.	***Kharif***				
1.	Up land paddy(paddy-I)	x_{11}	2.6	6.5	-
2.	Medium land paddy(paddy-II)	x_{12}	3.4	7.1	-
3.	Low land paddy(paddy-III)	x_{13}	3.9	8.0	-
4.	Maize	x_{14}	3.6	2.0	-
5.	Arhar	x_{15}	1.6	-	-
6.	Hybrid Napierbajra grass	x_{16}	-	30.0	-
7.	Subabool	x_{17}	-	7.0	15.0
B.	***Rabi***				
1.	Low land paddy(paddy-III)	x_{21}	4.2	8.5	-
2.	Mustard	x_{22}	1.3	-	-
3.	Mung	x_{23}	1.0	-	-
4.	Hybrid Napierbajra grass	x_{24}	-	20.0	-
5.	Subabool	x_{25}	-	3.0	15.0

IV. Net income generation from field crops

Most of the people of the watershed are small and marginal farmers. They depend on agriculture for their livelihood. So to improve the economic condition of the farmers, it is necessary to get more income from the field crops. This will be achieved by providing basic agricultural inputs such as seed, fertilizer, pesticides and loans at a reasonable interest rate (12%) to the farmers. The details of the inputs, investment cost, net return per hectare of land is presented in Appendix I to X. The net return from various crops and plantation activities are given in Table 4.7. The objective function for maximization of net income generation from field crops is given by:

$$Z_4(x) = 7051x_{11}+13822x_{12}+14458x_{13}+10378x_{14}+40073x_{15}+10800x_{16}$$
$$+12444x_{17}+15606x_{21}+9108\ x_{22}+24263x_{23}+7200\ x_{24}+9516\ x_{25}$$

$$(48)$$

TABLE 4.7   Net Income and Labor Requirement of Field Crops

S. No.	Crop/plantation	Variable name	Net income (Rs./ha)	Labor required (Man-days/ha)
A. Kharif				
1.	Up land paddy(paddy-I)	x_{11}	7051	102
2.	Medium land paddy(paddy-II)	x_{12}	13,822	132
3.	Low land paddy(paddy-III)	x_{13}	14,458	165
4.	Maize	x_{14}	10,378	160
5.	Arhar	x_{15}	40,073	93
6.	Hybrid Napierbajra grass	x_{16}	10,800	60
7.	Subabool	x_{17}	12,444	185
B. Rabi				
1.	Low land paddy(paddy-III)	x_{21}	15,606	165
2.	Mustard	x_{22}	9108	100
3.	Mung	x_{23}	24,263	65
4.	Hybrid Napierbajra grass	x_{24}	7200	90
5.	Subabool	x_{25}	9516	185

V. Labor employment generation

The objective of maximization of labor employment is considered in the plan to ensure livelihood security to the weaker section of the rural population like small and marginal farmers and landless laborers. So the aim is to utilize the existing human resources and provide employment to maximum number of people that may be possible.

The total labor need for different crops and plantation for the entire season are given in Table 4.7. So taking into consideration the total labor required by crops in man days, the objective function for maximization of labor employment generation is given by

$$Z_5(x)= 102x_{11} + 132x_{12} + 165x_{13} + 160x_{14} + 93x_{15} + 60x_{16} \qquad (49)$$
$$+ 185x_{17} + 165x_{21} + 100x_{22} + 65x_{23} + 90\,x_{24} + 185x_{25}$$

VI. Run off water augmentation

To meet the water requirements of the crops and plantation, it is essential to store as much as water possible for use in rabi season. So the objective for maximization of the stored volume of water in ponds are considered and mathematically expressed as below:

$$Z_6(x) = 2.1\ X_3 \tag{50}$$

4.4.3.2 Constraints of Multi Objective Model

I. Land area

The total watershed area is 506.67 ha, out of which 49.67 ha comes under homestead land, nallahs, roads, etc. So the net area allocated to different crops/ plantation activities should not exceed 457 ha, i.e., the treatable area of the watershed. Therefore:

$$x_{11} + x_{12} + x_{13} + x_{14} + x_{15} + x_{16} + x_{17} + x_3 \le 234 \tag{51}$$

$$x_{21} + x_{22} + x_{23} + x_{24} + x_{25} + x_3 \le 234 \tag{52}$$

From the land capability classification it is shown that 282.22 ha land is suitable for agriculture. So the land allocated to field crops should be less than or equal to 170.66. Therefore:

$$x_{11} + x_{12} + x_{13} + x_{14} + x_{15} \le 170.66 \tag{53}$$

Similarly in rabi season total land cultivable is limited to sum of medium land and low land, i.e., 148.17

$$x_{21} + x_{22} + x_{23} \le 148.47 \tag{54}$$

Out of 170.66 ha of arable land, 22.19 ha is under up land, 130.19 ha is medium land and 18.28 ha is low land. In upland the crops suggested is up land paddy, maize, arhar and in medium land the crops suggested are medium land paddy and arhar. Similarly in low land, low land paddy is suggested. So the respective constraints equations are as below:

$$x_{11} + x_{14} + x_{15} \le 22.19 \tag{55}$$

$$x_{12} + x_{15} \le 130.19 \tag{56}$$

$$x_{13} \le 18.28 \tag{57}$$

Similarly in rabi season, the total medium land is distributed between mustard and mung and the low land is to be cultivated with low land paddy. The respective constraints are shown below:

$$x_{22} + x_{23} = 130.19 \tag{58}$$

$$x_{22} + x_{23} \leq 130.19 \tag{59}$$

$$x_{21} \leq 18.28 \tag{60}$$

The total land under degraded forest, gochar, culturable and unculturable wasteland which can be bought under plantation, fodder and also for making the water storage structures is 174.78. So the constraints equations are

$$x_{16} + x_{17} + x3 \leq 63.34 \tag{61}$$

$$x_{24} + x_{25} + x3 \leq 63.34 \tag{62}$$

As subabool and hybrid Napierbajra grass is grown in both seasons, so total area cultivated in kharif season must be equal to the total area in rabi season.

$$x_{16} - x_{24} = 0 \tag{63}$$

$$x_{17} - x_{25} = 0 \tag{64}$$

II. Water quantity

The total quantity of water needed by different crops/plantation should be less than or equal to the total volume of water available from rainfall in kharif season at the desired probability level and from different storage structures in rabi season. The water requirement of the crops/plantation is given in Table 4.8.

The required constraint equations are as below:

a. $1.2x_{11} + 1.2x_{12} + x_{13} + 0.45x_{14} + 0.35x_{15} + 0.4x_{16} + 0.35x_{17} \leq 149.76$ ha-m (65)

(70% probability level)*

b. $x_{21} + 0.4x_{22} + 0.4x_{23} + 0.2x_{24} + 0.1x_{25} \leq 79.85$ ha-m (66)

A rough guide for estimating effective rainfall has been developed by U.S. Bureau of reclamation for arid and semi-arid regions (1969) in which the mean seasonal precipitation of transpiration/precipitation ratio method based on extensive field basis has been used for the determination. Rainfall values of 70% probability levels are considered in calculation for irrigation planning.

TABLE 4.8   Water Requirement (WR, m) of Different Crops

S. No.	Crop/plantation	Variable name	WR (m)
A. Kharif			
1.	Up land paddy (paddy-I)	x_{11}	1.20
2.	Medium land paddy (paddy-II)	x_{12}	1.20
3.	Low land paddy (paddy-III)	x_{13}	1.00
4.	Maize	x_{14}	0.45
5.	Arhar	x_{15}	0.35
6.	Hybrid Napierbajra grass	x_{16}	0.40
7.	Subabool	x_{17}	0.35
B. Rabi			
1.	Low land paddy (paddy-III)	x_{21}	1.00
2.	Mustard	x_{22}	0.40
3.	Mung	x_{23}	0.40
4.	Hybrid Napierbajra grass	x_{24}	0.20
5.	Subabool	x_{25}	0.10

III. Nutritional requirement

Protein and calorie are two main nutrients required by a man for better health. The quantity of food produced from different crop should meet the minimum nutritional requirements of the people. The recommended nutritional requirements per day per person of different age groups and the protein and calorie content of the crops are given in Tables 4.9 and 4.10.

A. Protein constraint

The total protein need of the people of the watershed is 51391 Kg. for one year. Therefore, the total protein available from different crops must be greater than or equal to the total requirement.

$$163x_{11}+231x_{12}+265x_{13}+306x_{14}+392x_{15}+286x_{21}+260x_{22}+240x_{23} \geq 51391$$

(67)

TABLE 4.9   Nutritional Requirements of Different Age Groups/Day

S. No.	Age group	Protein (g)	Calories (Cal.)
1.	Male	70	3000
2.	Female	55	2400
3.	Children	40	2200

TABLE 4.10   Protein and Calorie Contents of the Crops

S. No.	Crop	Protein (gms/kg)	Calories (Cals/kg)
1.	Paddy	68	3450
2.	Maize	85	3430
3.	Arhar	245	3480
4.	Mustard	200	2917
5.	Mung	240	3240

B. Calorie constraint

The total calorie need of the people of the watershed for one year is 2250955.6 Kcal. Therefore the total calorie available from different crops must be greater than or equal to the total requirement.

$$8280x_{11}+11730x_{12}+13455x_{13}+112348x_{14}+5568x_{15}+14490x_{21}$$
$$+ 3792x_{22} + 3240x_{23} \geq 2250955.6$$

$$(68)$$

IV. Labor requirement

The *kharif* season is taken from June to October and *rabi* from December to April. The labor requirement by the plantation and fodder crops is taken throughout the year. The labor required by different crops in a particular month should be less than or equal to the available labor of the watershed in that month. Assume that maximum 30% of the population can be engaged as laborers for agricultural practices. So the number of people available as laborers in a day is 690.

Total man-days available in a month= $690 \times 30 = 20,700$

Taking into consideration the labor requirement by different crops and total labor available for the watershed, the constraint equations are given below:

Jun $30x_{11}+30x_{12}+30x_{13}+35x_{14}+15x_{15}+25x_{16}+20x_{17} \leq 20700$ (69)

Jul $20x_{11}+35x_{12}+35x_{13}+30x_{14}+20x_{15}+10x_{16}+10x_{17} \leq 20700$ (70)

Aug $15x_{11}+20x_{12}+30x_{13}+15x_{14}+10x_{15}+10x_{16}+30x_{17} \leq 20700$ (71)

Sep $7x_{11}+12x_{12}+30x_{13}+20x_{14}+23x_{15}+5x_{16}+25x_{17} \leq 20700$ (72)

Oct $30x_{11}+35x_{12}+40x_{13}+60x_{14}+25x_{15}+10x_{17} \leq 20700$ (73)

$$\text{Nov} \quad 10x_{16}+110x_{17} \leq 20700 \tag{74}$$

$$\text{Dec} \quad 30x_{21}+20x_{22}+10x_{23}+10x_{24}+20x_{25} \leq 20700 \tag{75}$$

$$\text{Jan} \quad 35x_{21}+15x_{22}+10x_{23}+10x_{24}+20x_{25} \leq 20700 \tag{76}$$

$$\text{Feb} \quad 30x_{21}+10x_{22}+10x_{23}+15x_{24}+15x_{25} \leq 20700 \tag{77}$$

$$\text{Mar} \quad 30x_{21}+25x_{22}+10x_{23}+15x_{24}+10x_{25} \leq 20700 \tag{78}$$

$$\text{Apr} \quad 40x_{21}+30x_{22}+25x_{23}+10x_{24}+10x_{25} \leq 20700 \tag{79}$$

$$\text{May} \quad 30x_{24}+90x_{25} \leq 20700 \tag{80}$$

V. Bulk Requirement

The total quantity of food, fodder and fuel wood required per year by the people and livestock of the watershed is given in Table 4.11.

VI. Food Requirement

Food produced under each category of food should be more than the food requirement.

a. **Paddy**

The total requirement of the paddy for the watershed in one year is estimated as 309.78 tons. So yield of paddy is greater than or equal to the need.

$$2.4x_{11}+3.4x_{12}+3.9x_{13}+4.2x_{21} \geq 309.78 \tag{81}$$

TABLE 4.11 Estimated Food, Fodder and Fuel Wood Requirements of the Watershed

S. No.	Items	Estimated Quantity (tons)
1.	Food requirements	
	Paddy	309.78
	Maize	143.28
	Arhar	36.72
	Mustard	34.57
	Mung	30.47
2.	Fodder requirement	1420
3.	Fuel wood requirement	1680

b. **Maize**
   The total requirement of maize is 143.28 tons. So the total yield of maize should be greater than or equal to 143.28 tons

$$3.6x_{14} \geq 73.45 \tag{82}$$

c. **Arhar**
   The total requirement of arhar is estimated as 36.72 tons. So the total yield of arhar should be more than or equal to the estimated value.

$$1.6x_{15} \geq 36.72 \tag{83}$$

d. **Mustard**
   The total requirement of mustard is estimated as 34.57 tons. So the total yield of mustard should be greater than or equal to 36.96 tons.

$$1.3x_{22} \geq 34.57 \tag{84}$$

e. **Mung**
   The total requirement of mung is estimated as 30.47 tons. So the total yield of mung should be greater than or equal to 32.28 tons.

$$x_{23} \geq 30.47 \tag{85}$$

VII. Fodder Requirement
The total quantity of fodder available from different crops and plantation should be greater than or equal to the total requirement of the livestock of the watershed, i.e., 1420 tons each green and dry fodder as given in Appendix-XV.

Dry fodder

$$[6.5x_{11}+7.1x_{12}+8x_{13}+2x_{14}+8.5x_{21}] \geq 1420 \tag{86}$$

Green fodder

$$[30x_{16}+7x_{17}+20x_{24}+3x_{25}] \geq 1420 \tag{87}$$

8. Fuel wood Requirement
The total fuel wood requirement of the watershed is 1680 tons. Therefore, the total fuel wood available from plantation crops should be greater than or equal to 1680 tons.

$$[15x_{17}+15x_{25}] \geq 1680 \tag{88}$$

VIII. Area required for ponds

The total area needed for constructing the required number of ponds should be greater than or equal to 27.54 ha: The area required to augment the quantity of water needed for irrigation purpose in *rabi* season, which is:

$$X_3 \geq 27.54 \tag{89}$$

IX. Non negative constraints

All the variables should be either greater than or equal to zero. (Should be non-negative). Therefore, the constraint equations are:

$$x_{11} \geq 0, \quad x_{12} \geq 0, \quad x_{13} \geq 0, \quad x_{14} \geq 0, \quad x_{15} \geq 0, \quad x_{16} \geq 0,$$

$$x_{17} \geq 0, \quad x_{21} \geq 0, \quad x_{22} \geq 0, \quad x_{23} \geq 0, \quad x_{24} \geq 0, \quad x_{25} \geq 0, \quad x3 \geq 0 \tag{90}$$

These six objective functions along with thirty eight constraints are entered into the *Quick Statistical Business* (QSB) software to find out the required solutions.

4.5   RESULTS AND DISCUSSIONS

4.5.1   RAINFALL ANALYSIS

In the proposed multi-objective model, the assumption was that water requirement in kharif was met from rainfall and water requirement in rabi season was met from water harvesting structures. So the water available from the rainfall at different probability level was estimated. The monthly rainfall data of 34 years were gathered from the historical records. The collected data was tested with different probability distribution functions, in order to find out the best-fit probability distribution function. The probability distribution functions considered were: Normal distribution, Log normal distribution, Gumbel maximum, Log Pearson type-III and Weibull distribution. The rainfall data was analyzed using 'FLOOD' software.

The best-fit function was determined using Chi-square goodness of fit test. The probability function, which has least mean error, was selected as best fit distribution. The distribution function best fit for different months along with their RMSE value and mean error are given in Table 4.12. After that, the rainfall values at different probability level were calculated, using the same 'FLOOD' software. Monthly variations of rainfall at different probability levels were estimated and are given in Table 4.13. The rainfall values at different probability level are shown graphically in Figures 4.3–4.5.

TABLE 4.12 Best Fit Probability Distribution Function for Different Months

Month	Best fit distribution	Corresponding RMSE value	Corresponding Mean error
January	Log Pearson	0.0335	0.0268
February	Generalized Extreme value	0.04358	0.03258
March	Gamma	0.03929	0.03331
April	Exponential	0.03241	0.02619
May	Lognormal	0.03179	0.02743
June	Extreme value type III	0.02022	0.017
July	Gamma	0.02479	0.01951
August	Weibull	0.02919	0.0239
September	Gumbel maximum	0.03968	0.03299
October	Lognormal 3 parameter	0.03898	0.03187
November	Log Pearson	0.03858	0.03137
December	Weibull	0.04382	0.03384

TABLE 4.13 Expected Month Wise Rainfall (mm) at Different Probability Levels

Month	Probability level (%)								
	90	80	70	60	50	40	30	20	10
Jan	0	0	0	0	1.339	7.817	15.62	26.688	45.952
Feb	0	0	0	2.263	14.403	24.746	35.492	48.456	67.982
Mar	0	0	1.03	5.186	11.12	19.167	30.41	47.707	83.018
Apr	1.908	6.742	12.221	18.546	26.028	35.184	46.989	63.627	92.069
May	22.838	32.916	42.844	53.667	66.245	81.772	102.44	133.36	192.32
Jun	88.379	114.156	136.81	158.73	181.2	205.45	233.19	267.78	319.11
Jul	150.707	192.602	227.43	260.42	294.06	330.49	372.67	426.35	508.77
Aug	187.918	239.894	279.45	314.1	346.89	379.87	415.17	456.29	512.69
Sep	124.998	153.67	176.91	198.77	221.11	245.54	274.3	311.85	371.92
Oct	29.587	51.824	72.362	93.648	117.29	145.23	180.75	231.26	321.35
Nov	0	0	2.023	5.919	12.167	21.893	37.441	64.43	121.79
Dec	0	0	0	0	0	0	1.874	7.804	21.36

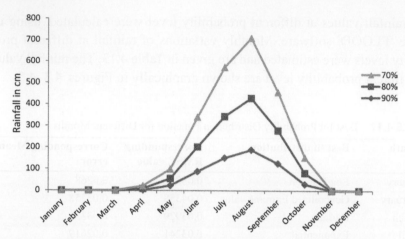

FIGURE 4.3   Monthly variation of rainfall at 90%, 80% and 70% probability levels.

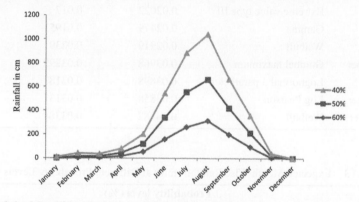

FIGURE 4.4   Monthly variation of rainfall at 60%, 50% and 40% probability levels.

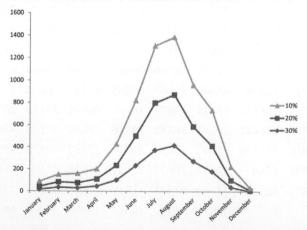

FIGURE 4.5   Monthly variation of rainfall at 30%, 20% and 10% probability levels.

4.5.2   WATERSHED TREATMENTS

4.5.2.1   Cultivation Practices for Different Slope Groups

The slope of the culturable lands of watershed varies between 0 and 5%. The degraded forests need renovation and afforestation, in wastelands and gochars. The grasslands needs controlled grazing. In the culturable waste-land some portion may be under plantation and some portion under fodder crop. The proposed cultivation practices are presented in Table 4.14.

Along with these measures, crop rotation and improved varieties of different crops should be grown in that watershed. The improved varieties grown in that watershed should be suitable for the climate of the watershed. The improved varieties not only increase the yield but also farmers will be more benefited from the crops. In the watershed management program some crop rotation and improved varieties of the crops are taken. The suggested crops along with their varieties are given in Table 4.15.

4.5.3   LAND ALLOCATION PLAN USING MULTI-OBJECTIVE APPROACH

The multi-objective functions along with the various constraints were formulated. The equations were solved and analyzed by an interactive technique, i.e., step method. The optimal solution was obtained for each objective function considering the resource constraints and requirement of the people. From the sets of optimal solution, a compromise solution for allocation of land under different crops and plantation was found out in *kharif* and *rabi* seasons.

TABLE 4.14   Suggested Cultivation Practices

Slope group (%)	Proposed cultivation practices
0–1	Any crop with proper crop rotation and green manuring to maintain soil fertility.
1–3	Some specified low duty crops with agronomical measures such as strip cropping and contour farming.
3–5	Grassland and forestry with controlled grazing and limited cutting of forestry trees.

TABLE 4.15 Suggested Crop Varieties for Different Types of Land

Type of land	Crops suggested	Varieties	Follow-up crop
Upland (Plain)	Paddy	Hira	Niger
		Parijat	Mustard
		Kalinga-III	
	Maize	Ganga-5	Niger
		Deccan-103	Mustard
	Maize + Arhar	Ganga-5	Mustard
		UPAS-120	Sesamum
Sloppy agricultural land (contoured lands)	Mixed cropping	Maize/jawar	Mustard
		Arhar + cowpea	Sesamum
		Arhar + Biri + Ragi	
Medium land	Paddy	Annapurna	Mung
		Cavery	Mustard
		Parijat	
	Arhar + Ragi	UPAS-120	Mung
		Dibyasinga	Mustard
Low land	Paddy	Swarna	Sesamum
		Jagannath	Mung
		Jajati	

4.5.3.1   Optimal Solutions for the Objective Functions

The six objective functions along with 38 constraints were solved for maximization separately using Quick Statistical Business (QSB) software. There is a feasible solution for each objective function. From the feasible solution the optimal solution for each function is obtained considering the other alternative optima, which are non-inferior. The set of optimal solution is given in Table 4.16.

4.5.3.2   Pay-Off Table (Matrix)

After finding out six set of optimal solution they are to be put in the expression for the objective functions. In the construction of pay-off table (Table 4.17), the rows correspond to the optimal solution and the columns are labeled by the objectives.

TABLE 4.16 Optimal Value Under Different Objective Functions

Crop variables	Area allocation under different objectives, ha					
	Z_1	Z_2	Z_3	Z_4	Z_5	Z_6
x_{11}	0	0	0	0	0	0
x_{12}	64.732	87.012	81.64	86.78	64.74	91.11
x_{13}	18.28	18.28	18.28	18.28	18.28	18.28
x_{14}	42.30	22.19	22.19	22.19	42.30	22.19
x_{15}	22.95	22.95	22.95	22.95	22.95	22.95
x_{16}	28.40	28.40	28.40	26.55	28.40	28.40
x_{17}	38.38	58.13	58.13	58.13	58.13	38.38
x_{21}	18.28	18.28	18.28	18.28	18.28	18.28
x_{22}	26.59	26.59	26.59	26.59	26.59	26.59
x_{23}	32.28	32.28	32.28	32.28	32.28	32.28
x_{24}	28.40	28.40	28.40	26.55	28.40	28.40
x_{25}	38.38	58.13	58.13	58.13	58.13	38.38
x_8	27.54	27.54	27.54	27.54	27.54	27.54
Optimal value of objective	Z_1= 624 tons	Z_2=2965.08 tons	Z_3= 1743.9 Tons	Z_4=Rs. 5678832	Z_5= 54005 Man-days	Z_6= 57.84 ha-m

TABLE 4.17 Pay-Off Table (Matrix)

Optimal Solution[*]	$Z_1(x)$ Ton	$Z_2(x)$ Ton	$Z_3(x)$ ton	$Z_4(x)$ Rs.	$Z_5(x)$ man-days	$Z_6(x)$ ha-m
			Objective functions			
X_1	624	2649.62	1151.4	5182377	46697	57.84
X_2	627.36	2965.08	1743.9	5715339	53728	57.84
X_3	609.10	2926.95	1743.9	5641087	53018	57.84
X_4	626.57	2870.94	1743.9	5678832	53419	57.84
X_5	624.03	2847.17	1743.9	5616197	54006	57.84
X_6	641.30	2796.68	1151.4	5338272	46961	57.84

Note: [*]optimal solution set for objective function Z_1, i.e., (x_{11}, x_{12}, x_{13}, x_{14}, x_{15}, x_{16}, x_{17}, x_{21}, x_{22}, x_{23}, x_{24}, x_{25}, x_8.)

The coefficients of different objective functions are selected. From the coefficient of objective functions, the value of $[\sum(C_j^k)^2]^{\frac{-1}{2}}$ (normalizing term) was calculated and given in Table 4.18.

TABLE 4.18   Calculation of Normalizing Term

Parameter	Objective functions					
	Z_1	Z_2	Z_3	Z_4	Z_5	Z_6
X_{11}	2.4	6.5	0	7051	102	0
X_{12}	3.4	7.1	0	13822	132	0
X_{13}	3.9	8	0	14458	165	0
X_{14}	3.6	2	0	10378	160	0
X_{15}	1.6	0	0	40073	93	0
X_{16}	0	30	0	10800	60	0
X_{17}	0	7	15	12444	185	0
X_{21}	4.2	8.5	0	15606	165	0
X_{22}	1.3	0	0	9108	100	0
X_{23}	1	0	0	24263	65	0
X_{24}	0	20	0	7200	90	0
X_{25}	0	3	15	9516	185	0
X_3	0	0	0	0	0	2.1
$\Sigma (Cj^k)^2$	68.38	1590.9	450	349,242,9723	210,902	4.41
$[\Sigma (Cj^k)^2]^{\frac{-1}{2}}$	0.121	0.025	0.047	0.00001692	0.0022	0.4762

After calculation of the normalizing term the value of scaling term $\left[\dfrac{M_k - n_k}{M_k} \right]$ is calculated, and the multiplication of the normalizing term with scaling term will give the value of 'α_k'.

The calculation of 'α_k' is given in Table 4.19. From the values of 'α_k' values of initial set of weights (W_k) are calculated and tabulated in Table 4.20. The above values of weights are used to derive the first compromise solution, by minimizing the deviation from the optimal solution. The objectives having largest difference between maximum and minimum value of objective functions are assigned with larger weights. From the above calculation the sixth objective having weight equal to zero, has already attain the optimal value. There is no necessity for calculation of compromise solution for this objective.

4.5.3.3 First Compromise Solution

Simultaneous optimization of all the six objectives is not possible, that is why the compromise solution is obtained by the initial set of weights as

TABLE 4.19 Calculation of 'α_y'

Objective function	Maximum value of objective function M_k	Minimum value of objective function n_k	$\dfrac{M_k - n_k}{M_k}$	$[\Sigma(C_j^k)^2]^{\frac{-1}{2}}$	α_k
$Z_1(x)$	641.29	609.10	0.050	0.12093	0.006
$Z_2(x)$	2965.08	2649.62	0.106	0.02507	0.0026
$Z_3(x)$	1743.9	1151.4	0.339	0.04714	0.0159
$Z_4(x)$	5715339	5182377	0.093	0.00001692	0.0000015
$Z_5(x)$	54006	46697	0.135	0.002177	0.0002293
$Z_6(x)$	57.84	57.84	0	0.47619	0

TABLE 4.20 Calculation of 'W_k' (Initial Set of Weights)

Objective function	Value of α_k	Value of W_k $\left(W_k = \dfrac{\alpha k}{\sum \alpha k}\right)$
$Z_1(x)$	0.006	0.2429
$Z_2(x)$	0.0026	0.1052
$Z_3(x)$	0.0159	0.6437
$Z_4(x)$	0.0000015	0.00000607
$Z_5(x)$	0.0002293	0.00928
$Z_6(x)$	0	0
	$\sum \alpha_k = 0.0247$	$\sum W_k = 0.99 \approx 1.00$

calculated in Table 4.20. From Table 4.20, it is clear that the objectives Z_1, Z_2 and Z_3 exhibit larger relative variations in their maximum and minimum values. Therefore, larger weights are assigned to these objective functions. The compromise solution is obtained for objectives one to five except objective No. 6. The first compromise solution will be obtained by solving the following linear programming:

Minimize the deviation 'd' subjected to: (91)

$$W_k [M_k - Z_k(x)] - d \leq 0, \text{ or}$$

$$W_k M_k - W_k Z_k(x) - d \leq 0, \text{ or}$$

$$W_k Z_k(x) + d - W_k M_k \geq 0, \text{ or}$$

$$W_k Z_k(x) + d \geq W_k M_k, \text{(This is the required constraint equation)} \quad (92)$$

$$d \geq 0 \quad (93)$$

Now the requisite equations are:

$$\text{Minimize 'd', subjected to:} \quad (94)$$

$$0.2429 \, (2.4 \, x_{11} + 3.4 \, x_{12} + 3.9 \, x_{13} + 3.6 \, x_{14} + 1.6 \, x_{15} + 4.2 \, x_{21}$$
$$+ 1.3 \, x_{22} + x_{23}) + d \geq 0.2429 \times 641.29, \text{ or}$$

$$0.582 \, x_{11} + 0.826 \, x_{12} + 0.947 \, x_{13} + 0.875 \, x_{14} + 0.389 \, x_{15}$$
$$+ 1.02 \, x_{21} + 0.315 \, x_{22} + 0.2429 \, x_{23} + d \geq 155.77 \quad (95)$$

$$0.1052 \, (6.5 \, x_{11} + 7.1 \, x_{12} + 8 \, x_{13} + 2 \, x_{14} + 30 \, x_{16} + 7 \, x_{17}$$
$$+ 8.5 \, x_{21} + 20 x_{24} + 3 \, x_{25}) + d \geq 0.1052 \times 2965.08, \text{ or}$$

$$0.684 \, x_{11} + 0.746 \, x_{12} + 0.841 \, x_{13} + 0.210 \, x_{14} + 3.156 \, x_{16}$$
$$+ 0.736 \, x_{17} + 0.894 \, x_{21} + 2.104 \, x_{24} + 0.315 \, x_{25} + d \geq 311.92 \quad (96)$$

$$0.6437 \, (15 \, x_{17} + 15 \, x_{25}) + d \geq 0.6437 \times 1743.9, \text{ or}$$
$$9.65 \, x_{17} + 9.65 \, x_{25} + d \geq 1122.54 \quad (97)$$

$$[0.00000607 \, (7051 \, x_{11} + 13822 \, x_{12} + 14458 \, x_{13} + 10378 \, x_{14}$$
$$+ 40073 \, x_{15} + 10800 \, x_{16} + 12444 \, x_{17} + 15606 \, x_{21} + 9108 \, x_{22}$$
$$+ 24263 \, x_{23} + 7200 \, x_{24} + 9516 \, x_{25})] + d \geq 0.00000607 \times 5715339, \text{ or}$$

$$0.0428 \, x_{11} + 0.0838 \, x_{12} + 0.0877 \, x_{13} + 0.063 \, x_{14} + 0.2432 \, x_{15}$$
$$+ 0.0655 \, x_{16} + 0.0755 \, x_{17} + 0.0947 \, x_{21} + 0.0552 \, x_{22} + 0.1472 \, x_{23}$$
$$+ 0.0437 \, x_{24} + 0.0577 \, x_{25} + d \geq 34.692 \quad (98)$$

$$[0.00928 \, (102 \, x_{11} + 132 \, x_{12} + 165 \, x_{13} + 160 \, x_{14} + 93 \, x_{15}$$
$$+ 60 \, x_{16} + 185 x_{17} + 165 \, x_{21} + 100 \, x_{22} + 65 \, x_{23} + 90 \, x_{24}$$
$$+ 185 \, x_{25})] + d \geq 0.00928 \times 54005, \text{ or}$$

$$0.946 \, x_{11} + 1.224 \, x_{12} + 1.531 \, x_{13} + 1.484 \, x_{14} + 0.863 \, x_{15}$$
$$+ 0.556 \, x_{16} + 1.716 \, x_{17} + 1.531 \, x_{21} + 0.928 \, x_{22} + 0.603 \, x_{23}$$
$$+ 0.835 \, x_{24} + 1.716 \, x_{25} + d \geq 501.166 \quad \text{for} \quad (99)$$

$$d \geq 0 \quad (100)$$

Solving equations 94 to 100, the first compromise solution set is obtained. These compromise solution set gives new function values. The percentage difference between these compromise value and optimum value is calculated. These values will help in deciding the achievement levels and trade-off between different objectives. The solution of the above equation gives the value of the variable as follows:

$$x_{11} = 0 \qquad x_{21} = 18.88$$

$$x_{12} = 81.64 \quad x_{22} = 26.59$$

$$x_{13} = 18.88 \quad x_{23} = 32.28$$

$$x_{14} = 22.79 \quad x_{24} = 28.80$$

$$x_{15} = 22.95 \quad x_{25} = 58.93$$

$$x_{16} = 28.80 \quad x_{8} = 27.54$$

$$x_{17} = 58.93 \quad d = 3.4$$

Putting these values, 1st compromise solution for different objectives is obtained. Differences in percentage of 1st compromise solution from the maximum value are given in Table 4.21.

From Table 4.21, it is observed that the objectives 2, 3, 4, and 6 have been satisfactorily achieved the maximum value, whereas objectives 1 and 5 are within the permissible range. Decision maker (DM: here the author) is

TABLE 4.21 First Compromise Solution

Objective function	Maximum value	First compromise solution	% difference from maximum value
$Z_1(x)$	641.29	616.30*	3.89
$Z_2(x)$	2965.08	2964.37*	0.24
$Z_3(x)$	1743.9	1743.9*	0
$Z_4(x)$	5715339	5690950*	0.04
$Z_5(x)$	54006	53676*	0.7
$Z_6(x)$	57.84	57.84*	0

*Satisfactorily achieved objectives (if percentage difference between first compromise solution and maximum value is within 5% then the objective function is considered to be satisfactorily achieved).

satisfied with this solution, so there is no need of further iteration. So from the above solution, the DM believed that if the land is put under the following crops and plantation activities taken with desired inputs, the achievement levels will be satisfactory (Table 4.22).

Achievement level of the selected objectives

1. Food production   = 616.30 ton
2. Fodder production   = 2964.37 ton
3. Fuel wood production   = 1743.90 ton
4. Net income from field crops   = Rs. 56,90,950.00
5. Labor employment generation   = 53676 man-days
6. Run off volume augmentation   = 57.84 ha-m

4.5.3.4   Food Grain Production

Food grain production from different crops based on the feasible land allocation plan is given in Table 4.23. It is found that the percent increase of food

TABLE 4.22   Land Allocation Proposals for the Watershed

Crop	Land allocation, ha
Kharif	
1. Paddy (medium land)	81.64
2. Paddy(low land)	18.88
3. Maize	22.79
4. Arhar	22.95
5. Hybrid Napierbajra grass	28.80
6. Subabool	58.93
Subtotal	**234 = A**
Rabi	
1. Paddy(low land)	18.88
2. Mustard	26.59
3. Mung	32.28
4. Hybrid Napierbajra grass	28.80
5. Subabool	58.93
Subtotal	**165.48 = B**
Grand total, A + B =	**399.48**

TABLE 4.23 Food Grain Productions in the Watershed

Crop	Food grain requirement in the watershed(ton)	Food grain achieved in the watershed(ton)	Percent increase over requirement
Paddy	325.54	420.65	29.22
Maize	73.45	79.89	8.77
Arhar	36.72	36.72	0
Mustard	34.57	34.57	0
Mung	30.47	32.28	5.94
Total	**500.75**	**604.11**	**20.64**

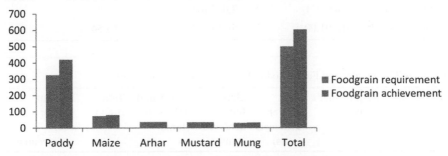

FIGURE 4.6   Food grain requirements vs. Food grain achievement.

grains of paddy and maize as per land allocation plan is more than 100% and the highest being obtained from the mustard (1510.46%), whereas the other crops just meet the requirement of the people of the watershed. The result of requirement versus achievement of different food grains inside the watershed is shown graphically in Figure 4.6.

4.5.3.4.1   Achievement of Calorie and Protein

The total calorie and protein available from different field crops as per land allocation plan are estimated and the values are presented in Table 4.24. The table showed that the calorie and protein requirements are sufficiently more than the requirements for the people of the watershed.

4.5.3.5   Fodder and Fuel Wood Production in the Watershed

Fodder and fuel wood productions from different crop activities are presented in Table 25.

4.5.3.6 Economics of the Proposed Planning

The compromised land allocation plan obtained for different crops was found out. The economics of the suggested crops in compromised land allocation planning is given in Table 4.26.

TABLE 4.24 Achievement of Calorie and Protein

Nutrients	Nutrient requirement in the watershed	Total nutrient achievement in the watershed	Percent increase over requirement
Calorie, Kcal	2,000,711	2,075,676	3.74
Protein, kg	51,391.00	59,378.26	15.54

TABLE 4.25 Fodder and Fuel Wood Achievement in the Watershed

Materials	Requirement (tons)	Dry fodder achieved (tons)	Green fodder achieved (tons)	Fuel wood achieved (tons)	Total (tons)	Percent increase over requirement
Fodder	1420	925.64	2001.30		2926.94	106.12
Fuel wood	1680	-	-	1743.9	1743.9	3.80

TABLE 4.26 Economics of the Suggested Crops in Compromised Land Allocation Planning

Crop	Area allocated (ha)	Cost of produce Rs./ha	Total cost of produce (Rs.)	Cost per ha(Rs.)	Total cost of production (Rs.)	Net benefit	B-C ratio
Kharif season – A							
Medium land paddy	81.64	51640	4215889.6	37818	3087461.52	1128428.08	
Low land paddy	18.88	59090	1115619.2	44632	842652.16	272967.04	
Maize	22.79	48160	1097566.4	37782	861051.78	236514.62	
Arhar	22.95	68800	1578960	28727	659284.65	919675.35	
Hybrid Napierbajra grass	28.80	45000	1296000	34200	984960	311040	
Subabool	58.93	52250	3079092.5	40172	2367335.96	711756.54	
Total = A			12383127.7		8802746.07	3580381.63	1.40

TABLE 4.26 Continued

Crop	Area allocated (ha)	Cost of produce Rs./ha	Total cost of produce (Rs.)	Cost per ha(Rs.)	Total cost of production (Rs.)	Net benefit	B-C ratio
Rabi season – B							
Low land paddy	18.88	63520	1199257.6	47914	904616.32	294641.28	
Mustard	26.59	39000	1037010	29787	792036.33	244973.67	
Mung	32.28	45000	1452600	20260	653992.8	798607.2	
Hybrid Napierbajra grass	28.80	30000	864000	22800	656640	207360	
Subabool	58.93	42750	2519257.5	32868	1936911.24	582346.26	
Total = B			**7072125.1**		**4944196.69**	**2127928.41**	**1.43**
Grand Total, A + B =			**19455252.8**		**13746942.76**	**5708310.04**	**1.42**

From the results obtained for maximization of different objectives, maximization of production of food, fodder, fuel wood is more compromising to the policies of the government, because government is interested in maximum production. Similarly the maximization of benefit is beneficial to the farmers. Maximization of labor employment is best suited to the unemployment status of the area and economic status of the watershed. It is left to the Govt. and the people of the watershed either to adopt compromise solution or one of the optimal solutions described earlier, as per their own option depending upon the situation.

4.6 CONCLUSIONS

1. Among different methods and interactive techniques, the step method is found to be best suited method to derive the compromise solution in watershed management program, because it can accommodate the problems encountered and is easy to understand.

2. As watershed management involves multi-disciplinary activities with special objectives like food, fodder, fuel wood production, maximization of net income from field crops, etc. Multi objective approach is found to be suitable approach for watershed management program.

3. The interactive method gave the best result for the planning of the watershed from the farmers social and economical point of view.
4. Various biological measures have been suggested for the watershed depending upon the soil, slope and land capability of the area.
5. The compromise solution developed will help the decision maker to generate new compromise solution by judging satisfactory objectives.
6. The optimal land allocation plan in ha for kharif and rabi season obtained was as follows:

Optimal land allocation, ha			
Kharif		*Rabi*	
Paddy (medium land)	81.64	Paddy(low land)	18.88
Paddy(low land)	18.88	Mustard	26.59
Maize	22.79	Mung	32.28
Arhar	22.95	Hybrid Napierbajra grass	28.80
Hybrid Napierbajra grass	28.80	Subabool	58.93
Subabool	58.93		

7. The optimal value of different objectives obtained in the model is given below:

     a. Food production   616.30 ton
     b. Fodder production   2964.37 ton
     c. Fuel wood production   1743.90 ton
     d. Net income from field crops   Rs. 56,90,950.00
     e. Labor employment generation   53676 man-days
     f. Run off volume augmentation   57.84 ha-m

8. The cropping intensity of the proposed planning is found to be 170.71% against the existing cropping intensity of 104%
9. The cost economics of the crops suggested in the proposed planning has been calculated and it has been found that the total benefit is Rs. 5,690,950.00
10. The benefit-cost ratio of the proposed planning in kharif season is 1.40 and in rabi season, it is 1.43. The resultant benefit-cost ratio of the planning is 1.42, which shows the economic viability of the proposed program.

11. The planning shows that, total land resources utilized in both the season is 399.48 ha and total water resources utilized in both the season is 229.61 ha-m.
12. The results obtained from maximization of production and maximization of income appears to be more compromising and beneficial to government and to the farmers as the farmer is interested in maximum income from his crops and government has the aim for maximum production of food grains and fodder from unit land area.

If the proposed planning with the suggested measures is taken, then the socio-economic standard of the watershed inhabitants can be increased to a reasonable stage. It is hoped that this compromise land allocation plan along with suggested land treatment measures, if implemented will help to boost the standard of living of the people of the watershed.

4.7 SUMMARY

In this chapter, the main objective was to come out with a solution to react against the different problems in the watershed considering various requirements and resources available for planning. The planning for proper management of the Badabandha Nala watershed was carried through interactive multi objective linear modeling considering six basic objectives. The interactive multi objective model comprises objectives like maximization of food, fodder, fuel wood, net income generation from field crops, labor employment generation and runoff water augmentation. These objectives were solved under a set of 38 resource constraints and requirement of the people and the livestock in the watershed. The management plan also includes some biological measures proposed for future development of the watershed. In biological measures, the management plan consists of proper agricultural inputs, improved varieties of crops, crop rotation and fertilizer management.

The multi objective mathematical model was formulated and the solution was obtained by one computer software package Quick Statistical Business software (QSB). The solution was analyzed with an interactive technique known as step method. By this method a compromised land allocation plan under different crops and plantation activities was obtained.

KEYWORDS

- **Agricultural inputs**
- **Agricultural practices**
- **Arhar**
- **Benefit-cost-ratio**
- **Blackgram**
- **Constraints**
- **Cost economics**
- **Crop rotation**
- **Cropping intensity**
- **Demographic pressure**
- **Effective rainfall**
- **Effective rainfall**
- **Fertilizer management**
- **Fodder**
- **Food grain**
- **Fuel**
- **Green gram**
- **Gross domestic product**
- **Hybrid Napier bajra grass**
- **Interactive technique**
- **Kharif**
- **Labor employment**
- **Land resources**
- **Low land**
- **Maize**
- **Mathematical model**
- **Maximization**
- **Medium land**
- **Minimization**
- **Mung**
- **Mustard**
- **Non-interactive technique**

- **Optimum solution**
- **Paddy**
- **Payoff table**
- **Payoff table**
- **Pond**
- **Probability distribution function**
- **Quick Statistical Business software**
- **Rabi**
- **Rainfall**
- **Runoff**
- **Subabool**
- **Upland**
- **Water Resources**
- **Watershed**

REFERENCES

1. Agnihotri, V., Sharma, J. S., and Mittal, S. P. (1996). Boosting hill resource economy through watershed management in Hoshiarpur, Shiwaliks. *Indian Journal of Soil Conservation,* 24(3), 248–252.
2. Alemayehu, F., Taha, N., and Nyseen, J. (2009). The impacts of watershed management on land use and land cover dynamics in Eastern Tigray (Ethiopia). *Journal Resource Conservation and Recycling,* 53(4), 192–198.
3. Beura, P. K., (1998). *Multi objective approach in optimum land and water resource planning for Mahanadi delta command.* Unpublished MTech thesis, CAET, OUAT, Bhubaneswar, India.
4. Chulsang Y., Wang, M., and Yang, M. (2005). Effect of global warming on rainfall distribution. *International Journal of Irrigation Planning,* 49(1), 149–160
5. Dmitri, M., Chulsang, M., and Maurizio, S. (2006). Application of Markov decision process in resource utilization. *Journal of Ecohydrology and Hydrobiology,* Vol. 49, 203–213.
6. Elsebaie, I. H. (2012). Developing rainfall intensity duration frequency relationship for two regions in Saudiarabia. *Journal of King South University Engineering Sciences,* Vol. 24, 131–141.
7. German, L., Mansoor, H., Alemu, G., Mazengia, T., Amade, T. and Storud, A. (2007). Participatory integrated watershed management: Evolution of concepts and methods in an ecoregional program of the eastern African highland. *Agricultural Systems,* Vol. 94, 189–204.

8. Ghimire, S., and Johston, J. (2013). Impact of domestic and agricultural rainwater harvest systems on watershed hydrology: A case study in the Alemarle-Pamlicoriver Basins (USA). *Journal of Ecohydrology and Hydrobiology*, Vol.13, 159–171.

9. Jayashree, G. S. (1994). *Linear programming approach to Agricultural resources management considering price fluctuations- a case study,* Unpublished MTech thesis, Department. of Agril. Engg., Indian Institute of Technology, Kharagpur, India.

10. Kaur, R., Srivastava, R., Betne, R., Mishra, K., and Dutta, D. (2004). Integration of linear programming and a watershed scale hydrologic model for proposing and optimized land –use plan and assessing its impact on soil conservation: A case study of the Nagwan watershed in the Hazaribagh district of Jharkhand, India. *International Journal of Geographical Information Science*, 18 (1), 73–98.

11. Khatik, G. L., Kurithe, R. S., and Sing, S. B. (1997). Impact of operational research project on agricultural production through integrated watershed management. *Indian Journal of Soil Conservation*, Vol. 25(2), 157–161.

12. Latinopoulos, D., and Mylopoulos, Y. (2005). Optimal allocation of land and water resources irrigated agriculture by means of goal programming: Application in Laudious River *Basin. Global NEST Journal*, 7 (3), 264–273.

13. Marineseu, M. V., Sowlati, T., and Maness T. C. (2005). The development of a timber allocation model using data envelopment analysis, *Canadian Journal of Forest Research*, 35 (10), 2304–2315.

14. Massimo, B., and Maurizio, B. (2006). A combined goal programming AHP approach to maintenance selection problem. *Reliability Engineering and System Safety*, Vol. 91 (7), 839–848.

15. Mishra, S., Prakash, S., Tiwari, M. K., and Lashkari, R. S.(2006). A fuzzy goal programming model of machine tool selection and operation quotation problem in FMS. *International Journal of Production Research*, Vol. 44 (1), 43–76.

16. Mohan, S., and Jothiprakash, V., 2000. Fuzzy system modeling for optimal crop planning, *Journal of the Institution of Engineers*, 81:9–16.

17. Mohanty, S., Maratha, R. A., and Singh, S. (1999). Probability analysis of annual maximum daily rainfall for Amravati. *Journal of Soil and Water Conservation*, Vol. 43, 15–17.

18. Nyonand, A. K. S., Sing, J., Yadav, K. S., and Singh, P. (1997) .Rehabilitation of degraded ecosystems through watershed management: A success story. *Indian Journal of Soil Conservation*, Vol. 25 (2), 162–166.

19. Pande, V. C., Kurothe, R. S., Sing, H. B., and Nambiar, K. T. N., 1998. Farm resources development: a case study of watershed management in semi arid tropics of Gujarat, *Indian Journal of Soil Conservation* 26 (1): 52–56.

20. Paul, J. C., Mishra, S. B., and Dhal, G. C. (2004). Multi Objective Programming Approach for Watershed Planning- A case study. *Journal of Soil and Water Conservation*, Vol. 3(1&2),73–79.

21. Phadnawis, A. N., Birajdar, J. M., Nangre, M. L., and Asmatoddin, M. D. (1993). Resource management in rainfed watershed- padalsi Maharashtra. *Indian Journal of Soil Conservation* 25(1): 40–45.

22. Prasad, S. N., Singh, R., Prakash, C., Rao, D. H., Gadekar, H., Katiyar V. S., Pasan, A., and Raghupathy, R. (1997). Impact of watershed management on runoff watershed development and productivity of arable lands in south Eastern Rajasthan. *Indian Journal of Soil Conservation, Vol.* 25 (1), 68–72.

23. Samaras A., and Koutitas, C. (2012). Integrated approach to quantity the impact of watershed management on coastal morphology. *Ocean & Costal Management*, Vol. 69, 68–77.
24. Senapati, P. C., Sharma, S. D., and Lal, R. (1979). Probability analysis of annual rainfall data at Bhubaneswar, Orissa. *Journal of Agril. Engg., ISAE,* Vol. 16(4), 131–139
25. Senapati, P. C. (1988). *Water Resources Planning for Canal Command Area of Orissa,* Unpublished PhD thesis, CAET, OUAT, Bhubaneswar.
26. Sheik, A., and Khan, S. A. (2005). Integral programming approach for optimal resource allocation work flow automation design. *9th international multi topic conference ILEE INMIC,* 1–5.
27. Sheng, A. (1972). Soil conservation classification in Taiwan. *International Journal of Soil Conservation, Vol.* 43(2), 41–45.
28. Sheng Y. (2000). Rainfall analysis at Niggata meteorological station, Japan. *International Journal of Irrigation Planning*, Vol. 44(2), 132–140.
29. Tefera, B., and Geert, S. (2010). Land management erosion problems and soil and water conservation in fincha watershed, western Ethiopia. *Land use Policy*, Vol. 27, 1027–1037.
30. Wang, L., Meng, W., Guo, H., Zhang, Z., Liu, Y., and Fan, Y. (2006). An interval fuzzy multi objective watershed management model for the lake Qionghai watershed, China. *Water Resource Management*, Vol. 20, 701–721.
31. Williaim, H., Helen, E., and Nyseen, J. (2007). An integrated multiple criteria decision making approach for resource allocation in higher education. *International Journal of Innovation and Learning*, Vol. 4 (1), 49–62.

APPENDIX I COST OF CULTIVATION AND NET RETURN OF PADDY (UP LAND) AND *KHARIF* (BROADCASTED): RS./HA

S, No.	Components	Unit	Rate (Rs.)	Total (Rs.)
1.	Cost of seed	80 kg/ha	Rs. 15/kg	1200.00
2.	Cost of seed treating chemical	2 gm/kg	Rs. 1.60/gm	256.00
3.	Farm yard manure	5MT	Rs. 600/MT	3000.00
4.	Chemical Fertilizer			
	N	60 kg	Rs. 12.70/kg	762.00
	P_2O_5	30 kg	Rs. 37.50/kg	1125.00
	K_2O	30 kg	Rs. 20.80/kg	624.00
5.	Micro nutrient			500.00
6.	Cost of PP chemicals	2 Nos.	Rs. 400/No.	800.00
7.	Weedicide	1 kg	Rs. 400/kg	400.00
8.	Tractor operation	11 hrs.	Rs. 400/hr	4400.00
9.	Bullock labor	5 Nos.	Rs. 150/Nos.	750.00

Continued

S, No.	Components	Unit	Rate (Rs.)	Total (Rs.)
10.	Human labor	102 MD	Rs. 150/MD	15300.00
11.	Irrigation	-	-	-
12.	Total investment			29117.00
13.	Interest on investment	6 month	12%	1747.00
14.	Land revenue			25.00
15.	Total cost of cultivation			30889.00
16.	Total yield			
	Grain	24 qt	Rs. 1310/qt	31440.00
	Straw	65 qt	Rs. 100/qt	6500.00
17.	Gross income			37940.00
18.	Net return = S. No. 17 – S. No. 15 =			**7051.00**

APPENDIX II　COST OF CULTIVATION AND NET RETURN OF PADDY (MEDIUM LAND) – *KHARIF* (BROADCASTED): RS./HA

S. No.	Components	Unit	Rate (Rs.)	Total (Rs.)
1.	Cost of seed	80 kg/ha	Rs. 15/kg	1200.00
2.	Cost of seed treating chemical	2 gm/kg	Rs. 1.60/gm	256.00
3.	Farm yard manure	5MT	Rs. 600/MT	3000.00
4.	Chemical Fertilizer			
	N	80 kg	Rs. 12.70/kg	1016.00
	P_2O_5	40 kg	Rs. 37.50/kg	1500.00
	K_2O	40 kg	Rs. 20.80/kg	832.00
5.	Micro nutrient			500.00
6.	Cost of PP chemicals	2 Nos.	Rs. 400/No.	800.00
7.	Weedicide	1 kg	Rs. 400/kg	400.00
8.	Tractor operation	14 hrs.	Rs. 400/hr	5600.00
9.	Bullock labor	5 Nos.	Rs. 150/Nos.	750.00
10.	Human labor	132 MD	Rs. 150/MD	19800.00
11.	Irrigation	-	-	-
12.	Total investment			35654.00
13.	Interest on investment	6 month	12%	2139.00
14.	Land revenue			25.00
15.	Total cost of cultivation			37818.00

Continued

S. No.	Components	Unit	Rate (Rs.)	Total (Rs.)
16.	Total yield			
	Grain	34 qt	Rs. 1310/qt	44540
	Straw	71 qt	Rs. 100/qt	7100.00
17.	Gross income			51640
18.	Net return = (S. No. 17 – S. No. 15) =			**13822.00**

APPENDIX III   COST OF CULTIVATION AND NET RETURN OF PADDY (LOW LAND)- *KHARIF* (BROADCASTED), RS./HA

S. No.	Components	Unit	Rate (Rs.)	Total (Rs.)
1.	Cost of seed	60 kg/ha	Rs. 15/kg	900.00
2.	Cost of seed treating chemical	2 gm/kg	Rs. 1.60/gm	192.00
3.	Farm yard manure	5MT	Rs. 600/MT	3000.00
4.	Chemical Fertilizer			
	N	84 kg	Rs. 12.70/kg	1066.80
	P_2O_5	44 kg	Rs. 37.50/kg	1650.00
	K_2O	42 kg	Rs. 20.80/kg	873.00
5.	Micro nutrient			500.00
6.	Cost of PP chemicals	2 Nos.	Rs. 400/No.	800.00
7.	Weedicide	1 kg	Rs. 400/kg	400.00
8.	Tractor operation	18 hrs.	Rs. 400/hr	7200.00
9.	Bullock labor	5 Nos.	Rs. 150/Nos.	750.00
10.	Human labor	165 MD	Rs. 150/MD	24750.00
11.	Irrigation	-	-	-
12.	Total investment			42082.00
13.	Interest on investment	6 month	12%	2524.00
14.	Land revenue			25.00
15.	Total cost of cultivation			44632.00
16.	Total yield			
	Grain	39 qt	Rs. 1310/qt	51090.00
	Straw	80 qt	Rs. 100/qt	8000.00
17.	Gross income			59090.00
18.	Net return = (S. No. 17 – S. No. 15) =			**14458.00**

APPENDIX IV   COST OF CULTIVATION AND NET RETURN OF MAIZE *(KHARIF), RS./HA*

S. No.	Components	Unit	Rate (Rs.)	Total (Rs.)
1.	Cost of seed	15 kg/ha	Rs. 40/kg	600.00
2.	Cost of seed treating chemical	2 gm/kg	Rs. 1.60/gm	72.00
3.	Farm yard manure	5MT	Rs. 600/MT	3000.00
4.	Chemical Fertilizer			
	N	80 kg	Rs. 12.70/kg	1016.00
	P_2O_5	40 kg	Rs. 37.50/kg	1500.00
	K_2O	40 kg	Rs. 20.80/kg	832.00
5.	Micro nutrient	-	-	-
6.	Cost of PP chemicals	2 Nos.	Rs. 400/No.	800.00
7.	Weedicide	1 kg	Rs. 600/kg	600.00
8.	Tractor operation	8 hrs.	Rs. 400/hr	3200.00
9.	Bullock labor	-	-	-
10.	Human labor	160 MD	Rs. 150/MD	24000.00
11.	Irrigation	-	-	-
12.	Total investment			35620.00
13.	Interest on investment	6 month	12%	2137.00
14.	Land revenue			25.00
15.	Total cost of cultivation			37782.00
16.	Total yield			
	Grain	36 qt	Rs. 1310/qt	47160.00
	Straw	20 qt	Rs. 50/qt	1000.00
17.	Gross income			48160.00
18.	Net return = (S. No. 17 – S. No. 15) =			**10378.00**

APPENDIX V   COST OF CULTIVATION AND NET RETURN OF ARHAR (KHARIF), RS./HA

S. No.	Components	Unit	Rate (Rs.)	Total (Rs.)
1.	Cost of seed	20 kg/ha	Rs. 25/kg	500.00
2.	Cost of seed treating chemical	3 gm/kg	Rs. 1.60/gm	96.00
3.	Farm yard manure	5 MT	Rs. 600/MT	3000.00

S. No.	Components	Unit	Rate (Rs.)	Total (Rs.)
4.	Chemical Fertilizer			
	N	20 kg	Rs. 12.70/kg	254.00
	P_2O_5	40 kg	Rs. 37.50/kg	1500.00
	K_2O	20 kg	Rs. 20.80/kg	416.00
5.	Micro nutrient			1000.00
6.	Cost of PP chemicals	2 Nos.	Rs. 500/No.	1000.00
7.	Weedicide	0.75 kg	Rs. 600/kg	450.00
8.	Tractor operation	5 hrs.	Rs. 400/hr	2000.00
9.	Bullock labor	3 Nos.	Rs. 150/Nos.	450.00
10.	Human labor	93 MD	Rs. 150/MD	13950.00
11.	Irrigation	-	-	-
12.	Total investment			24616.00
13.	Miscellaneous		10%	2461.60
14.	Interest on investment	6 month	12%	1264.65
15.	Land revenue			25.00
16.	Total cost of cultivation			28727.00
17.	Total yield			
	Grain	16 qt	Rs. 4300/qt	68800.00
18.	Gross income			68800.00
19.	Net return = (S. No. 18 – S. No. 16) =			**40,073.00**

APPENDIX VI   COST OF CULTIVATION AND NET RETURN OF PADDY (LOW LAND)-*RABI* (TRANSPLANTED): RS./HA

S. No.	Components	Unit	Rate (Rs.)	Total (Rs.)
1.	Cost of seed	60 kg/ha	Rs. 15/kg	900.00
2.	Cost of seed treating chemical	3 gm/kg	Rs. 1.60/gm	288.00
3.	Farm yard manure	5MT	Rs. 600/MT	3000.00
4.	Chemical Fertilizer			
	N	84 kg	Rs. 12.70/kg	1066.80
	P_2O_5	44 kg	Rs. 37.50/kg	1650.00
	K_2O	42 kg	Rs. 20.80/kg	873.00
5.	Micro nutrient			500.00
6.	Cost of PP chemicals	2 Nos.	Rs. 400/No.	800.00
7.	Weedicide	1 kg	Rs. 400/kg	400.00

Continued

S. No.	Components	Unit	Rate (Rs.)	Total (Rs.)
8.	Tractor operation	18 hrs.	Rs. 400/hr	7200.00
9.	Bullock labor	5 Nos.	Rs. 150/No.	750.00
10.	Human labor	165 MD	Rs. 150/MD	24750.00
11.	Irrigation	15 Nos.	Rs. 200/No.	3000.00
12.	Total investment			45178.00
13.	Interest on investment	6 month	12%	2710.70
14.	Land revenue			25.00
15.	Total cost of cultivation			47914.00
16.	Total yield			
	Grain	42 qt	Rs. 1310/qt	55020.00
	Straw	85 qt	Rs. 100/qt	8500.00
17.	Gross income			63520.00
18.	Net return = (S No. 17 - S No. 15) =			**15606.00**

APPENDIX VII   COST OF CULTIVATION AND NET RETURN OF MUSTARD *(RABI)*: RS./HA

S. No.	Components	Unit	Rate (Rs.)	Total (Rs.)
1.	Cost of seed	10 kg/ha	Rs. 30/kg	300.00
2.	Cost of seed treating chemical	3 gm/kg	Rs. 1.60/gm	48.00
3.	Farm yard manure	5MT	Rs. 600/MT	3000.00
4.	Bio-fertilizer			200.00
5.	Chemical Fertilizer			
	N	60 kg	Rs. 12.70/kg	762.00
	P_2O_5	30 kg	Rs. 37.50/kg	1125.00
	K_2O	30 kg	Rs. 20.80/kg	624.00
6.	Micro nutrient			1000.00
7.	Cost of PP chemicals	2 Nos.	Rs. 500/No.	1000.00
8.	Weedicide	0.78 kg	Rs. 600/kg	468.00
9.	Tractor operation	8 hrs.	Rs. 400/hr	3200.00
10.	Bullock labor	3 Nos.	Rs. 150/Nos.	450.00
11.	Human labor	100 MD	Rs. 150/MD	15000.00
12.	Irrigation	3 Nos.	Rs. 300/Nos.	900.00
13.	Total investment			28077.00
14.	Interest on investment	6 month	12%	1684.62

Continued

S. No.	Components	Unit	Rate (Rs.)	Total (Rs.)
15.	Land revenue			25.00
16.	Total cost of cultivation			29787.00
17.	Total yield			
	Grain	13 qt	Rs. 3000/qt	39000.00
18.	Gross income			39000.00
19.	Net return = (S No. 18 - S No. 16) =			**9108.00**

APPENDIX VIII   COST OF CULTIVATION AND NET RETURN OF MUNG *(RABI)*: RS./HA

S. No.	Components	Unit	Rate (Rs.)	Total (Rs.)
1.	Cost of seed	25 kg/ha	Rs. 30/kg	750.00
2.	Cost of seed treating chemical	3 gm/kg	Rs. 1.60/gm	120.00
3.	Farm yard manure	3 MT	Rs. 600/MT	1800.00
4.	Chemical Fertilizer			
	N	20 kg	Rs. 12.70/kg	254.00
	P_2O_5	40 kg	Rs. 37.50/kg	1500.00
	K_2O	20 kg	Rs. 20.80/kg	416.00
5.	Micro nutrient	-	-	-
6.	Cost of PP chemicals	2 Nos.	Rs. 500/No.	1000.00
7.	Weedicide	0.75 kg	Rs. 600/kg	450.00
8.	Tractor operation	5 hrs.	Rs. 400/hr	2000.00
9.	Bullock labor	3 Nos.	Rs. 150/Nos.	450.00
10.	Human labor	65 MD	Rs. 150/MD	9750.00
11.	Irrigation	2 Nos.	Rs. 300/No.	600.00
12.	Total investment			19090.00
13.	Interest on investment	6 month	12%	1145.40
14.	Land revenue			25.00
15.	Total cost of cultivation			20260.00
16.	Total yield			
	Grain	10 qt	Rs. 4500/qt	45000.00
	Straw	-	-	-
17.	Gross income			45000.00
18.	Net return = (S. No. 17 – S. No. 15) =			**24263.00**

APPENDIX IX   COST OF CULTIVATION AND NET RETURN OF HYBRID NAPIER BAJRA GRASS THROUGHOUT THE YEAR-*(KHARIF & RABI)*: RS./HA

S. No.	Components	Unit	Rate (Rs.)	Total (Rs.)
1.	Cost of planting	40000 slips/ha	Rs. 0.40/slips	16000.00
2.	Farm yard manure	8 MT	Rs. 600/MT	4800.00
3.	Chemical Fertilizer			4000.00
4.	Tractor operation	7 hrs.	Rs. 400/hr	2800.00
5.	Bullock labor	2 Nos.	Rs. 200/Nos.	400.00
6.	Plant protection	4 Nos.	Rs. 500/No.	2000.00
7.	Human labor	150 MD	Rs. 150/MD	22500.00
8.	Irrigation			4500.00
9.	Total investment			57000.00
10.	Total yield			
	Fodder	500 qt	Rs. 150/qt	75000.00
11.	Gross income			75000.00
12.	Net return = (S No. 11- S No. 9) =			**18000.00**

Season wise Net return from the grass

Season	Yield	Price
Kharif	30	10800
Rabi	20	7200
Total		**18000**

APPENDIX X   COST OF CULTIVATION AND NET RETURN FROM THE SUBABOOL PLANTATION THROUGHOUT THE YEAR *(KHARIF & RABI)*: RS./HA

S. No.	Components	Unit	Rate (Rs.)	Total (Rs.)
1.	Site preparation including soil work	80 MD	Rs. 150/MD	12000.00
2.	Alignment & staking	60 MD	Rs. 150/MD	9000.00
3.	Cost of plants	5500 Nos.	Rs. 3/No.	16500.00
4.	Planting and gap filling	70 MD	Rs. 150/MD	10500.00
5.	Weeding, mulching, etc.	80 MD	Rs. 150/MD	12000.00

Continued

S. No.	Components	Unit	Rate (Rs.)	Total (Rs.)
6.	Cost of insecticides			440.00
7.	Protection and fencing	80 MD	Rs. 150/MD	12000.00
8.	Irrigation	2 Nos.	Rs. 300/No.	600.00
9.	Total investment			73040.00
10.	Total yield			
	Fuel wood	30 t/yr	Rs. 3000/t	90000.00
	Leafs	100 qt/yr	Rs. 50/qt	5000.00
11.	Gross income			95000.00
12.	Net return = (S.No. 11 − S.No. 9) =			**21,960.00**

Season wise Net return from the Subabool plantation

S. No.	Season	Yield (t/ha)		Total
		Fodder	Fuel wood	
1.	Kharif	7	15	12444
2.	Rabi	3	15	9516
Total				**21960**

APPENDIX XI   CALCULATION OF FOOD REQUIREMENT

Per capita food requirement

$$F_n = \frac{MF_n N_m + FF_n N_f + CF_n N_c}{\text{Total population}}, \text{ where:}$$

MF_n, FF_n, CF_n are the food requirements of male, female and children respectively in gms/day. N_m, N_f, N_c are the number of male, female and children respectively.

Total food requirement in quintals = $F_n \times$ total population $\times$ 365

1. Paddy, $F_n = \dfrac{400 \times 1124 + 350 \times 995 + 275 \times 185}{2304} = 368.37$ gms/person/day

   Paddy requirement for the total watershed = $\dfrac{368.37 \times 2304 \times 365}{1000000}$

   $= 309.78 \approx 310$ ton

2. Maize, $F_n = \dfrac{200 \times 1124 + 150 \times 995 + 100 \times 185}{2304} = 170.38$ gms/person/day

    Maize requirement of the watershed $= \dfrac{170.38 \times 2304 \times 365}{1000000} = 143.28$ tons

3. Arhar, $F_n = \dfrac{50 \times 1124 + 40 \times 995 + 25 \times 185}{2304} = 43.67$ gms/person/day

    Arhar requirement of the watershed $= \dfrac{43.67 \times 2304 \times 365}{1000000} = 36.72$ tons

4. Mustard, $F_n = \dfrac{50 \times 1124 + 35 \times 995 + 20 \times 185}{2304} = 41.11$ gms/person/day

    Mustard requirement of the watershed $= \dfrac{41.11 \times 2304 \times 365}{1000000} = 34.57$ tons

5. Mung, $F_n = \dfrac{40 \times 1124 + 35 \times 995 + 20 \times 185}{2304} = 36.23$ gms/person/day

6. Mung requirement of the watershed $= \dfrac{36.23 \times 2304 \times 365}{1000000} = 30.47$ tons

SUMMARY

S. No.	Food requirement	Estimated quantity (tons)
1.	Paddy	309.78
2.	Maize	143.28
3.	Arhar	36.72
4.	Mustard	34.57
5.	Mung	30.47

APPENDIX XII   CALCULATION OF NUTRITIONAL REQUIREMENT

1. Calorie Need:

Calorie need/person/day(C)

$$C = \dfrac{C_m \times N_m + C_f \times N_f + C_c \times N_c}{\text{Total population}}$$

where, C_m, C_f and C_c are the calorie need of the male, female and children respectively. Total calorie need of the people (CP)

$$CP = C \times \text{population} \times \text{days, where:}$$

$$C = \frac{2600 \times 1124 + 2200 \times 995 + 2000 \times 185}{2304} = 2379.08 \text{ cal/person/day}$$

Total calorie need of the people of the watershed = (2379.08×2304× 365)/1000 = 2000711 Kcal

2. Protein need:

Protein need/person/day (P)

$$P = \frac{P_m \times N_m + P_f \times N_f + P_c \times N_c}{\text{Total population}}$$

Where: P_m, P_f and P_c are the protein need of the male, female and children respectively. Total protein need of the people (PP)

$$PP = P \times \text{population} \times \text{days}; P = \frac{70 \times 1124 + 55 \times 995 + 40 \times 185}{2304} = 61.11 \text{ gms/person/day}$$

Total protein need of the people of the watershed = (61.11 × 2304 × 365)/1000=25556kg

APPENDIX XIII MONTH WISE LABOR REQUIREMENT OF DIFFERENT CROPS PROPOSED IN THE WATERSHED

Crop	Jan	Feb	Mar	Apr	May	Jun	Jul	Aug	Sep	Oct	Nov	Dec
1. Kharif												
Paddy-I	-	-	-	-	-	30	20	15	7	30	-	-
Paddy-II	-	-	-	-	-	30	35	20	12	35	-	-
Paddy-III	-	-	-	-	-	30	35	30	30	40	-	-
Maize	-	-	-	-	-	35	30	15	20	60	-	-
Arhar	-	-	-	-	-	15	20	10	23	25	-	-
2. Rabi												
Paddy-III	35	30	30	40	-	-	-	-	-	-	-	30
Mustard	15	10	25	30	-	-	-	-	-	-	-	20
Mung	10	10	10	25	-	-	-	-	-	-	-	10

Continued

Crop	Jan	Feb	Mar	Apr	May	Jun	Jul	Aug	Sep	Oct	Nov	Dec
Hybrid Napierbajra grass	10	15	15	10	30	25	10	10	5	-	10	10
Subabool	20	15	10	10	90	20	10	30	25	10	110	20

Total labor requirement of different crops

Paddy-I	Paddy-II	Paddy-III	Maize	Arhar	Paddy-III	Mustard	Mung	Hybrid Napier Bajra grass	Subabool
102	132	165	160	93	165	100	65	150	370

APPENDIX XIV   CALCULATION OF FODDER REQUIREMENT OF LIVESTOCK IN THE WATERSHED

S. No.	Animals	Body weight of animal (qt)	No. of animals	Feed rate per qt of body weight (kg/day)	Fodder requirement (kg/day)
1.	Bullock	4.5	82	3.0	1107
2.	Cow	4.0	200	2.7	2160
3.	Buffaloes	6.0	20	3.5	420
4.	Sheep	0.7	58	2.5	102
5.	Goat	0.45	112	2.0	100.8
Total					**3889.80 kg/day**

Total fodder requirement in a year = 3889.80 × 365/1000 = 1419.77 ≈1420 ton (each green and dry fodder)

APPENDIX XV   CALCULATION OF FUEL WOOD REQUIREMENT

S. No.	Name of the village	No. of families per village
1.	Jadupur	0
2.	Ghasedihi	208
3.	Barapatna	111

Continued

S. No.	Name of the village	No. of families per village
4.	Panchugaon	230
5.	Nachhipur	61
Total		**610**

By analyzing the data it is found that the no of house hold, which are well-off according to their economic and social scenario are 210. Assuming the average family consumption of fuel wood as 350 kg/family/month, the total annual fuel wood requirement is $= 350 \times 400 \times 12/1000 = 1680$ ton.

CHAPTER 5

SIMULATION OF OPTIMUM DIKE HEIGHT IN RAINFED RICE FIELD

TRUSHNAMAYEE NANDA,[1] BANAMALI PANIGRAHI,[2] and BALRAM PANIGRAHI[3]

[1]*Research Scholar, Department of Agricultural and Food Engineering, Indian Institute of Technology, Kharagpur, West Bengal, India, 721302, Tel.: +91-8895705688, E-mail: nanda.trushnamayee@yahoo.com*

[2]*Research Associate, Department of Regulatory Sciences, GVK Informatics Pvt. Ltd., Hyderabad, Telangana, India, Tel.: +91-9440942065, E-mail: banamali.panigrahi25@gmail.com*

[3]*Professor and Head, Department of Soil and Water Conservation Engineering, College of Agricultural Engineering and Technology, OUAT, Bhubaneswar, Odisha, India, Mobile: +91-9437882699, E-mail: kajal_bp@yahoo.co.in*

CONTENTS

5.1   INTRODUCTION

Paddy (*Oryza sativa*) is grown over a total area of 159 million ha [5]. It is the largest consumer of irrigation water [20]. The contribution of paddy yield as a percentage of global paddy yield from irrigated, rainfed, upland and food prone ecosystems is 76%, 17%, 4% and 3%, respectively [9]. About 25–33% of world's fresh water is used for irrigation exclusively in paddy cultivation. In Asia, more than 80% of the developed freshwater resources are used for irrigation of which, paddy cultivation consumes more than 90% of irrigation water. By 2025, it is expected that 2 million ha of Asia's irrigated dry-season rice and 13 million ha of its irrigated wet season rice will experience *physical water scarcity*, and most of the approximately 22 million ha of irrigated dry season rice in South and Southeast Asia will suffer *economic water scarcity* [19]. Drought is one of the main constraints to high yield in rainfed rice production systems in both the lowlands and the uplands. In India, paddy is grown over an area of 43 million ha with an annual production of 124 million tons [10] and average productivity is only 2–3.5 tons/ha [24].

As paddy is one of the high water requiring crops, which likes to grow "with the feet in the water," the large volume of water should be applied efficiently minimizing loss constraints. However, poor water use efficiency (WUE) is a major management constraint in lowland paddy production systems. The WUE for transplanted paddy is only 20–30% [19, 21]. Generally, a substantial amount of applied water is lost during land preparation of soil from bypass flow through cracks [4], by deep percolation from root zone, seepage through bunds [11–13] and evapotranspiration [7, 8, 17]. Wopereis et al. [22] estimated cumulative seepage and percolation (SP) losses during a crop cycle for a well-puddled paddy field alone to be as high as 350 cm. Paddy crop has some sensitivity to water stress and some tolerance to water excess. Therefore, water condition of paddy field is to be controlled to keep adequate water supply under submergence condition, usually eliminating the risk of water deficit.

Field experiment was conducted in a site in Bhubaneswar [15] for water balance study in order to estimate optimum dike height in an irrigated paddy. Proper irrigation management demands application of water at the time of actual need of the crop with just enough water to wet the effective root zone soil. The interval between two irrigations should be as wide as possible to save irrigation water without adversely affecting the growth and yield. Hence, water saving irrigation practices are being promoted during the last two decades of research on enhancing rice productivity [1, 2, 14, 24]. It is important to calculate the irrigation water requirement and other water balance parameters of paddy grown under various water management practices in the area of interest. Moreover, typical water-saving schemes in paddy production systems involve alternate wetting and drying conditions in low-land paddy soils [3, 18] leading to unsaturated soil water regimes. Recently, Yang et al. [23] showed 7–11% increase in yield with up to 38% reduction in irrigation water by maintaining critical soil water potential (SWP) at 15 kPa. This will help in assessing how yield will be affected by different water management practices saving different amount of irrigation water.

Because of the intensification of agriculture, per capita availability of water resources is declining day by day in many Asian countries. It is estimated that in India by the year 2050, the share of water for irrigation in agriculture will dwindle to 70% from the present share of 80% now. The declining water resources and the reduced share of its availability for agriculture have affected all the rainfed rice farmers. It is high time now to save the costly irrigated water and economize its use in agriculture. Since rice is a major water-consuming crop, it is important to save the irrigation water in rice field with new and innovative techniques of irrigation and water management.

In rainfed ecosystem, there is no facility of providing supplemental irrigation to crop and rice farmers entirely depend on rainfall and its distribution for growth and production of crop. Since, rainfall is highly erratic and uneven in distribution and onset and withdrawal of monsoon in the study region are highly erratic, users prefer to store maximum amount of rainfall in the field (high ponded depth) by increasing dike height. This eliminates the chances of failure of the crop yield, minimizes irrigation frequencies and amounts. But they are not aware what should be the optimum dyke height in their rice fields. The study has revealed some farmers keep more than 60 cm dyke height, which is oversize and hence causes losses of crop field towards construction of dyke. Some keep as low as 10–15 cm dike height, which is undersize and cannot store maximum rainwater in the field. Hence, there is a need to study and find out what should be the optimum dyke height so that it will be economical and at the same time can store maximum rainwater in the rice field.

Ponding water depth is related to many hydrologic parameters like rainfall, evapotranspiration, seepage and percolation, surface runoff, and irrigation depth, etc. It is necessary to analyze the water balance for long term, which takes into account the variation of rainfall over several years in order to obtain the estimate of the ponding condition in the field. Hence the present study was undertaken with the following objectives to: (i) develop a water balance model for rice grown in *kharif* season under rainfed condition; and (ii) determine optimum dike height in rice field for conservation of maximum rain water.

5.2   MATERIALS AND METHODS

5.2.1   STUDY SITE AND BASIC DATA

The present study on simulation of dike height of paddy field was undertaken for the Bhubaneswar region of Odisha, Eastern India. Geographically, Bhubaneswar is situated in the eastern coastal plain of Odisha (Figure 5.1) and southwest of the Mahanadi River between 21°15' North latitude 85°15' East Longitude and at an altitude of 45 m above MSL. The study area has a tropical climate, specifically a tropical wet and dry climate. Average temperature ranges between a minimum of around 10°C in the winter to a maximum of 42–45°C in summer. Sudden afternoon thunderstorms are common in April and May. The southwest monsoon appears in June with an average annual rainfall is 154 cm, most of which is recorded between June and October. The mean relative humidity ranges from 15.5 to 90.5%. The dominant soil group in the study area is sandy loam, acid lateritic with pH ranging from 4.8 to 5.6, and poor in organic matter. The soil has very low water holding capacity and dries up quickly after cessation of rainfall. Hence, cultivation

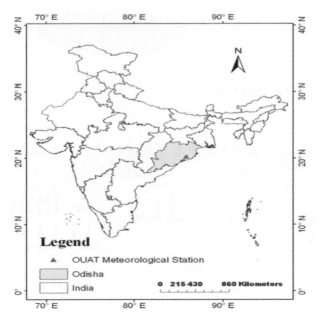

FIGURE 5.1   Index map of study area.

of crop on residual moisture is difficult. Values of field capacity, wilting point and saturation moisture content of rice field in the study area within 45 cm root zone depth are found to be 120, 42, and 170 mm, respectively. The meteorological data includes daily rainfall and pan evaporation collected for 22 years from 1992 to 2013 from the meteorological station of Orissa University of Agriculture and Technology (OUAT), Bhubaneswar.

In the present study, dry seeded rice (DSR) grown in upland topo sequence in wet season/kharif (rainy) without any provision of supplemental irrigation (SI, i.e., rainfed) forms the basis of the modeling. The dike heights of the field are adequate to check any inflow to and outflow from the field. The rice fields are considered as leveled fields. The model uses daily rainfall and other climatological data, soil, and crop data of the study area. Effective root zone depth of rice (DSR or broadcast rice) is taken as 45 cm.

5.3   DEVELOPMENT OF WATER BALANCE MODEL

The various water balance parameters considered in the model are shown in Figure 5.2. The inflow to the field consists of total water supplied from

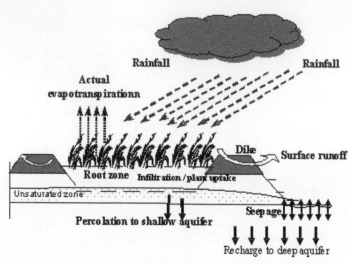

FIGURE 5.2   Water balance of rice field.

rainfall and out flow from the field consists of actual evapotranspiration, seepage and percolation, and surface runoff. Considering the effective root zone of rice as a single layer and neglecting the capillary rise of groundwater in upland topo sequence where groundwater lies more than 1.5 m below crop effective root zone, the generalized water balance model is given as:

$$S_i = S_{i-1} + R_i - AET_i - SP_i - SR_i \tag{1}$$

where, S = soil moisture content, R = rainfall, SP = seepage and percolation loss, AET = actual evapotranspiration, SR = surface runoff from the field and sub-script i = time index taken as 1 day in the study. In the above water balance model, all parameters are considered in mm.

If soil moisture content in the effective root zone of rice is more than saturated moisture content (SAT), then ponding will occur in field. Under the ponding phase, water balance in rice field is given as:

$$PD_i = PD_{i-1} + R_i - AET_i - SP_i - SR_i \tag{2}$$

where, PD is the ponding depth, mm and other terms are defined as above and are given in mm.

5.3.1 WATER MANAGEMENT PRACTICE

The crop is grown under rainfed condition. Hence, there is no provision of supplemental irrigation to crop at any time during its growth period. The dike height is so constructed that it can store maximum rainwater in the field. However, during the initial stage when the rice seeds are broadcast in field and the seedlings are small, the ponded (standing) water in the field is harmful since it may rot the seedlings. Similarly, during the ripening/maturity stage of rice, no standing water is allowed in the field, which will delay the harvest and interfere in crop reaping. In this study, no ponded water was allowed in the field during first 15 and the last 15 days of the crop growth period. Any ponding water in the field during these periods was drained out as surface runoff.

5.3.2 ACTUAL EVAPOTRANSPIRATION

The actual evapotranspiration (AET_i) on any day 'i' is expressed as:

$$AET_i = Kc_i Ks_i ETo_i \qquad (3)$$

where, Kc = crop coefficient that depends on growth stage of the crop; Ks = crop stress coefficient that is a function of the relative available soil moisture content in the field; and ETo = reference crop evapotranspiration.

Daily ET_o was estimated by Pan evaporation method for the simulation period. In pan evaporation method, evaporation pan provides a measurement of the combined effect of temperature, humidity, wind speed and sunshine on the reference crop evapotranspiration ET_o. The value of ET_o can be calculated as:

$$ET_o = Kpan. Epan \qquad (4)$$

where, ET_o = reference crop evapotranspiration; $Kpan$ = pan coefficient; $Epan$ = pan evaporation.

Different types of evaporation pans are being used. The best known pans are the class A evaporation pan (circular pan) and the Sunken Colorado pan (Square pan) [6]. If the water depth in the pan drops too much (due to lack of rain), water is added and the water depth is measured before and after the water is added. If the water level rises too much (due to rain) water is taken out of the pan and the water depths before and after is measured. For the class A evaporation pan, the Kpan varies between 0.35 and 0.85, average Kpan = 0.70 [6]. For the sunken Colorado pan, the Kpan varies between 0.45 and 1.10, average Kpan =0.80 [6]. The value of pan coefficient was taken as 0.8 for this study.

Values of K_c for rice are assumed as 1.05 during crop establishment (CE), 1.10 during both (crop development) CD and Reproductive stage (RS), and 0.95 for maturity stage (MS) [6]. The value of K_s in Eq. (3) is 1.0 under no water stress condition. But as the ponding water vanishes from the rice fields, soil moisture stress occurs that is usually provided by K_s, which consequently decreases the value of AET. In the present study, under unsaturation case, K_s is assumed to vary linearly with the ratio of soil moisture content (S_i) to saturation moisture content (SAT) that is termed as relative available soil moisture content [1, 17] and under ponded case it is assumed as 1. Value of SAT is 170 mm in 45 cm effective root zone depth of rice.

$$K_{si} = S_i/SAT \qquad (5)$$

5.3.3 SEEPAGE AND PERCOLATION

Water loss due to seepage and percolation (SP) in rice fields is often inseparable and so both the terms are considered as single component [21]. The value of SP in the rice field is an extremely variable factor depending on soil and drainage condition. Under different cultural and water management practices, the values of SP are reported to vary from one to 25 mm/day [9]. For rainfed upland rice when most of the time soil in the effective root zone depth remains under unsaturated, SP is estimated [17] as below:

$$SP_i = -16.45 + 0.145 \, (S_{i-1} + R_i - AET_i - SRi) \qquad (6)$$

Under ponding stage, SP is estimated [17] as below:

$$SP_i = -16.45 + 0.145 \, (PD_{i-1} + SAT + R_i - AET_i - SRi) \qquad (7)$$

All terms in Eqs. (6) and (7) are expressed in mm. The value of SP in the model is computed at the end of each day whereas R and SR if any are assumed to occur at the beginning of the day. Water balance model for rainfed rice is run using Eqs. 1 to 6 mentioned as above on daily basis for all 22 years of simulation from June 16 to October 3 of each year.

5.3.4 SIMULATION OF RICE YIELD

The yields of rice are affected by soil water deficits that might occur at any point in the growing season due to differential irrigation. The authors have used the rice yield model as developed by Panigrahi [16]:

$$Y_{ar} = -13.06 + 0.05\ AET_{a1} + 0.07\ AET_{a2} + 0.75\ AET_{a3} + 0.45\ AET_{a4} \quad (8)$$

where, Y_{ar} = actual yield of rice (1000 kg/ha); and AET_{a1}, AET_{a2}, AET_{a3}, and AET_{a4} = values of AET (cm) during CE, CD, RS and MS, respectively.

The authors have taken a short duration rice of 110 days for simulation of water balance model. The duration of growth stages of CE, CD, RS, and MS of the rice from the day of germination till harvest, and these stages were assumed as 25, 25, 35, and 25 days, respectively.

A computer program was written in C^{++} language to estimate the water balance model parameters. The model parameters were obtained on daily basis from which total values in each stage as well as that of the whole season were calculated. Seasonal values of water balance parameters for 22 years are presented in Table 5.1.

The simulated water balance model parameters and the yield data are found to be stochastic varying from year to year. Hence, they were subjected

TABLE 5.1 Seasonal Values of Water Balance Parameters (cm) and Yield Obtained During Simulation Period

Year	Rainfall	Actual evaporation	Seepage and percolation	Surface runoff	Maximum ponded depth	Yield, kg /ha
1992	104.4	38.5	56.8	9.1	19.9	1717.0
1993	100.9	37.7	54.4	8.8	19.0	1422.4
1994	110.2	40.1	58.8	11.3	23.3	2298.7
1995	85.7	31.2	48.9	5.6	17.2	1019.4
1996	90.7	33.5	51.2	6.0	16.5	1438.7
1997	99.5	36.8	56.6	6.1	16.2	1091.1
1998	105.6	38.9	57.2	9.5	22.4	1864.3
1999	112.7	41.0	59.6	12.1	25.6	2637.4
2000	94.6	34.4	55.6	4.6	16.9	1414.9
2001	114.7	41.1	60.8	12.8	26.1	2617.2

TABLE 5.1 Continued

Year	Rainfall	Actual evaporation	Seepage and percolation	Surface runoff	Maximum ponded depth	Yield, kg /ha
2002	100.4	36.9	55.5	8.0	21.4	1127.9
2003	111.0	40.3	63.9	6.8	22.2	2379.7
2004	103.8	38.2	58.5	7.1	18.7	1606.5
2005	78.3	29.7	43.8	4.8	16.8	919.2
2006	88.0	32.6	50.4	5	15.3	1315.9
2007	102.5	38.4	57.9	6.2	21.4	1680.1
2008	115.1	40	64.4	10.7	27.4	2269.3
2009	100.5	37.3	53.1	10.1	22.7	1275.2
2010	87.7	32	48.5	7.2	19.2	896.9
2011	90.4	33.2	51.2	6	20.1	1036.5
2012	100.4	37.8	55.5	7.1	22.6	1459.2
2013	96.6	38.6	53.3	4.7	20.1	1153.8

to probability analysis by fitting different probability distribution functions. In total, the following 12 distribution functions were studied through software "Flood." The distributions are (i) Normal (ii) Log Normal (2-p), (iii) Log Normal (3-p), (iv) Gamma (v) Extreme value (maximum), (vi) Extreme value (minimum), (vii) Exponential, (viii) Pearson, (ix) Log Pearson, (x) Extreme value type III, (xi) Generalized extreme value and (xii) Generalized Pareto distribution. Seasonal values of the water balance model parameters and the yield were predicted by different distributions at different probability levels (PE) from 10 to 90%. The predicted values of maximum ponding depth and yield are presented in Tables 5.2 and 5.3, respectively. The trend was same for other water balance parameters like ponding depth. Since at a given PE level, each distribution gives different values, it is suggested to use a best-fit distribution to predict the variates. In the present study five statistical criteria are used to find out the best-fit distribution. They are (i) Chi-square test, (ii) mean absolute relative error (MARE), (iii) model efficiency (ME), (iv) root mean square error (RMSE); and (v) coefficient of determination (CD). These tests are described as follows:

TABLE 5.2 Maximum Ponding Depth at Different Probability of Exceedance by Various Distributions

Types of distribution	Probability, %								
	10	20	30	40	50	60	70	80	90
Normal distribution	27.9	26.6	25.5	25.0	23.8	23.1	21.8	21.0	20.1
Log normal (2-p) distribution	28.1	27.2	26.2	25.8	24.3	24.0	22.9	22.0	20.4
Log normal (3-p) distribution	28.0	27.4	26.0	25.1	24.0	23.2	22.5	21.6	21.0
Gamma distribution	27.5	26.0	25.1	24.0	22.9	22.0	21.1	20.3	19.6
Extreme value (max) distribution	25.2	24.3	23.8	23.0	22.1	21.0	19.3	18.2	17.0
Extreme value (min.) distribution	26.0	25.3	24.5	23.7	22.0	20.3	19.0	17.9	16.4
Exponential distribution	26.6	25.2	24.3	23.0	22.1	20.7	19.1	18.0	16.8
Extreme value (Type III) dist.	26.9	25.6	24.1	22.7	21.0	19.1	17.4	16.1	15.8
Log Pearson distribution	28.0	26.5	25.1	23.8	22.4	21.3	20.7	19.1	17.2
Pearson distribution	28.2	27.0	25.3	23.7	22.8	21.6	20.1	19.2	18.0
GEV distribution	28.0	26.3	25.6	24.0	22.7	21.5	20.0	18.4	17.1
Generalized Pareto distribution	26.2	25.1	23.5	22.1	21.0	20.3	19.2	18.6	16.7

TABLE 5.3 Simulated Rice Yield (kg ha^{-1}) at Different Probability of Exceedance by Various Distributions

Types of distribution	Probability, %								
	10	20	30	40	50	60	70	80	90
Normal distribution	2488.3	2164.8	1890.3	1648.3	1441.6	1273.5	1138.9	1043.7	985.9
Log normal (2-p) distribution	2486.7	2167.8	1887.9	1647.0	1442.6	1275.2	1140.3	1044.6	986.2
Log normal (3-p) distribution	2480.2	2163.8	1885.3	1645.0	1440.2	1274.5	1140.0	1042.8	987.4
Gamma distribution	2482.4	2165.7	1886.4	1647.2	1441.5	1275.8	1142.5	1043.4	988.0
Extreme value (max) distribution	2468.8	2147.8	1867.8	1626.6	1421.6	1250.7	1123.5	1029.4	970.0
Extreme value (min.) distribution	2468.0	2146.5	1868.8	1625.6	1420.0	1250.1	1121.8	1028.8	968.9
Exponential distribution	2465.0	2144.3	1865.6	1624.7	1418.8	1248.7	1120.0	1027.5	968.0
Extreme value (Type III) dist.	2471.0	2148.9	1868.5	1627.3	1421.9	1250.2	1121.7	1028.9	970.1
Log Pearson distribution	2468.9	2147.9	1874.3	1630.2	1420.6	1254.9	1126.7	1028.5	970.0
Pearson distribution	2494.3	2167.8	1889.4	1652.1	1448.2	1273.8	1145.2	1055.9	991.8
GEV distribution	2498.1	2165.7	1890.2	1650.3	1445.7	1274.6	1143.7	1050.1	990.4
Generalized Pareto distribution	2470.3	2150.1	1870.0	1628.7	1422.8	1253.9	1124.5	1030.1	971.5

The value of the Chi-square is given as:

$$x^2 \sum_{i=1}^{9} \frac{(o_i - p_i)^2}{p_i} \tag{9}$$

where, x^2 is the value of chi-square, O is the observed value, and P is the predicted value, and summation is done from i equals to 1 to 9, i.e., 10 to 90% PE.

The value of mean absolute relative error ($MARE$) is given as:

$$MARE = \sum_{i=1}^{9} \frac{\left| O_i - P_i \right|}{P_i} \bigg/ n \tag{10}$$

where, O and P are as defined earlier. Summation is done from $i = 1$ to 9, i.e., 10 to 90% PE and n is number of data point, i.e., 9.

Model efficiency is defined as:

$$ME = 1 - \frac{\sum_{i=1}^{9}(O_i - P_i)^2}{\sum_{i=i}^{9}\left(O_i - \bar{O}\right)^2} \tag{11}$$

where, $\bar{O}$ is the mean of the observed data, and other parameters are as defined earlier.

Root mean square error is given as:

$$RMSE = \sqrt{\sum_{i=1}^{9} \frac{(O_i - P_i)^2}{n}} \tag{12}$$

Coefficient of determination (CD) is defined as:

$$CD = \frac{\sum_{i=1}^{9}\left(P_i - \bar{O}\right)^2}{\sum_{i=1}^{9}\left(O_i - \bar{O}\right)^2} \tag{13}$$

The *RMSE, MARE* and chi-square values indicate the extent to which the simulations are overestimating or underestimating the observed values. The smaller the *RMSE, MARE* and chi-square, the closer are the simulated values to the observed values. The *CD* statistics describe the ratio of the scatter of the simulated values to that of the observed values. *CD* value of 1 indicates that the simulated values perfectly match the observed values. The model efficiency (ME) can have the highest value of 1. A value closer to 1 indicates that *ME* is perfect and predictions are better. Observed values of water balance parameters and yield at 10 to 90% PE levels were predicted by Weibull's distribution. The value predicted at 50% PE level by Weibull's distribution is considered as the mean of the observed data. Values of Chi-square, *MARE, ME, RMSE* and *CD* were calculated for each distribution using above mentioned formulae. Tables 5.4 and 5.5 represent the values of statistical tests for ponding depth and yield, respectively. Statistical values of other parameters were worked out but not presented here in this chapter

TABLE 5.4   Statistical Parameters for Best-Fit Probability Distribution Function for Maximum Ponding Depth

Distribution	Chi-square	MARE	ME	RMSE	CD
Normal distribution	4.47	0.169	0.969	3.37	0.71
Log normal (2-p) distribution	6.28	0.204	0.980	4.03	0.80
Log normal (3-p) distribution	5.67	0.192	0.977	3.83	0.75
Gamma distribution	2.93	0.134	0.972	2.68	0.73
Extreme value (max) distribution	0.55	0.052	0.960	1.12	0.68
Extreme value (min.) distribution	0.59	0.054	0.919	1.22	0.69
Exponential distribution	0.62	0.059	0.953	1.24	0.65
Extreme value (Type III) dist.	0.42	0.048	0.980	1.05	0.86
Log Pearson distribution	1.84	0.104	0.965	2.18	0.74
Pearson distribution	2.17	0.114	0.934	2.37	0.81
GEV distribution	1.75	0.099	0.946	2.15	0.76
Generalized Pareto distribution	0.37	0.044	0.988	0.93	0.95

TABLE 5.5 Statistical Parameters for Best-fit Probability Distribution Function for Yield Data

Distribution	Chi-square	MARE	ME	RMSE	CD
Normal distribution	0.57	0.0065	0.985	5.49	0.78
Log normal (2-p) distribution	0.62	0.0069	0.976	5.09	0.76
Log normal (3-p) distribution	0.46	0.0058	0.987	4.38	0.80
Gamma distribution	0.63	0.0068	0.971	6.05	0.77
Extreme value (max) distribution	0.78	0.0071	0.969	7.67	0.68
Extreme value (min.) distribution	0.83	0.0074	0.959	10.4	0.70
Exponential distribution	0.95	0.0093	0.940	8.21	0.64
Extreme value (Type III) dist.	0.53	0.0063	0.986	4.76	0.77
Log Pearson distribution	0.89	0.0089	0.948	5.36	0.70
Pearson distribution	0.82	0.0085	0.951	3.74	0.72
GEV distribution	0.97	0.0099	0.926	4.76	0.69
Generalized Pareto distribution	0.38	0.0024	0.992	2.97	0.93

to save the space. The observed values of ponding depth and yield (worked out by Weibull's distribution) and those simulated/predicted by the best-fit distribution as decided by the statistical tests at different PE levels are also shown in Tables 5.4 and 5.5, respectively.

5.4 RESULTS AND DISCUSSION

5.4.1 WATER BALANCE MODEL PARAMETERS

The water balance parameters were SMC, AET, SP, SR and ponded depth (PD); and these were simulated by the developed water balance model for *kharif* season for 22 years on daily basis. The daily values were used to compute the seasonal values for each year. The computed seasonal values were found to be stochastic (Table 5.1) and so they were predicted by different probability density functions by using software *Flood*. Values of seasonal water balance parameters were predicted at 10–90% probability levels by different distributions. The values for the ponding depth and yield are presented in Tables 5.2 and 5.3, respectively.

5.4.1.1   Actual Evapotranspiration (AET)

Values of AET ranged from 29.7 to 41.1 cm during 22 years of simulation with a mean value of 36.73 cm and standard deviation of 3.37 cm. Value of AET was maximum for the year 2001 (41.1 cm) having experienced a seasonal rainfall of 114.7 cm, which is the second highest (highest rainfall was 115.1 cm in 2008). The lowest value of AET was associated for the year 2005 with a value of 29.7 cm and the rainfall in that year was the minimum of all the years (78.3 cm). Thus, the study concludes that values of AET of rice are dependent on rainfall/irrigation and since the rainfall in 2001 was maximum, AET was maximum for that year and similarly since the rainfall in 2005 was minimum, AET for 2005 was minimum (Table 5.1). The average value of AET was observed to be 36.8% of the total average losses (total losses = AET + SP + SR). Values of AET were predicted by different probability distribution functions. It was observed that as the probability of exceedance (PE) was increased, values of AET were found to decrease. At 10% PE level, values of AET were maximum for all the distributions and at 90% level, they were minimum. From the statistical tests, it was revealed that *Generalized Pareto distribution* gave the best-fit value at different PE levels.

5.4.1.2   Seepage and Percolation

Values of seepage and percolation (SP) ranged from 43.8 to 64.4 cm with a mean value of 55.26 cm and standard deviation of 5.01 cm. The value of SP was minimum for the year 2005 (43.8 cm) since the seasonal rainfall in that year was also minimum (78.3 cm). Similarly the year 2008 had a maximum seasonal rainfall of 115.1 cm (highest of all 22 years) and this was consequential to the highest SP value of 64.4 cm in 2008 (Table 5.1). Like AET, SP values were found to decrease when the PE levels were increased from 10 to 90%, and it was highest at 10% PE level for all the probability distribution functions. However, like AET, *Generalized Pareto distribution* was observed to be the best-fit distribution of all the 12 distributions from different statistical tests. Contribution of SP was observed to be the highest of all water balance parameters with an average value of 55.4% of total losses in the rice field.

5.4.1.3   Surface Runoff

In the rainfed ecosystem, farmers depend only on rainfall for cultivation of rice. So, they want to store as much water as possible in the field through raised dikes and therefore they do not allow any surface runoff. But to save the crop from rotting when they are just sown or are small in height, they dispose off all the surface runoff from the field up to 15 days of sowing the seeds in the field. Similarly, during the harvest stage (15 days before harvest) they dispose off all the surface runoff from the field. In other periods, they store entire runoff in the diked field. Values of surface runoff (SR) ranged from a minimum of 4.7–12.8 cm (Table 5.1) with an average value of 7.7 cm and standard deviation of 2.49 cm. Value of SR was maximum for the year 2001 (12.8 cm) and minimum for the year 2013 (4.7 cm). Values of SP were observed to vary with rainfall. The year with high rainfall is associated with high value of SR. Like AET and SP, SR was also found to vary with different PE levels, and was highest at 10% PE level for all the probability levels and lowest at 90% PE levels for all distributions. However, from various statistical tests, it was noted that *Generalized Pareto distribution* was observed to be the best-fit distribution of all the 12 distributions. Average contribution of SR to the total seasonal losses was 7.8%. The study reveals that the average value of SR of 7.7 cm if stored in an on-farm reservoir, can provide at least one supplemental irrigation to rice during the dry spell and thus can save the crop form drought.

5.4.1.4   Ponding Depth

Ponding depth is also called as standing water. Ponding depth in field is an important water balance parameter, which happens when the soil moisture content is above the saturation level. In the rainfed situation, farmers prefer to maintain all rainwater as ponding depth (also called as ponded depth, PD) in their rice field except the first and last 15 days of crop growth period as mentioned above. This is possible by constructing higher dikes around the field. In the present study, simulation of maximum ponded depth is determined by water balance model on early basis. The study reveals that the ponded depth depends on many factors like AET, SP and SR along with the pre-assigned water management practice. Values of PD ranged from a minimum of 15.3 cm to a maximum of 27.4 cm during the 22 years of simulation (Table 5.1). Maximum PD of 27.4 cm occurred in 2008 when there was the highest seasonal rainfall of 115.1 cm. Similarly the year 2006 has very little rainfall (88.0 cm), which resulted in obtaining lowest PD of 15.3 cm. The average value of PD during the simulation period was 20.5 cm with a standard deviation of 3.36 cm.

Like other water balance parameters, PD was found to be stochastic. Hence, the PD data were fitted to different probability distribution functions through the software "Flood" and the values were predicted at different PE levels. Table 5.2 represents the values of PD by different distributions at 10 to 90% PE levels. From these data of Table 5.2, it is noted that as PE increases from 10 to 90%, the values of PD decreases for all the distributions.

In order to find out the best-fit distribution out of tested 12 ones, statistical tests like Chi-square, MARE, ME, CD and RMSE were conducted with the observed and predicted data. Weibul's distribution was taken as the observed data. Values of different statistical parameters obtained for various distributions are presented in Table 5.4. The statistical tests indicated that *Generalized Pareto distribution* gave the lowest values of Chi-square (0.37), MARE (0.044) and RMSE (0.93) and highest values of ME (0.988) and CD (0.95) and thus is found to be the best-fit distribution amongst all other distributions. The next best-fit distribution was Extreme Value Type III with values of chi-square (0.42), MARE (0.048) and RMSE (1.05) and highest values of ME (0.980) and CD (0.86) (Table 5.4). Values of PD at 10 to 90% PE levels were therefore calculated by *Generalized Pareto distribution* and at an average 50% PE level, PD was obtained as 21.0 cm. Since, small soil water conservation structures are designed at 5 years return period (20% PE level), value of PD was worked out to be 25.1 cm (Table 5.2). With consideration of 20% free board, the dike height should be 30.1 cm. Hence, it is recommended that in rainfed ecosystem to conserve maximum rainwater, the dike heights in the rice field should be 30.1 cm. Presently, the farmers have dike heights in their fields ranging from 15 to 60 cm.

5.4.2 SIMULATION OF RICE YIELD

Rice yield was simulated by Eq. (8). The yield of rice ranged from 896.91 to 2637.45 kg/ha with a mean value of 1574.6 kg/ha and standard deviation of 548.7 kg/ha. The yield data are found to have large deviation and dispersion with a large value of standard deviation. The yields are highly variable depending on the magnitude and distribution of rainfall. The yield was maximum for the year 1999 (2637.45 kg/ha) and minimum for the year 2010 with a value of 896.91 kg/ha. Though the rainfall in 1999 (112.7 cm) was not the maximum as compared to 2005, which had a rainfall of 115.1 cm, but its distribution was more uniform in 1999 than 2008 and so, the yield was maximum in 1999. Similarly, the year 2010 had a rainfall of 87.7 cm, which is more than the drought year 2005 (rainfall = 78.3 cm). But the distribution of rainfall in 2010 was less uniform than 2005 and therefore, the yield was the lowest in this year (Table 5.1).

Since rice yields were stochastic, they were fitted to different probability distribution functions through the software "*Flood*" and the values were predicted at different PE levels. Values of rice yields by different distributions at 10 to 90% PE levels were worked out as shown in Table 5.3. Table 5.3 indicates that as PE increases from 10 to 90%, the values of rice yields decreases for all the distributions. In order to find out the best-fit distribution, statistical tests like Chi-square, MARE, ME, CD and RMSE were conducted with the observed and predicted data. Just like ponding depth, Weibul's distribution was taken as the observed data in the statistical tests.

Values of different statistical parameters obtained for various distributions are presented in Table 5.5. The statistical tests indicated that like the case of PD, *General Pareto distribution* gave the lowest values of Chi-square (0.38), MARE (0.0024) and RMSE (2.97) and highest values of ME (0.992) and CD (0.93) and thus is found to be the best-fit distribution amongst all other distributions. The next best-fit distribution was Log Normal (3-p) distribution with values of chi-square (0.46), MARE (0.0058) and RMSE (4.38) and highest values of ME (0.987) and CD (0.80) (Table 5.5). Values of PD at 10 to 90% PE levels were therefore calculated by *Generalized Pareto distribution* and at an average 50% PE level, yield was 1422.8 kg/ha. The yield data at 10% PE was the highest (2470.3 kg/ha) and at 90% PE, it was the lowest (971.5 kg/ha) (Table 5.3).

5.5 CONCLUSIONS

The study of simulation of water balance model parameters in rainfed rice and yield indicates that yield of rice is very sensitive depending on the magnitude and distribution of rainfall. The rainfed rice yields varied from 896.91 to 2637.45 kg/ha with a mean value of 1574.6 kg/ha. From different statistical tests, it was observed that the *Generalized Pareto distribution* was the best-fit distribution and hence the yield were forecasted by this distribution. At 10% PE level, the yield was predicted as 2470.3 kg/ha and as PE level increased to 90%, the yield was found to decrease to 971.5 kg/ha. The simulation of water balance parameters indicated that seepage and percolation is the major loss in rice field. The loss due to SP accounts for 53% of total losses in the fields.

All the water balance parameters were found to be stochastic and so they were fitted to 12 different probability distribution functions through a software "Flood" and with statistical tests the *Generalized Pareto distribution* was found to be the best-fit one. The value of ponding depth at 20% PE level by this best-fit distribution was predicted to be 25.1 cm. Hence a dike height of 25.1 cm can store this maximum ponding depth in the rice field. With consideration of 20% free board, the maximum dike height is worked out to be 30.1 cm. Hence, the study suggests that in rainfed rice fields in eastern region of India, farmers can go for construction of optimum dike height of 30.1 cm so that maximum rainwater can be conserved in the field which will enhance the yield. Presently the rainfed farmers in the region construct dike heights varying from 15 to 60 cm, which is highly variable.

5.6 SUMMARY

The present study investigates to find out the optimum height of dike for broadcast rice fields grown in *kharif* (rainy) season in eastern region of India. Simulation of water balance model parameters was done for rainfed rice field without provision of any supplemental irrigation. However, during the initial stage when the rice seeds are broadcast in field and the seedlings are small, the ponded (standing) water in the field is harmful. Similarly, during the ripening/maturity stage of rice, no standing water is allowed in the field. In this study, no ponded water was allowed in the field during first 15 and the last 15 days of the crop growth stages. In all other stages, all the rainwater was allowed to remain present in the field which is the standard water management practice of the rainfed rice farmers in the region.

In the present study, a water balance model of rainfed rice was developed and the water balance model parameters were simulated for 22 years (1992–2013) starting from the day of onset of monsoon (rainy season) to withdrawal of monsoon, which is June 16 to October 3 (110 days).

Various water balance model parameters like seepage and percolation (SP), actual evapotranspiration (AET), surface runoff (SR) and ponded depth (PD) were simulated for each year during the simulation period on daily basis from which seasonal values were computed. An available simulated rice yield model was used to simulate the rice yield for each year using the data of AET.

The study revealed that the seasonal values of the model parameters including yield were stochastic and so these values were fitted to probability distribution functions. Twelve different probability distribution functions were considered in this study. By statistical tests, *Generalized Pareto distribution* was found to be the best-fit distribution for all the model parameters including yield. Hence, the values of different model parameters and yield were simulated at different probability levels by this best-fit distribution. The optimum dike (which is decided by ponding depth) height at 5 years return period (20% PE level) by *Generalized Pareto distribution* was obtained as 25.1 cm and with 20% free board, the optimum dike height suggested is 30.1 cm. So, the rainfed farmers in the eastern region of the country can go for construction of a maximum dike height of 30.1 cm, which will conserve maximum rainwater in the field, decrease the requirement of supplemental irrigation and hence can enhance the rice yield.

KEYWORDS

- actual evapotranspiration
- Chi-square
- coefficient of determination
- deviation
- dike
- farmers
- mean
- mean absolute relative error
- model efficiency
- optimum
- percent

- **probability distribution function**
- **probability**
- **rainfed**
- **return period**
- **rice**
- **root mean square error, RMSE**
- **seepage and percolation**
- **simulation**
- **standard deviation**
- **statistical tests**
- **supplemental irrigation**
- **surface runoff**
- **water balance model**
- **water balance parameters**
- **yield**

REFERENCES

1. Agrawal, M. K., Panda, S. N., and Panigrahi, B., (2004). Modeling Water balance parameters for rainfed rice. *Journal of Irrigation and Drainage Engineering,* ASCE, 130 (2):134–148.
2. Arora, V. K., (2006). Application of a rice growth and water balance model in an irrigated semi-arid subtropical environment. *Agr. Water Manage.,* 83:51–57.
3. Belder, P., Bouman, B. A. M., and Spiertz, J. H. J., (2007). Exploring options for water savings in lowland rice using a modeling approach. *Agric. Syst.,* 92: 91–114.
4. Cabangon, R. J., and Tuong, T. P., (2000). Management of cracked soils for water saving during land preparation for rice cultivation. *Soil Till. Res.,* 56:105–116.
5. Food and Agriculture Organization of the United Nations-FAO., (2010). *FAOSTAT,* Rome, Italy, December.
6. Food and Agriculture Organization of the United Nations-FAO, (1996). Modernization of irrigation schemes: past experiences and future. *Proceedings of the Expert Consultation,* Bangkok, Thailand.
7. Hardjoamidjojo, S., (1992). The effect of flooding and method of water application on water requirements and yield of wet land paddy. In: Murty, V. V. N., Koga, K. (Eds.), *Soil and Water Engineering for Paddy Field Management. Proc. Int. Workshop on Soil and Water Engineering for Paddy Field Management,* Asian Inst. Technol., Bangkok, Thailand, 28–30 January, pp. 63–71.

8. Humphreys, L., Muirhead, W., Fawcett, B. J., and Townsend, J., (1992). Minimizing deep percolation from rice. *Farmers Newsletter (Griffith, NSW, Australia)*, 172:41–43.
9. IRRI, (1993). Rice research in a time of change. *IRRI's medium-term plan for 1994±1998*, International Rice Research Institute, Los Baños, Philippines, pages 79.
10. IRRI, (2004). *Annual Report*, IRRI, Los Banos, Philippines.
11. Janssen, M., and Lennartz, B. B., (2007). Horizontal and vertical water and solute fluxes in paddy rice fields, *Soil Till. Res.*, 94:133–141.
12. Janssen, M., and Lennartz, B. B., (2008). Characterization of preferential flow pathways through paddy bunds with dye tracer tests. *Soil Sci. Soc. Am. J.*, 72:1756–1766.
13. Janssen, M., and Lennartz, B. B., (2009). Water losses through paddy bunds: methods, experimental data, and simulation studies. *J. Hydrol.*, 369:142–153.
14. Khepar, S. D., Yadav, A. K., and Sondhi, S. K., (2000). Water balance models for paddy field for intermittent irrigation practices. *Irrig Sci.*, 19:199–208.
15. Mishra, A.,(1999). Irrigation and drainage needs of transplanted rice in diked rice fields of rainfed medium lands. *Irrig Sci.*, 19:47–56.
16. Panigrahi, B., (2001). *Water Balance Simulation for optimum design of on-farm Reservoir in Rain fed Farming System*. PhD thesis, IIT, Kharagpur.
17. Sharma, P. K., and De Datta, S. K., (1985). Puddling influence on soil, rice development, and yield. *Soil Science Society of America Journal*, 49:1451–1457.
18. Tabbal, D. F., Bouman, B. A. M., Bhuiyan, S. I., Sibayan, E. B., and Sattar, M. A., 2002. On-farm strategies for reducing water input in irrigated rice: case studies in the Philippines. *Agric. Water Manage.*, 56:93–112.
19. Tuong, T. P., and Bouman, B. A. M., (2002). Rice production in water scarce environments. In: *J. W. Kijne, R. Barker & D. Molden, eds. Water productivity in agriculture: Limits and opportunities for improvement*. The Comprehensive Assessment of Water Management in Agriculture Series, 1:13–42.
20. Tuong, T. P., Bouman, B. A. M., and Mortimer, M., (2005). More rice, less water – Integrated approaches for increasing water productivity in irrigated rice-based systems in Asia. *Plant Prod. Sci.*, 8:231–241.
21. Walker, S. H., and Rushton, K. R., 1984. Verification of lateral percolation losses from irrigated rice fields by a numerical model. J. *Hydrol.*, 71:335–351.
22. Wopereis, M. C. S., Bouman, B. A. M., Kropff, M. J., Berge, H. F. M. T., and Maligaya, A. R., (1994). Water use efficiency of flooded rice fields, I. Validation of the soil–water balance model SAWAH. *Agric. Water Manage.*, 26:277–289.
23. Yang, J., Liu, K., Wang, Z., Du, Y., and Zhang, J., (2007). Water-saving and high-yielding irrigation for lowland rice by controlling limiting values of soil water potential. *J. Integr. Plant Biol.*, 49:1445–1454.
24. Yoon, G. H., and Jeon, J., (2003). Mass balance analysis in Korean paddy rice culture. *Paddy Water Environ.*, 1:99–106.

PART II

MODELING IRRIGATION SYSTEMS IN CANAL COMMAND

PERFORMANCE ASSESSMENT OF RICE IRRIGATION PROJECT USING REMOTE SENSING AND GEOGRAPHICAL INFORMATION SYSTEM

G. CHANDRA MOULI,[1] SUDHINDRA N. PANDA,[2] and V. M. CHOWDARY[3]

[1]*Principal Investigator, Precision Farming Development Centre, Institute of Agricultural Engineering and Technology, PJTS Agricultural University, Rajendra Nagar, Hyderabad–500030, India; Tel.: +91-9437667879; E-mail: gaddamchandramouli@gmail.com*

[2]*Department of Agricultural and Food Engineering, Indian Institute of Technology, Kharagpur–721302, West Bengal, India; Tel.: +91-3222-283140 (Work); E-mail: snp@iitkgp.ac.in, snp@agfe.iitkgp.ernet.in, sudhindra.n.panda@gmail.com*

[3]*Scientist, Regional Remote Sensing Centre, National Remote Sensing Centre, Kolkata–700156, West Bengal, India; E-mail: muthayya.chowdary@gmail.com*

Edited version of "*Gaddam Chandra Mouli, 2010. Performance Assessment of a Major Irrigation Project using Remote Sensing and Geographical Information System, Unpublished PhD Thesis, Agricultural and Food Engineering Department, Indian Institute of Technology, Kharagpur, 721302, India.*"

CONTENTS

6.1   INTRODUCTION

Irrigated agriculture is the largest water user at global level, consuming about 80% of the world's developed water resources [62]. In India, the gross irrigated area is likely to reach 107 million-ha by 2025 from the existing 79 million-ha [36]. Though, India accounts for the highest percentage of cultivated area in the world, where 75% of the population relying on agriculture, yet its crop productivity is invariably very low as compared to other agriculturally advanced countries. Low productivity is due to inefficient irrigation management practices being followed across the country. Moreover, due to growing population of India, which will be expected to reach 1395 million by 2025, the share of water for agriculture will go down with increasing demand for water in industry, hydropower, and domestic sectors [57]. Thus, in future, the irrigation water management needs to be more efficient with the production of more crops per unit of water.

Irrigation is the only option for accomplishing crop productivity in arid and semiarid regions due to scanty and erratic rainfall. Major canal irrigation projects in the developing countries often suffer from inequitable distribution of water due to excess use by the upper reach farmers to grow water intensive crops like rice [9]. Lack of sufficient information on water distribution within an irrigation system is still a major limitation for efficient management of water.

The study in this chapter was carried out in the Addanki branch canal command of the Nagarjuna Sagar right main canal, where canal water is the only source of irrigation. The Nagarjuna Sagar is a major irrigation project, located in the lower part of Krishna basin of Andhra Pradesh, India. There exists stiff competition for the distribution of irrigation water among different canal reaches. Upper reach farmers normally overdraw canal water for irrigating rice. Some of the tail reach farmers, who are uncertain of getting canal water supply; have replaced rice by alternate rain-fed crops like Bengal gram and maize.

Rice requires adequate supply of canal water so as to meet its demand. Under persisting competition for water, one of the greatest challenges for water managers is to match the demand and supply as it largely influences the variability in productivity [39]. Hence, a comprehensive study needs to be done to fully understand the water use, surplus and water deficit, and productivity for estimating the performance of an irrigation system. Performance assessment is considered to be one of the most critical elements for improving irrigation system management [2] as water deficit phenomenon is becoming common in many irrigation projects of the country.

In semiarid regions, a small variability in rainfall and land parameters can have profound effect on irrigation demand and crop productivity. Distributary-wise interpolated values of aforementioned parameters are the guiding factors for assessing irrigation system performance. The prevailing reservoir operational policy of the irrigation project is guided by water availability in the reservoir and cropping pattern framed at the initial days of project formulation. Further, without consideration of actual crop area and crop growth stages, water is being released from the outlets at a constant pre-decided rate, which is a great loss of precious water resources.

Based on aforementioned issues associated with the irrigation project, the following objectives were formulated to conduct the present research in this chapter:

- to predict rainfall, soil salinity, and rice productivity at unmeasured locations in the irrigated command area using Kriging technique.
- to develop land use/land cover map of the study area using satellite imagery.
- to assess the performance of the irrigated command using remote sensing and GIS.
- to develop and validate regression models for the rice productivity for the irrigated command.

6.2 GEOGRAPHICAL INFORMATION SYSTEM (GIS) – GEOSTATISTICAL ANALYSIS

Spatial data management tools of Geographic information system (GIS) can analyze spatial variability in soil, crop, and water supply in a canal command, while dealing with complex problems of water resources management [44]. GIS has the capability of generating and overlay various data layers in order to relate them over space and time. Crop yield is an outcome of many complex soil and climate factors, and their effect on yield might be better interpreted through the use of GIS [18].

Spatial and temporal analysis of actual water supply in different parts of the irrigation project identifies how and where to improve the performance of the irrigation scheme [22]. Although documentation of variability in soil properties is important, it is also important to ascertain if the variability falls in a range where it may limit crop growth and crop productivity. For estimation of soil properties at unsampled points, Kriging technique is normally used with different variogram models, and Kriged maps clearly showed the presence of both large- and small scale variability in a small field, where it is expected to have uniform trend [16].

GIS provides a set of powerful statistical tools for analyzing the spatial variability of the parameters using a number of conventional interpolation techniques. It generates surfaces of interesting phenomena by using measured sample points.

Geostatistical methods, based on the theory of describing the relationship between the spatially random variables, are increasingly utilizing the spatial correlation between neighboring observations to predict the attributes at unmeasured locations [5, 29, 32, 59]. Several authors [6, 12, 15, 26] have shown that geostatistics provide a far better estimate of the attributes than any other conventional methods. For low-density networks of rain gauges, geostatistical interpolation outperforms techniques of inverse square distance or Thiessen polygon that ignore the pattern of spatial dependence, which is usually observed for rainfall data prediction [25].

The present investigation is undertaken to develop the spatial variability prediction maps for rice productivity and monthly rainfall data during the crop growth period by using appropriate semi-variogram model based upon the cross-validation error statistics.

6.2.1 GENERATION OF LAND USE AND LAND COVER FROM REMOTE SENSING IMAGES

Manual integration of information of the large irrigation project (>10,000 ha) requires huge expenditure on manpower and time. Analysis of remote sensing data helps to study the different land features in a large project by its capability of synoptic view. Further, working on the GIS platform is faster, more accurate and, therefore, cost-effective. Hence use of the satellite imagery and GIS can ease the data integration and analysis of very large data sets.

Remote sensing data acquired from space-borne platforms, owing to their wide synosivity and multispectral acquisition, offer unique opportunities for the study of soils, LULC, and other parameters required for hydrologic modeling of large areas. Combining information obtained from satellite remote sensing with ground data in a GIS format proved to be efficient in identifying major crops and their condition and determining area and yield of wheat crop [49]. Remote sensing, with varying degrees of accuracy, has been able to provide information on land use, irrigated area, crop type, biomass development, crop yield, crop water requirements, crop evapotranspiration, salinity and waterlogging [8].

The spatial distribution of changes in cropping pattern was mapped using multi-temporal imagery from the moderate resolution imaging spectrometer, which could identify areas in single, double, or continuous cropping. With the use of these images, the study-identified areas affected by low canal releases and showed a widespread shift from double to single cropping scenario [10, 22].

6.2.2 PERFORMANCE ASSESSMENT BASED ON REMOTE SENSING AND GIS

Various workers used remote sensing as a tool for assessment of irrigation performance by processing satellite images. Performance indicators from RS algorithms supplemented by ground data have been suggested by [7, 35]. Topographical maps are required for geo-referencing of satellite data, subsequently for locating the sample areas/points on the ground during field check and lastly for transferring thematic details [43]. Performance indicators are necessary for assessing and analyzing water delivery systems. Without considering the uniformity of the spatial and temporal variability of water delivery at different levels, proper assessment of variability of performance is not possible [30].

On developing the AREASUM indices, adequacy was computed in terms of relative water supply (RWS) for Tarafeni South Main Canal command, using ET_C (SAVI based), which was found to be 0.44, that falls under high water deficit category (RWS < 0.5). However, the RWS for combined water supply from canal and ponds is improved to 0.66, which falls under water deficit category (0.5 <RWS< 0.9) [23].

6.2.3 ASSESSMENT OF IRRIGATION DEMAND–SUPPLY

The irrigation simulation model, CRIWAR, has been found to be helpful software in estimation of crop water demand, taking into consideration different cropping pattern, crop areas and respective sowing dates and crop duration [11]. The estimation of crop evapotranspiration as well as net irrigation requirement for each soil type at a given meteorological station is calculated using CRIWAR model.

Multidate RS data of WiFS (wide field sensor) in IRS-IC satellite was used for crop classification and computation of crop coefficient from NDVI. The sensor, WiFS, has 188 m ground resolution and two spectral bands in red (620–680 nm) and near-infrared (770–860 nm) region. One scene for each month was used for generating temporal NDVI distribution. Using monthly climatic data, ET_o values were estimated by the modified Blaney-Criddle method because of its simplicity and normal availability of data. Monthly estimates of ET_o by aforesaid method were observed fairly close to the estimates of a more rigorous Penman method [46].

The irrigation water supply and demand were analyzed for Tarafeni South main canal command of Kangsabati project in Paschim Midnapur district, West Bengal, India, using tools of RS and GIS. Reference crop evapotranspiration was estimated using FAO Penman-Monteith equation. The study concluded that the crop water demand in the command area was more than the supply during the months of *Rabi* season from December to April [37].

Crop identification and water requirement at distributary level was calculated for the Pehure High Level Canal and the Upper Swat Canal system in the North Western Frontier Province of Pakistan [52]. Unsupervised classification of multi temporal satellite images were used to identify various crops and cropping pattern in the area. These calculated areas were compared with the seasonal data recorded by the irrigation department. ET was calculated using CROPWAT model by the Penman-Monteith method for calculating reference crop evapotranspiration at various stages of crop growth. Then water required for each individual crop was calculated. The results were found very encouraging. It was observed that results of this study could be useful for water managers to release the canal supplies based on crop water requirement.

Gaur et al. [22] used integrated approach to assess how cropping patterns and the spatial equity of canal flow changed with the water supply variations in the left canal command area of Nagarjuna Sagar. The integrated approach was used to assess changes in the spatial equity of canal flow and land use with water supply variations in the head, middle, and tail reaches of the left main canal command of Nagarjuna Sagar during water surplus (2000–2001), normal (2001–2002), and deficit (2002–2003) years. A study conducted in the Hirakud canal command on irrigation water supply and demand estimation suggested a reduction of areas under rice and replacing the rice area with crops of low-water demand [45].

6.2.4   REGRESSION EQUATION BETWEEN RICE PRODUCTIVITY AND NDVI

NDVI correlates well with the spatial and temporal changes of crop conditions. Many attempts have been made to estimate crop yields from satellite data [27, 42] recommended the use of NDVI at crop heading stage for estimating potential harvestable yield. Murthy et al. [38] observed that in case of non-availability of crop yield data, simplified statistical relationships could be used for estimating the yield from NDVI taken from single date image acquired during the heading stage of grain crops.

Because rice transplantation is staggered across the command area, satellite data from any one date do not represent the same growth stage at all locations. Consequently, an innovative approach of time composition was attempted, using co-registered multi-date satellite data. The maximum value of NDVI for each rice pixel was picked from among the satellite overpasses encompassing the period of rice at heading across the command area [55]. However, data from crop cutting experiments are necessary to validate these types of statistical relationships [4].

6.2.5   NON-LINEAR MULTIPLE REGRESSION EQUATIONS BETWEEN RICE PRODUCTIVITY AND AGRICULTURAL INPUTS

The agricultural inputs applied for increasing the crop productivity in irrigated commands are namely water and nutrients. The farmers are always under the impression, that if these inputs are applied in maximum dosages, the crop productivity increases considerably. But this might not be true, the productivity is likely to show a nonlinear declining trend after reaching a certain level to input (s). Many of the water production functions presented in the literature were developed relating crop yield (Y) to applied water, which usually includes irrigation water to satisfy crop water requirements. Higher application of fertilizers would lead to soil and groundwater pollution along with negative effect on the produce also. Chemical fertilizer can be reduced significantly without yield reduction by applying with agricultural waste material [54]. The research conducted on crop yield seeks to find a model that describes the data well and aids in defining reasonable fertilization recommendations that result in optimum crop yield [48].

6.3 THEORETICAL CONSIDERATIONS

This section deals with theoretical concepts and error statistics of Geostatistical analysis, development of vegetation indices from remote sensing imagery, and performance assessment of irrigation system using remote sensing and GIS. Besides, the expressions for performance assessment indicators, the theory for estimation of evapotranspiration, and the water supply-demand estimation are covered in this chapter. The linear and nonlinear mathematical expressions developed for prediction of rice productivity are also incorporated in this chapter.

6.3.1 GEOSTATISTICAL ANALYSIS

Ordinary Kriging technique of a single variable is most robust and is frequently used for accounting of data fluctuations and considerations of a global trend over the study region [24, 61]. In general, things that are close together tend to be more alike than the things farther apart, the same is reflected in the semi-variogram cloud obtained. It can be observed that as the distances between the stations remain small, the semi-variogram is also small, meaning that the attribute values are very similar and, therefore, highly dependent on one another because of their close spatial proximity. But, as the distance (lag) between the stations increases, a rapid increase in the semi-variance is observed, meaning that the spatial dependency of the attributes drops rapidly. With further increase in distance between the stations, as the cloud flattens it exhibits that the rainfall data can now be no longer correlated. Eventually a critical value of lag known as the range occurs, at which the variance levels off and stays essentially flat [26]. It is this range within which the data can be said to be spatially dependent on one another and beyond which, the distance between the stations make no difference, they remain totally unrelated at any of the larger distances. The maximum value that the semi-variogram attains at the range is called the sill.

6.3.1.1   Framework of Calculations

Kriging technique basically comprises of various semi-variogram models such as Circular, Spherical, Tetraspherical, Pentaspherical, Exponential, Gaussian, Rational quadratic, Hole effect, K-Bessel, J-Bessel, and Stable functions for fitting the semi-variogram[31]. In this chapter using geostatistical analysis spatial interpolation (ordinary Kriging) are conducted to characterize the spatial distributions of rainfall, rice productivity, and soil salinity parameters.

The aim of using the Kriging technique for spatially interpolating soil and crop yield parameters is to predict the parameter values at unmeasured locations (x_0) within the system domain (D) using information available elsewhere in $D(x_1, x_2.................., x_n)$. The first step in estimating or "Kriging" the value of (A_0) at position (x_0) is to assume that its value is a linear function of the unknown values $A_i(x_i)$

$$A_0^*(x_0) = \sum_{i=1}^{n} \lambda_i A_i(x_i) \tag{1}$$

where, $A_i(x_i)$ and $A_i^*(x_i)$ are the observed and estimated values, respectively, at location x_i; λ_i denotes the Kriging weight of the parameter $A_0(x_0)$ for "n" number of nearby sample points to be used in estimation and should meet the condition:

$$\sum_{i=1}^{n} \lambda_i = 1 \tag{2}$$

The optimal weight λ_i is calculated such that the estimation of $A_0^*(x_0)$ by $A_0(x_0)$ is unbiased and the variance of $[A_0^*(x_0) - A_0(x_0)]$ is minimized. The equation for the experimental semi-variogram is given by:

$$\gamma^*(h) = \frac{1}{2n(h)} \sum_{i=1}^{n(h)} [A(x_i) - A(x_i + h)]^2 \tag{3}$$

where, $\gamma^*(h)$ represents the estimated value of semi-variance for lag h; $n(h)$ is the number of experimental pairs separated by vector h.

6.3.1.2 Error Statistics

An appropriate semi-variogram model need to be identified in case of multi-year data and determination of spatial distribution of parameters considered. The model that yields the minimum standard error has to be chosen for further analysis [31, 32, 58]. The mean standardized error, i.e., reduced mean error (RME) is used to test the predictability of the developed models and should be close to zero for the model to be acceptable. RME is expressed as:

$$\text{RME} = \frac{1}{n}\sum_{i=1}^{n}[\frac{(A^*(x_i) - A(x_i))}{\sigma_{ki}}] \cong 0 \qquad (4)$$

The root mean square error (RMSE) value should be within the range $1 \pm [2(2/n)^{1/2}]$ for the model to be acceptable [51] and should be close to one. RMSE is used to check the consistency between the estimation errors and the standard deviation of the observed values and RMSE is expressed as:

$$\text{RMSE} = (\frac{1}{n}\sum_{i=1}^{n}[\frac{(A^*(x_i) - A(x_i))}{\sigma}]^2)^{1/2} \cong 1 \qquad (5)$$

6.3.2 REMOTE SENSING APPLICATION

The classification of remote sensing data of LISS III (IRS P6) provides information on crops, cropping pattern and hence serves as an appropriate input for crop water requirement of the command area. Similarly the development of vegetation indices from reflected radiation with AWIFS data can be helpful for crop yield modeling and assessing the crop condition in the command area. Hence, the cloud free data of LISS III (IRS P6) of 24 days of receptivity is selected for generation of LULC and AWiFS (IRS P6) data available with 5 days frequency of receptivity are selected for covering different stages of crop growth, because lack of frequent coverage with LISS III data.

6.3.2.1   Relation Between Remote Sensing Indices and Crop Parameters

Normalized Difference Vegetation Index (NDVI) as suggested by Tucker [56] is well accepted index that influences crop productivity and can be used because of its linear relation with the crop yield. The crop/vegetation reflects high in near infrared radiance of the electromagnetic spectrum due to its canopy geometry, the health of the standing crop and absorbs high in the red reflected radiance due to its biomass and accumulated photosynthesis. The NDVI represents the integrated effect of various factors that influence crop production. In the present study, NDVI is estimated from series of AWiFS images covering crop growth period to assess the performance of irrigation command areas in terms of crop yield and crop condition.

6.3.2.2   Estimation of Normalized Difference Vegetation Index (NDVI)

The satellite derived NDVI can be used as an index to assess the crop productivity and condition of crops across the large command area. Hence the index is estimated by using reflected radiation in red (0.6 μm to 0.7 μm) and near-infrared (0.7 μm to 1.1 μm) wave length bands and is represented in the expression as:

$$NDVI = \frac{NIR - R}{NIR + R} \tag{6}$$

where, R and NIR are reflectance in red and near-infrared wave length regions.

Index values can range from −1.0 to 1.0, but vegetation values typically range between 0.1 and 0.7. Higher index values are associated with higher levels of healthy vegetation cover.

6.3.2.3   TCVI Generation From Series of Images

The ground truth from field visits in the study area revealed that there is considerable staggering in rice transplantation in the command area. The NDVI also changes with respect to crop stage and crop condition (health). For any particular crop, the NDVI extracted value keeps on increasing until heading stage. Since the use of single date satellite data cannot capture the critical crop growth (heading) in the entire command area, series of data is used. Hence, Time Composited Vegetation Index (TCVI), the maximum NDVI value for each pixel, is generated from series of data, in this case using eight images of AWiFS [40]. Thus, TCVI, which represents the NDVI value corresponding to the heading stage of crop, is generated for crop yield prediction.

6.3.2.4   NDVI and Crop Productivity

The relationship between NDVI at heading stage of cereal crops (TCVI) with its yield can be linearly correlated [55]. A robust linear yield model has been developed correlating TCVI (max NDVI) with yield data observed in the crop cutting experimental plots of the study area. Considering the image acquisition dates that coincide with the crop duration, the crop yield of mid-rice and late-rice have been related to NDVI to develop a linear regression equation.

6.3.2.5   Crop Condition Assessment in the Command Area

The crop condition at a given time during its growth period is influenced by complex interactions between soil-water-plant and atmosphere. Depending on the availability water and other inputs, crop condition varies. The maximum NDVI in the entire command area is assessed in order to observe the crop condition spatially. In general, crops conditions are categorized based on NDVI values as very good (> 0.5), good (0.4–0.5) and average (< 0.40) [40].

6.3.3   ESTIMATION OF CROP WATER DEMAND

6.3.3.1   Evapotranspiration

Using CRIWAR simulation model, the irrigation water requirement at different time periods (daily, weekly, and monthly) is estimated for different cropping pattern in a growing season [11]. The model initially estimates reference evapotranspiration rate (ET_o) and then the potential evapotranspiration (ET_p) based on the crop coefficient, K_c, input in crop factor file of the model. In this study, the crop coefficient values of FAO 56, adjusted to local conditions [53] are considered in CRIWAR simulation model. The sequence of computations performed by the model is represented in Eqs. (7)–(11). The reference evapotranspiration is calculated by the Penman-Monteith equation [3]:

$$ET_o = \frac{0.408\,\Delta\,(R_n - G) + \dfrac{890\,\gamma\,U_2\,(e_a - e_d)}{T_m + 273}}{\Delta + \gamma\,(1 + 0.339\,U_2)} \tag{7}$$

where, ET_o is the reference evapotranspiration, mm/day; Δ is the slope of the saturation vapor pressure temperature curve, kPa/°C; γ is the psychometric constant, kPa/°C; R_n is net solar radiation, MJ/m²/day; T_m is the mean daily air temperature, °C; $(e_a - e_d)$ is the vapor pressure deficit of air, kPa; G is the soil heat flux density, MJ/m²/day; and U_2 is the wind velocity at 2 m height, m/s;

The wind velocity adjustment is made by the following expression [3]:

$$U_2 = \frac{4.852 U_z}{\ln\left(\dfrac{z - 0.08}{0.015}\right)} \tag{8}$$

where, U_z is measured wind velocity at height of z (m) above the ground surface, m/s.

Potential evapotranspiration is calculated based on the values of crop coefficient, K_c, generated from the guidelines [3, 20] using expression as:

$$ET_p = K_c \times ET_o. \tag{9}$$

6.3.3.2   Crop Water Requirement

The monthly crop water requirements (CWR) are estimated by subtracting the effective rainfall (P_e) from potential evapotranspiration, ET_p as:

$$CWR = ET_p - P_e \tag{10}$$

CRIWAR uses the following semi-empirical formula to calculate effective rainfall per month, P_e

$$P_e = f(1.253P^{0.824} - 2.935) \ 10^{0.001ET_p} \tag{11}$$

where, P_e is the effective precipitation in mm/month; P is the total precipitation in mm/month; ET_p is the total crop evapotranspiration in mm/month; and f is the correction factor depending on depth of irrigation application. f is estimated a depth of irrigation water application, D_a.

If $D_a < 75$ mm/turn, then $f = 0.133 + 0.201\ln D_a$ or
If $D_a \geq 75$ mm/turn, then $f = 0.946 + 7.3 \times 10^{-4} D_a$

6.3.3.3   Gross Irrigation Water Requirement

The gross irrigation requirement (GIR) or irrigation demand is computed from the crop water requirement (CWR) with irrigation system efficiency (accounting for losses during conveyance, distribution, application, and special purposes) [22]:

$$GIR = \frac{CWR}{Irrigation\ Efficiency} \tag{12}$$

6.3.4   PERFORMANCE ASSESSMENT INDICATORS

Irrigation system performance is assessed with three broad indicators namely, productivity, equity and adequacy; and after assessing these indicators, they are compared among different distributaries commands for categorizing the performance.

6.3.4.1   Productivity

An attempt is made to estimate rice productivity at unmeasured locations by Kriging technique using measured data obtained from crop cutting experiments. The productivity of rice for different distributaries commands is retrieved from the Kriged surfaces.

6.3.4.2   Equity (NDVI Based Equity)

Any irrigation distribution system, which practices equity in water allocation and distribution, will have uniformity in cropped area and crop vigor.

However, if there is a large and consistent variation in cropped area and vigor between the distributaries, the distribution system cannot be considered to be practicing equity. To assess equity, maximum and average NDVI of inter- and intra-distributary (head, middle, and tail reaches) commands belonging to a branch canal were compared [47].

6.3.4.3 Adequacy (Relative Water Supply)

The adequacy answers to what level is the quantity of water supplied is sufficient to meet the crops growth requirements. The most comprehensive measure of adequacy in terms of relative water supply (*RWS*) is computed by the following expression [33] as:

$$RWS = \frac{IR + RN}{GIR} \tag{13}$$

where, *IR* represents the irrigation water supply; *RN* the rainfall; and *GIR* the gross irrigation requirement.

Gross irrigation requirement is the total amount of water including losses that must be applied by irrigation such that evapotranspiration may occur at the potential rate and optimal crop productivity may be achieved. Only part of the applied water is actually 'used' by the plant to meet evapotranspiration requirement, which is called as net irrigation requirement [47].

If *RWS is > 3 then* water surplus is excessive,

RWS within 2 to 3 represents high water surplus,

1.1 to 2.0 represents moderate water surplus,

0.9 to 1.1 represents adequate water surplus

0.5 to 0.9 represents water deficit.

6.3.5 *PREDICTION OF RICE PRODUCTIVITY*

6.3.5.1 Simple Linear Regression Model

In the entire rice growth period, maximum NDVI (TCVI) from RS and corresponding rice productivity from crop cutting experiments (CCE) is recorded at different locations of the command area. A simple linear regression has been developed between NDVI and rice productivity. The developed relationship is used for predicting rice for the entire command area.

6.3.5.2   Multiple Nonlinear Regression Model

A multiple nonlinear regression equation has been developed between crops input (total depth of water and nutrients application in a season) and output parameter(s) (crop productivity from CCE). The quadratic form of equation for two variables in the production surface is used for relating rice productivity with application of input parameters [28]:

$$Y = a + b_1 X_1 + b_2 X_2 - b_3 X_1^2 - b_4 X_2^2 + b_5 X_1 X_2 \qquad (14)$$

where, Y is the rice productivity; X_1 is the total depth of water applied in a season; X_2 is the quantity of total nutrients (NPK) applied to rice in a season; and $a, b_1,......,b_5$ are regression coefficients.

From the developed regression coefficients, one can get an overall idea about the sensitivity of the input variable (s) on rice productivity.

6.4   STUDY AREA AND DATA COLLECTION

This section deals with acquisition of data namely, irrigation water supply, crop parameters, and weather data. The remote sensing images are used for identification of crops and vegetation indices with the help of software (ERDAS Imagine and ArcGIS).

6.4.1   LOCATION OF THE STUDY AREA

The study area is situated in Prakasam district of Andhra Pradesh State of India and lies between latitudes of 15° 40′ 48″ to 16° 1′ 12″ North and longitudes of 79° 56′ 24″ to 80° 22′ 48″ East. The main source of irrigation is water from the Addanki Branch Canal and some portion of water received through precipitation from both the South-West and North-East monsoon seasons.

6.4.2   DATA COLLECTION

The information on canal supplies, crops, and soils is obtained from different sources. The general tendency of upper reach farmers is extraction of maximum amount of canal water as compared to the lower reach. The soil and land use data were procured from the Department of Agriculture. Information on crop acreages, administrative block level precipitation, and changes in cropping pattern in the command area were obtained from the Handbook of Prakasam district statistics [14]. The groundwater level data for period from 1990 to 2004 covering certain points in the command area was also collected from the Deputy Director, State Groundwater Department, Prakasam district.

6.4.3 IRRIGATION SUPPLY

The data pertaining to water supply to Addanki branch canal and to different distributaries is collected from the Deputy Executive Engineer office, Addanki. In view of poor water level in the Nagarjuna Sagar reservoir during 2004–2005 due to prevailing drought situation in the region, irrigation water was supplied very late to the command area.

6.4.4 DISTRIBUTARY NETWORK

The index sketch of the study area is procured from I&CAD office. The toposheets in 1:50000 scale and district map is procured from the Survey of India (SOI), Hyderabad. These maps have been used to identify the command area network. The important features of the study area and constituent distributaries and their locations with respect with branch canal are shown in Figure 6.1. The length and *culturable* command area (CCA) of five distributaries under the study area are presented in Table 6.1. Rajupalem distributary has been observed to have smaller CCA and length as compared to all other distributaries. Rajupalem and Nutalapadu distributaries are located at upper and lower reaches of the command area.

6.4.5 CROP PRODUCTION DATA

General soil, available water and crop features of the command area at three reaches (upper, middle, and lower) have been procured from various agencies and shown in Table 6.2 [14, 19]. In spite of control measures put forth by authorities of Department of Agricultural (DOA) on aerial extent for growing of rice, farmer's top preference continued to be on their staple

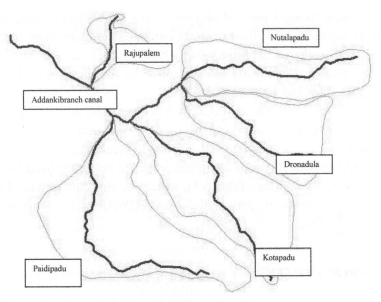

FIGURE 6.1   Location of distributaries in the command area.

TABLE 6.1   Significant Features of Distributaries in Command Area

Name of the distributary	Length (m)	Culturable command area (ha)
Rajupalem	8,000	2,434
Paidipadu	28,000	15,005
Kotapadu	24,400	11,008
Dronadula	17,000	7,565
Nutalapadu	19,300	7,925
Culturable command area = CCA.		

TABLE 6.2   General Features of Three Reaches of the Command Area

Parameter	Upper reach	Middle reach	Lower reach
Available water capacity (mm/m)	High 150–200	Medium 100–150	Low 50–100
Principal crops in *Kharif* season	Red gram, Cotton, Chillies	Red gram, Cotton, Chillies	Cotton, Red gram, Chillies
Principal crops in *Rabi* season	Rice, Bengal gram, Tobacco, Maize	Rice, Bengal gram, Maize, Tobacco	Bengal gram, Tobacco, Rice, Maize
Soil type	Deep, well-drained, red clayey, calcareous soils.	Moderately deep, well-drained, red clayey soils.	Deep, well-drained, red coastal loamy soils.

food, rice only. Bengal gram (Chickpea, *Cicer arietinum*) became highly attractive crop in the area so much so the crop area increased from 20,000 to 80,000 ha during the period from 2004 to 2008. Maize (*Zea mays*) is another alternative crop that is already being adopted by few farmers in the lower reach of the command area.

Among these crops, rice is water intensive wet crop and all other crops are supplied with intermittent irrigation. The identification of crops grown in the command area is carried out by remote sensing images. The durations of crop growth stages of identified crops are collected from the State Department of Agriculture (Table 6.3), which are subsequently used while incorporating the cropping pattern file in the CRIWAR model for crop water demand simulation.

6.4.5   CROP COEFFICIENT

The information about duration of irrigated wet and dry crops is generated out of discussions with the Department of Agriculture, and I & CAD authorities. Similarly, FAO crop coefficients of irrigated wet and dry crops are adjusted for the local area [50, 53], which is furnished in Table 6.4. The ground truth information shows that in upper-reach mostly rice is grown along with some ID crops, whereas in lower-reach mostly ID crops along with small area under rice is grown. Crop coefficient (K_c) information so generated is used by incorporating into the CRIWAR model in crop factor file. Consideration of K_c value plays an important role in estimation of crop water demand for irrigated crops.

TABLE 6.3   Duration of Crop Growth Stages

Crop	Duration of different crop growth stages (days)				Total crop duration (days)
	Initial stage	Development stage	Mid season	Late season	
Bengal gram	20	30	25	20	95
Chillies	45	60	70	35	210
Cotton	30	50	50	45	175
Maize	20	25	35	20	100
Red gram	35	60	65	40	200
Rice	20	40	45	15	120
Tobacco	20	30	40	30	120

TABLE 6.4   Crop Coefficient (FAO 56) Adjusted for the Study Area

Crop	Initial stage	Development stage	Mid season	Late season
Chillies	0.55	0.78	1.00	0.71
Cotton	0.80	0.91	1.01	0.70
Red gram	0. 84	0.93	1.02	0.56
Rice	1.10	1.15	1.20	0.90

6.4.6   WEATHER DATA

The overall climatic condition of the command area resembles with the semiarid tropics with prolonged dry spells, which are critical for the survival of crops. The *Kharif* (monsoon) and *Rabi* (winter) cropping season extends from June–October and November–February, respectively. The meteorological data consists of monthly rainfall, maximum and minimum temperatures, relative humidity, wind speed, and sunshine hours for the period from 1990 to 2005, was procured from Indian Meteorological Department (IMD), Hyderabad. The data collected from the nearby weather station namely, Ongole in Prakasam district was used in the analysis.

The analysis of 15-year rainfall data, shows that the annual average rainfall of the command area is 849 mm, which is received almost in equal proportion during south-west (June–September) and north-east (October–December) monsoons. Monthly rainfall is recorded at different administrative block headquarters within the command area [14]. Using rainfall of measured locations, distributary-wise rainfall for different months in 2004–2005 was predicted from the spatially interpolated Kriged surfaces.

6.4.7   SOIL SALINITY AND NUTRIENTS

Soil salinity is a problem of arid and semiarid regions referring to the amount of soluble salt present in the soil. In general, EC of soil varies depending on the amount of moisture held by the soil particles, their size and texture. The electrical conductivity (EC), pH, phosphorous, potash, and organic carbon status at 30 locations of the command area are collected at 30 cm soil depth to assess soil salinity and nutrient status. While testing soil samples in the command area, it is ascertained that alkalinity problem is persistent (pH = 7.9 to 8.3) as compared to salinity level (EC<1.0 dS/m), which is under the safe limit. The residual amounts of fertilizers observed in the samples indicated high level of application of nutrients.

6.4.8 GROUND WATER LEVEL

The information obtained from Prakasam district map has indicated that very deep water table conditions are prevalent in the command area. Water table is located as deep as 100–150 m at upper reach to 20–50 m at lower reach of the command area, which is 10 km away from the sea shore. Hence, due to deep table condition and huge cost of pumping, farmers of upper reach are not inclined to use of groundwater. During visit to the command area, it is noticed that farmers are mostly dependent on canal water for irrigation due to its low unit cost. However, the data obtained from the groundwater department confirmed deep water table conditions in the command area and at certain places it was observed that tubes wells have already dried up because of lowering of water table.

6.4.9 REMOTE SENSING (RS) IMAGES

The Resource-Sat-1 satellite, also known as IRS-P6, consists of LISS-III and AWiFS sensors, both containing four spectral bands is especially designed for land and water management and agricultural applications. These four spectral bands are band 2 (green), band 3 (red), band 4 (near infrared, NIR) and band 5 (short wave infrared, SWIR). The radiometric characteristics of these four bands are 0.52 to 0.59, 0.62 to 0.68, 0.77 to 0.86 and 1.55 to 1.70 μm, respectively. The LISS-III sensor has radiometric quantization of 7 bits and provides a combined swath of 141 km with a spatial resolution of 23.5 m. The AWiFS sensor has radiometric quantization of 10 bits and provides a combined swath width of 740 km with a spatial resolution of 56 m at nadir.

6.4.10 ERDAS IMAGINE AND ARCGIS SOFTWARE

The ERDAS (Earth Resource Data Analysis System) integrated software is used by thousands of professionals worldwide for urban and regional planning, and natural resource management. The LISS III and AWiFS data of the study area are processed in ERDAS Imagine for making extensive analysis.

ArcGIS is an integrated collection of GIS software products that provides a standards-based platform for spatial analysis, data management, and mapping. Using Kriging analogy, the spatial interpolation is carried out in ArcMap of ArcGIS software for rainfall, soil and crop productivity parameters of the command area.

6.5 METHODOLOGY

This section includes methodology for analysis of command area data. Geostatistics tool is used to analyze soil, rainfall, and crop productivity parameters using ArcGIS software. ERDAS imagine software is used to process remote sensing images for generation of LULC.

6.5.1 FORMULATION OF CROP CALENDAR

The ground truth observed during personal visits and crop area information obtained from the Handbook of Prakasam district statistics [14] indicated that rice is a prominent crop in the upper reach of the study area. Crops like Bengal gram, maize and tobacco are predominantly grown along with other rain-fed crops at lower-reach. Along the aforementioned cropping scenario, some ID crops are also grown in the entire command area. Hence, numerous field visits were made in the command area for collecting ground truth. In addition, the Prakasam district atlas [19] has been used to identify some relevant information on crops and soils for making preliminary overview of the study area.

6.5.2 GEOSTATISTICAL AND REMOTE SENSING ANALYSIS

Availability of scanty rainfall plays an important role in crop productivity in semiarid climate. Similarly, salinity and nutrient status of soil that influence crop yield, are important parameters. Hence, geostatistical analysis (Kriging) of rainfall, salinity, and productivity is used to predict reliable estimates at unmeasured locations of the command area.

Crop areas/phenology, land features, and hydrological conditions of the canal commands is precisely identified by processing multi-satellite images in ERDAS Imagine software. Further, the water demand and supply relationship in the command area is assessed effectively by the use of remote sensing derived crop areas. Spatial variation in irrigation system performance is better assessed by integration of remote sensing, GIS, and simulation of crop water demand.

6.5.3   DATA PROCESSING IN ArcGIS

6.5.3.1   Geostatistical Analysis

The geostatistical (Kriging) analysis has been performed on rainfall, salinity, and rice productivity parameters of 2004–2005 for the study area using ArcMap in ArcGIS software. These parameters have shown a clear trend in interpolated surfaces. These afore mentioned parameters were transformed to the real world coordinate system using geoprocessing feature of ArcGIS followed by a complete analysis. Using Arc toolbox, the point, line, and polygon feature classes of the irrigation command were transformed from geographic to coordinate system of NAD 1983 projection datum of 44N grid under UTM zone. The transformation ensured that the point, line, and polygon feature classes represent uniform map scale and actual latitudes and longitudes over land surface, which were required for the geostatistical analysis. The layers are then added and viewed in ArcMap. The elementary step of geostatistical analysis is exploratory data analysis in which the histogram, normality, trend of data, voronoi mapping, semi-variogram cloud and cross covariance cloud of the raw data are observed. These steps ensure that the data points are normally distributed and there are no global or local outliers [51]. The best-fit model is confirmed only after performing the cross validation and verifying the error statistics.

6.5.3.2   Semi-Variogram Model

The mean square prediction error should be minimum for a model to provide accurate predictions. If the predictions are close to the measured values, the mean standardized prediction error should be close to 0, and the reduced-mean-square standardized error should be close to 1. When the average estimated prediction standard errors are close to the root-mean-square prediction standard errors from cross-validation, one can be sure that the prediction standard errors are appropriate. Thus, utilizing the spatial variance structure available in a semi-variogram, the most appropriate model was chosen to yield the best linear unbiased estimate of the parameter calculated from weighted values measured in its local neighborhood.

6.5.4   DATA PROCESSING WITH ERDAS IMAGINE

6.5.4.1   Generation of Base Map

The ERDAS Imagine software has been used for processing of toposheets and the Prakasam district administrative map. The procured toposheets (1:50,000 scale) and the index map of the study area have been scanned and geo-referenced for further analysis. The canal network and significant features of the irrigation command area have been developed by digitization using ERDAS imagine. Base map is thus extracted by clipping the mosaic of toposheets to boundary shape of the study area. All data of the study area is modeled keeping the base map as reference area.

6.5.4.2   Geo Referencing the Images

Remote sensing images are geo-referenced in ERDAS Imagine software. Initially, one of the IRS LISS III images (reference image) was geo-referenced by registering with GCP's (Ground Control Points) observed toposheets. Images of other dates were geo-referenced using image-to-image GCP's. The 3 images of LISS-III and the 8 images AWiFS of Resourcesat-1 (IRS P6) have been geo-corrected for carrying out further analysis.

6.5.5   GENERATION OF LAND USE/COVER MAP

In the image classification approach, all the images were registered to the same datum, projection, and coordinate system. The thematic maps in Prakasam district atlas [19] for the study area are also used to support the crop identification from the satellite images.

The aerial extent of any distributary could be extracted from the classified image of LULC of the command area by clipping to the respective distributary boundary. Crop area statistics, derived from LULC of each distributary, is used as an input for estimation of crop water requirement in CRIWAR model. In order to generate composite LULC map based on multi-images, it is necessary to delineate each image into two classes, i.e., vegetated areas (V), and non-vegetated areas (NV).

6.5.5.1　Delineation of Vegetated and Non-Vegetated Areas

Delineation of vegetated and non-vegetated areas could be made in a simple form by visually separating the areas that appear to have no-vegetation on each image and also verifying from toposheets. However, visually interpreting each image would be very time consuming and cumbersome, hence, an automated process to delineate the vegetated and non-vegetated areas for each satellite image was made by using the NDVI for three images in the season [56].

The derivation of NDVI is based on vegetation, which have a characteristic spectral response that is significantly different soil and other land targets. The spectral signature of vegetation in the electromagnetic spectrum is determined by plant pigments (chlorophyll), which preferentially absorbs blue and red light for photosynthesis, leading to a low reflectance in the visible wavelengths (0.4–0.7 μm). The spectral response becomes primarily controlled by the cellular microstructure of the leaf in near infrared (NIR) wavelengths (0.7–1.3 μm). Subsequently, unsupervised classification with ISODATA clustering was performed on each NDVI image and iterative labeling was done by examining the NDVI cluster image with FCC image having RGB combination. By labeling the NDVI cluster image, it was possible to create a binary map of vegetation (V) and non-vegetation (NV) areas.

6.5.5.2　Land Use/Land Cover

For identification of land cover dynamics, NDVI binary (vegetation, non-vegetation) images were generated from three time periods by examining their RASTER attributes with corresponding FCC images. A composite image that indicates land cover dynamics for three time periods was generated by adding all the NDVI binary images.

Pixels with vegetation are assigned the values of 1 and 2 for November 19, 2004 and December 13, 2004 images, respectively, and the overlapping pixels would have a value of 3 as a result of the image addition. The possible classes after addition of these two images are given in the Table 6.5. Further, this composite image is added to third image having vegetation with

TABLE 6.5   Land Cover Dynamics in Composite Image of Two Time Periods

Type of Vegetation	Time 1 image	Time 2 image	Time 1 + Time 2	
	November, 19 2004	December, 13 2004	Composite image of two time periods	
	Reclassed value	Reclassed value	Reclassed values after addition	Land cover dynamics
NV	0	0	0	NV-NV
V	1	2	1	V-NV
			2	NV-V
			3	V-V

Note: NV = non-vegetation; V = vegetation.

a pixel value of 4. Overall, the resultant composite layer will result in 8 possible combinations of land cover dynamics in the study area. Possible land cover dynamics in three time period composite image are presented in Table 6.6. Subsequently, all the reclassified NDVI images are overlaid by addition. Resultant classes were labeled as different LULC classes based on the GCPs, crop calendar, and NDVI. The decision rules framed for identifying different classes were discussed in subsequent sections.

TABLE 6.6   Land Cover Dynamics in Composite Image of Three Time Periods

Land cover dynamics	Time 1 + Time 2 images	Type of vegetation	Time 3 image	Time 1 + Time 2 + Time 3	
				Composite image of three time periods	
	Reclassed value		Reclassed value	Reclassed values after addition	Land cover dynamics
NV-NV	0	NV	0	0	NV-NV-NV
V-NV	1	V	4	1	V-NV-NV
NV V	2			2	NV-V-NV
V-V	3			3	V-V-NV
				4	NV-NV-V
				5	V-NV-V
				6	NV-V-V
				7	V-V-V

Note: NV = non-vegetation; V = vegetation.

6.5.5.3   Labeling of LULC Classes in the Composite Image

Classification of multi-date LISS III images based on some heuristic rules, is more useful in comparison to single date image for crop identification [41, 60]. Thus, in the present study, heuristic/decision rules and feedback knowledge were envisaged for labeling the combination of eight land cover dynamics into individual crops in the command area (Table 6.6). The feedback from local farmers and technical personnel of the State DOA and I & CAD enabled to formulate crop calendar (Table 6.7).

The decision rules used for image classification are based on planting, harvesting, and phenology of major crops in the command area (Table 6.8). An overlay process of the branch canal and distributaries was done for identification of rice area, knowing that rice needs a constant source of water and grown in close proximity to irrigation canals.

Vector point coverage of GCPs, indicating crop types in the study area, was overlaid on an image to get more accurate LULC map. The criteria of combining GCP'S along with land slope (usually rice areas are flat and adjacent to canal) was also used for labeling the clusters of composite image. Further, the job of labeling the classes was made easy by the use of March 19, 2005 image, because by this date majority of irrigated crops has been harvested.

TABLE 6.7   Crop Calendar for the Command Area During 2004–2005

Crop	Sowing Date	Harvest Date	Initial Stage (Days)	Development Stage (Days)	Mid Season (Days)	Late season (Days)	Crop duration (Days)
Early rice	15/Oct	15/Feb	20	40	45	15	120
Mid rice	10/Nov	10/Mar	20	40	45	15	120
Late rice	5/ Dec	5/Apr	20	40	45	15	120
Bengal gram	15/Nov	20/Feb	20	30	25	20	95
Chillies	1/Sep	30/Mar	45	60	70	35	210
Cotton	1/Sep	25/Feb	30	50	50	45	175
Early maize	10/Aug	20/Nov	20	25	35	20	100
Late maize	01/Jan	10/Apr	20	25	35	20	100
Red gram	15/Aug	5/Mar	35	60	65	40	200
Tobacco	20/Oct	20/Feb	20	30	40	30	120

TABLE 6.8 Decision Rules for Land Use Map Based on the Temporal Data

Land cover dynamic	Description of land feature	Possible class	Remarks
NV-NV-NV	(Non cropped area) uncultivable land or wasteland/water	Settlement/water bodies/bare land	Delineated into various individual classes using ISO clustering technique
V-NV-NV	Crop seen only in November image and not in subsequent images	Other crops	From crop calendar, crops like early maize are harvested by November 20 and not seen in subsequent images
NV-V-NV	Crop seen only in December image	Bengal gram	Crop is seen in December 13 image only and not in other images
V-V-NV	Early crops seen in November and December	Early rice & mid rice, red gram, cotton, tobacco	Further segregation of crops was presented in Section 6.5.6.3.1
NV-NV-V	Late crops seen only in March image.	Other crops	From farmers feedback and crop calendar information, classified as other crops
V-NV-V	Double cropped area	Maize	In this class, vegetation is seen two times, verifying with the ground truth and crop calendar, it is classified as maize
NV-V-V	Late crops seen in December and March images	Late rice	The late-rice, which starts from December 5 are seen in December and March images
V-V-V	Long duration crop	Chillies	Chillies with duration from 1 September to 30 March is covered in all the three images.

Note: NV = non-vegetation; V = vegetation.

6.5.5.4 Segregation of V-V-NV Class

Crop is seen both in November 19 and December 13 images and could not be seen in March 19 image in V-V-NV class. The crop calendar shows majority of crops, namely early rice, mid rice, cotton, red gram, and tobacco, are identified under the aforementioned class. Further identification of individual crops is carried out with decision rules mentioned in Table 6.9.

TABLE 6.9   Decision Rules for Labeling of V-V-NV Class

Crop cover dynamics	Possible crops	Landmark criteria.	Labeled crop	Heuristic knowledge	Identification of crop
V-V-NV	Early rice, mid rice, red gram, cotton, tobacco	Slope < 2% & proximity to canal	Rice	NDVI, ground truth	Early rice
				NDVI, crop calendar	Mid rice
		Slope > 2% and away from canal	Non- rice	Scattered ground truth, NDVI	Red gram
				Proximity to tobacco barns, ground truth, NDVI	Tobacco
				NDVI, ground truth	Cotton

Hence, identification of individual classes is carried out by the use of guiding rules, ground truth, and feedback knowledge. Other crops such as maize, grams, vegetables, fruit crops, plantation, etc., for which ground truth is not available have been classified into one group as 'other crops'. The sequence of methodology for LULC classification and identification of different crops is shown in the flow chart (Figure 6.2).

6.5.5.5   Segregation of NV–NV–NV Class

Iterative labeling was done to NDVI cluster images of November 2004 and March 2005 images. In order to identify water bodies, settlement, and other land features like hilly terrain, scrubs that are seen clearly in toposheets, were also digitized and used for labeling these non-crop classes. Similarly, water bodies, which were also labeled in this particular class, can be easily identified in the multispectral image. The objective of generating LULC is to depict the cropping pattern that will estimate the crop water demand of the canal command.

6.5.5.6   Accuracy Assessment

The accuracy of classification depends on several factors; such as spectral separability of the land cover classes, heterogeneity of the land surface, and

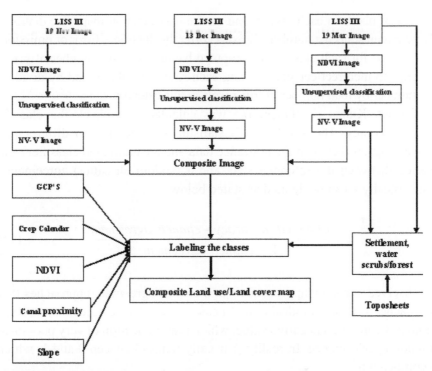

FIGURE 6.2   Methodology for land use/land cover classification.

extent of ground truth. Accuracy analysis was performed on the classified image with the accuracy assessment function of ERDAS Imagine software. Random method is selected for carrying out accuracy assessment. It lists two sets of class values for the randomly selected points. One set of class value is automatically assigned to these random points according to classification and the other set of classified value is input by the user. These reference values are based on ground truth data and information collected from field visits and available maps.

These class values are reported in a contingency table, where overall classification accuracy and misclassification between categories are identified. It usually takes the form of a ($m \times m$) confusion matrix, where m are the number of classes under investigation. The rows represent the assumed true classes, while the columns are associated with remote sensing derived land use classes. One of the most important characteristics of such matrix is their ability to summarize errors of omission and commission. Errors of the commission/omission occur when some of the pixels of the specified class actually belong to other category. The total column on the extreme right summarizes the assumed true number of pixels in each class found on the classification map. The diagonal element represents the number of pixels correctly classified. Error or omission/commission for each class were computed by summing the number of pixels assigned to incorrect categories along each row and column and by dividing this number by total number of true pixels in this category.

Further, Kappa coefficient ($\hat{k}$) was calculated to check the classification accuracy. The $\hat{k}$ ("KHAT") statistic is a measure to differentiate between actual (between the reference data and an automated classifier) and chance (between the reference data and a random classifier) agreements. Conceptually $\hat{k}$ can be defined as stated below

$$\hat{k} = \frac{observed \;\; accuracy - chance \;\; agreement}{1 - chance \;\; agreement} \tag{15}$$

As observed accuracy approaches 1 and chance agreement approaches 0, $\hat{k}$ approaches 1, this is an ideal case. In case where chance agreement is large enough, $\hat{k}$ can take a negative value, which is an indication of very poor classification performance. In reality, $\hat{k}$ usually ranges between 0 and 1, which is computed by:

$$\hat{k} = \frac{N\sum_{i=1}^{r} x_{ii} - \sum_{i=1}^{r}(x_{i+} . x_{+i})}{N^2 - \sum_{i=1}^{r}(x_{i+} . x_{+i})} \tag{16}$$

where, N is the total number of observations; r is the number of rows in the error matrix; x_{ii} is the number of observations in row i and column i (on the major diagonal); x_{i+} is the total observations in row i (shows as marginal total to right of the matrix); and x_{+i} is the total observations in column i (shown as marginal total at bottom of matrix).

In this chapter, the accuracy was estimated by selecting 500 random samples representing various LULC categories. The Kappa coefficient was originally developed to test classification accuracy. Kappa coefficient, $\hat{k} = 1$, indicates perfect agreement between the classification categories, while $\hat{k} = 0$, indicates the observed agreement equals the chance agreement [17]. The $\hat{k}$ value greater than 0.75 indicates very good to excellent agreement, while a value between 0.40 and 0.75 indicates fair to good agreement. A value of less than or equal to 0.4 indicates poor agreement between the classification categories [23, 34].

6.5.5.7   Estimation of Vegetation Indices

The vegetation index, NDVI (Normalized Difference Vegetation Index) shows an integrated effect of applied water and soil nutrients on crop condition/vigor. The NDVI is observed to vary spatially in distributaries along the length of branch canal [13]. The crop condition also exhibits expected crop yield.

During crop growth season of 2004–05, the NDVI has been estimated by using reflectance received in NIR and Red bands, for selected eight images of AWiFS. The eight NDVI images have been overlaid sequentially, one over the other, for extracting the maximum NDVI for all the pixels. These images provide critical crop condition spatially in the entire command.

6.5.6   ASSESSMENT OF PERFORMANCE INDICATORS

The objective of performance assessment of the irrigation project is to identify problem distributaries and analyze reasons for further improvement in the functioning of the irrigation project. The indicators considered in the performance assessment are crop productivity, equity, and adequacy.

6.5.6.1   Crop Productivity

The recorded values of rice productivity in kg/ha (Chief Planning Officer, 2006) were taken from CCE plots (5 m × 5 m size). The geometric position of CCE plots at different village locations was recorded using GPS (Global Positioning System) instrument. The location-wise measured rice productivity was used to predict the productivity at any unmeasured locations of the distributary by using Kriging interpolation techniques.

6.5.6.2   Equity

Equity is a measure of the uniformity of water allocation in the distribution system. The equity is estimated in terms of NDVI based equity, irrigation depth, and irrigation and cropping intensity.

6.5.6.3   Adequacy of Irrigation

The adequacy in the present study is estimated in terms of two parameters namely, relative water supply and water utilization index. The relative water supply (RWS) is estimated as the ratio of seasonal water supply to the irrigation water demand in any distributary. WUI is the area covered per one million m^3 of irrigation water.

6.5.6.4   CRIWAR for Irrigation Demand

CRIWAR 2.0 (Crop Irrigation Water Requirement) software was developed by ILRI, Wageningen. It is a useful tool in the operation of irrigation systems. Crop water requirement (CWR) is calculated by subtracting the effective rainfall from the ET_p. Gross irrigation requirement (GIR) for selected command is estimated by considering an irrigation system efficiency of 60% (Gaur et al., 2008), which accounts for conveyance and other losses. In CRIWAR model, crop factor (K_c) values are adjusted to local weather conditions of right canal command area of Nagarjuna Sagar project [53] in accordance with FAO guidelines [3].

6.5.6.5   Irrigation Supply Demand Gap in Crop Growth Season

The simulation by CRIWAR model is carried out for estimation of crop water demand on monthly basis. Monthly supply-demand gap is computed to study its impact on crop productivity. Steps used in water supply-demand relationship are shown in Figure 6.3.

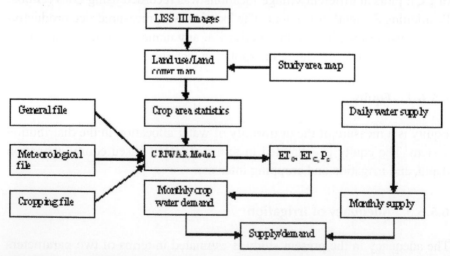

FIGURE 6.3   Irrigation supply-demand analysis in the command area.

6.5.7 PREDICTION OF RICE PRODUCTIVITY

Regression models have been developed to predict rice productivity in a cropping season using two approaches (NDVI and crop inputs).

6.5.7.1 Simple Linear Regression Model

The correlation between the NDVI and crop yield has been tested by some researchers. The correlation was reported to be highest at the heading stage of crop. Before and after heading stage, the correlation found weaker. In the present study, a simple regression model was developed between NDVI and the rice yield of 2004–05 in the selected command area.

The individual values of TCVI (maximum NDVI) and average NDVI values for different locations and the respective rice yields are plotted to get scatter diagram. The line of best-fit in each case is generated by discarding extreme points so as to keep the sufficient degree of coefficient of determination (R^2). The model, showing higher correlation of NDVI on crop yield, is selected for the prediction of rice productivity in the irrigation project.

6.5.7.2 Multiple Nonlinear Regression Model

Application of total depth of water and nutrients (NPK) are the two major factors for influencing crop productivity. A nonlinear second order multiple regression model has been developed using central composite design (CCD) after considering its parameters at three levels, where total seasonal water depth and nutrients (NPK) applied has been considered as input variables and crop productivity as output. The range of input variables has been chosen as 0.612–1.546 m and 284–660 kg/ha for water depth and nutrients (NPK), respectively. The significance of the parameters and their interaction terms on the response has been studied by conducting analysis of variance (ANOVA) to check the adequacy of the model. The effect of parameters and their interactions on the output was depicted through the surface plot using MINITAB 14 software.

6.6 RESULTS AND DISCUSSIONS

This section deals with the results related to spatial analysis of rainfall, land parameters and productivity. It also deals with depiction of land use and land cover classes, crop area statistics and vegetation indices from remote sensing images. The estimated values of performance indicators are compared among different distributary commands. Regression models for rice productivity are also dealt in this chapter.

6.6.1 VARIABILITY IN RAINFALL AND CANAL WATER SUPPLY

From the data of canal supply, it is noticed that upper reach distributary received fairly adequate amount of supply as compared to lower reaches. Supply-design ratio of lower reach distributary was observed to be 50% of the upper reaches during effective months of crop growth period from November 2004 to February 2005. It is also an indication of inequality in water distribution system of the command area.

6.6.2 VARIABILITY IN RICE PRODUCTIVITY

Rice is the prominent crop grown in the command area. Initial observation of rice productivity at point locations shows its variation from upper to lower reaches of the command area. The productivity trend noticed in the study area is in close agreement with the results observed by Ref. [1] that unequal distribution of water has a direct influence on productivity. Reduction in rice productivity in the lower reach of the command area could be attributed due to deficit supply of canal water during critical crop growth stage.

6.6.3 VARIABILITY IN SOIL MOISTURE

Spatial distribution of soil moisture content in the study area during irrigation period was retrieved (Figure 6.4) from the Prakasam district atlas [19]. High (> 150 mm/m), medium (100–150 mm/m), and low (50–100 mm/m) soil moisture content was observed in upper, middle, and low reaches, respectively. It is thus evident from Figure 6.4 that the soil moisture content during the cropping season has been observed to be in accordance with the canal water release pattern at different reaches, and not based on the ground water contribution, which is quite deep. For example, poor level of soil moisture content in lower reaches of the command area is due to low canal releases pattern.

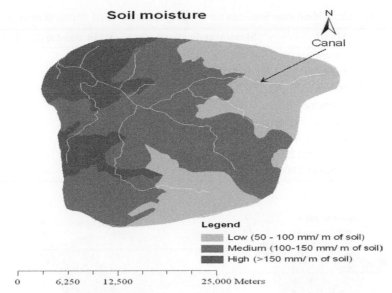

FIGURE 6.4   Soil moisture content in command area.

6.6.4 GEOSTATISTICAL ANALYSIS

The geostatistical analysis of rainfall, salinity, and rice productivity is performed using Kriging technique to compare interpolated parameters among different distributaries.

6.6.4.1 Rainfall

Kriging is performed for monthly rainfall data from October 2004 and March 2005. Cross validation statistics have been performed on monthly rainfall using different models (Table 6.10). The Gaussian model gave the desired error statistics as compared to other models tested. Hence, it is finally selected for spatial interpolation and correlation of monthly rainfall. The interpolated surfaces have been generated for October 2004, November 2004, January 2005, February 2005, and March 2005 (Figure 6.5). During December 2004 as there was no rainfall recorded in the entire command area, hence Kriged surface could not be generated.

During November 2004 to March 2005, scanty monthly rainfall below 35 mm is received that exhibits prominent dry spell condition in the

TABLE 6.10 Cross Validation Statistics for Monthly Rainfall Using Different models

Model	Month	Mean standardized error	RMS standardized error
Spherical	November 2004	−0.040	0.960
	February 2005	−0.002	0.960
Exponential	November 2004	−0.030	0.930
	February 2005	0.002	0.960
Gaussian	November 2004	−0.040	0.990
	February 2005	0.002	0.970
K-Bessel	November 2004	−0.040	0.980
	February 2005	0.003	0.960

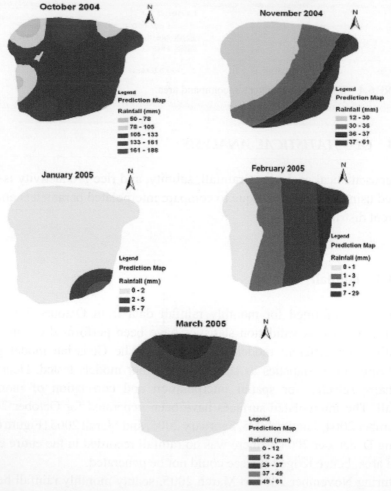

FIGURE 6.5 Rainfall prediction map in different months.

command area (Table 6.11). The total rainfall during the crop period from October 2004 to March 2005 has been recorded as 178.7 mm at upper distributary (Rajupalem) and 183.8 mm at lower distributary (Nutalapadu). Similarly, depth of canal water application for the same period in case of Rajupalem and Nutalapadu distributaries was 904 and 570 mm, respectively.

6.6.4.2 Rice Productivity

The spatial Kriged surface of interpolated value of rice productivity in the command area varied from 6277 kg/ha at upper distributary to 3386 kg/haat lower distributary (Figure 6.6). Based on water release pattern,

TABLE 6.11 Monthly Rainfall (mm) in Different Distributaries

Month	Rajupalem	Paidipadu	Kotapadu	Dronadula	Nutalapadu
Oct 04	125.1	125.0	111.5	125.0	131.5
Nov 04	26.0	31.2	33.0	32.5	31.2
Dec 04	0.0	0.0	0.0	0.0	0.0
Jan 05	0.4	1.4	0.4	0.4	0.4
Feb 05	1.1	2.2	4.5	5.5	7.5
Mar 05	26.1	12.0	12.5	14.0	13.2
Total	**178.7**	**171.8**	**161.9**	**177.4**	**183.8**

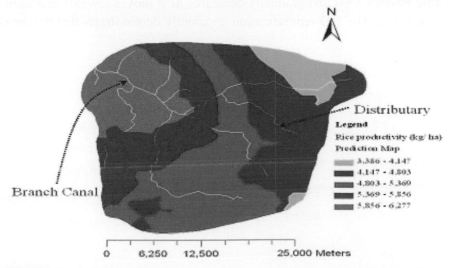

FIGURE 6.6 Spatial distribution of rice yield.

crop productivity has shown spatial variation in irrigation performance in the command area under scanty rainfall of 185 mm during the crop season of 2004–05. The adequate supply of canal water released in the upper reach resulted in higher productivity, whereas declined irrigation supply resulted in poor crop productivity at the tail reach of the command area.

Generally in an irrigation project, the upper reach farmers draw much more water as compared to the lower reach farmers, resulting in inequitable distribution of water across the region. The present case also revealed that there is a significant variation in the productivity of rice along the length of the branch canal and also along the length of all distributaries as well. Figure 6.6 shows spatial variability in rice productivity within the canal command. Poor crop productivity noticed in the lower reach could be attributed to inadequate canal water supply during critical crop growth period.

6.6.4.3   Trend Analysis of Rice Productivity

A 3-D representation of rice productivity demonstrates its spatial variability in the study area (Figure 6.7). Each vertical stick in the trend analysis plot represents the location and value (productivity) of each data point. The points are projected into the perpendicular planes, an east-west and a north-south plane. A best-fit curve is then placed through the projected points. Productivity curve shows that the value of productivity starts out with a high value towards west and gradually decreases as it moves towards east until it levels out. The 3-D representation apparently demonstrates that the data

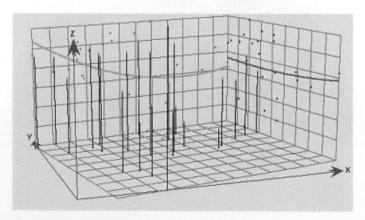

FIGURE 6.7   Trend analysis plot for rice productivity.

seems to exhibit a strong trend in the west-east direction and a weaker one in the north-south direction (Figure 6.7).

Hence, the present study validates that there is a considerable variation in canal water release pattern along its length showing a significant impact on rice productivity. Consequently, productivity declined gradually along the canal length from upper to lower reaches of the irrigation project

6.6.4.4   Soil Salinity on Crop Productivity

Salinity is a measure of concentration of soluble salts in the soil. Some salts are useful (many chemical fertilizers are in salt form), but too much salt of any kind is detrimental to plants and other organisms. Soil salinity levels are spatially interpolated for the entire command area, based on collected soil samples analyzed at the Soil Testing Laboratory of Prakasam district (Figure 6.8).

The salt affected soils are generally characterized by pH and EC (electrical conductivity). The pH readings of the study area varied from 7.7 to 8.5 shows soil alkalinity problem even though variation is insignificant. The crop yield decreased with the increase in salinity above 2.0 dS m^{-1} [63]. Hence, soil salinity status at any unmeasured locations of the study area, is assessed from the interpolated values measured data (Figure 6.8).

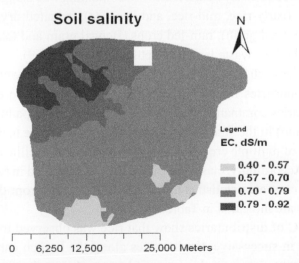

FIGURE 6.8   Soil salinity in the command area.

The prediction map was generated with ordinary Kriging method by inputting the core sample data along with corresponding coordinates of latitude and longitude for the study area. However, it is observed that soil salinity levels are not posing any threat to crop yield. The prediction map shows EC values of 0.7–0.9, 0.7–0.8, and 0.6–0.7 for upper, middle, and lower reaches, respectively. Hence, the predicted soil EC values in the entire command area are well below 2 dS/m (safe limit), making the soils classified as non-saline category.

6.6.4 ANALYSIS OF REMOTE SENSING IMAGES

The IRS P6 images of both LISS III and AWiFS, which are processed in the ERDAS Imagine 8.5 software, enabled to obtain reliable crop area statistics and vegetation indices that have been further used in estimation of crop water demand and performance assessment, respectively. Based on the crop water demand, supply-demand relationship of irrigation project has been analyzed and results are interpreted.

6.6.4.1 Generation of Land Use/Land Cover (LULC)

The unsupervised classification performed in ERDAS Imagine on LISS III data has resulted in twelve LULC classes, containing rice staggered in three sowing dates (early-rice, mid-rice, and late-rice), irrigated dry crops (chilies, cotton, and red gram), rain-fed crops (Bengal gram and tobacco), other crops, water bodies, waste land, and settlement (Figure 6.9).

Areas under each distributary commands are obtained by clipping to its respective boundaries from LULC map. The LULC maps so extracted for five distributaries commands (Rajupalem, Paidipadu, Kotapadu, Dronadula, and Nutalapadu) in the study area are represented in Figure 6.10. The study area consists of multiple crops being grown during the satellite data acquisition period. Accordingly, classified LULC map has resulted in twelve LULC classes. The crop area statistics that have been derived from the LULC of distributaries are indicated in Table 6.12.

The LULC of distributaries show that rice was observed to decrease in aerial extent in successive distributaries along the branch canal, whereas ID crops have successively increased towards its lower-reach. The

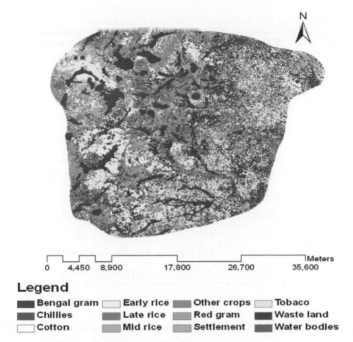

FIGURE 6.9 LULC of the command area.

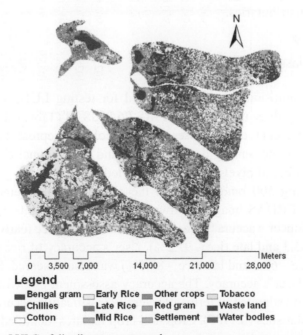

FIGURE 6.10 LULC of distributary commands.

TABLE 6.12　Crop Area (ha) Statistics of Different Distributaries

Land use/land cover	Rajupalem	Paidipadu	Kotapadu	Dronadula	Nutalapadu
Early rice	308.20	3386.50	1079.95	953.21	1319.75
Mid rice	184.62	426.22	367.07	121.19	103.31
Late rice	188.22	771.71	956.27	601.16	265.03
Bengal gram	157.06	1941.96	2592.36	1483.05	1386.77
Chillies	64.06	524.29	453.67	410.98	623.00
Cotton	103.90	1042.39	599.30	852.22	926.60
Other crops	795.77	3224.63	2146.42	1845.10	1949.37
Red gram	216.89	717.06	305.44	197.85	235.35
Settlement	28.30	109.61	183.56	95.38	147.43
Tobacco	3.60	1439.20	1703.70	759.38	396.01
Waste land	381.09	1394.60	558.04	215.88	537.78
Water bodies	2.49	26.93	62.16	29.78	34.81

crop area statistics obtained for different distributaries are represented in Table 6.12. Based on classified LULC map, the attribute values was prepared incorporating the areas occupied by different LULC classes in respective distributaries.

6.6.4.2　Classification Accuracy

The error (confusion) matrix is prepared for testing LULC classification accuracy using three time images of IRS P6 LISS III (Table 6.13). The rows of matrix represent the true classes while the columns represent classes associated with remote sensing derived LULC and the diagonal element represents the number of pixels correctly classified.

Considering 500 random points, the classification accuracy has been estimated in ERDAS Imagine software using random points method. High value of Producer's accuracy was observed in case of rice (early rice – 96.9, mid rice – 83.3 and late rice – 97.2%). User's accuracy of rice (early rice – 80.0, mid rice – 68.2 and late rice – 87.5%) was also observed. Bengal gram has highest User's accuracy. The accuracy assessment of LULC classification has been estimated in terms of overall classification accuracy (81.2%) and Kappa statistics (0.7787). Crop/non-crop features identified in the study area and corresponding accuracy statistics are represented in Table 6.14.

TABLE 6.13 Error (Confusion) Matrix of LULC Classification for Command Area

Classified Data	Reference Data												
	1	2	3	4	5	6	7	8	9	10	11	12	Total
1	2												2
2	1	15			2		2	1			1		22
3	1		35		1			1		1	1		40
4				5				1					6
5					81	2	2	2		1	1		89
6					1	21		4		1		3	30
7		1	1		4	53	4						63
8				3	5	2	9	100	2		4		125
9		2			2			9	64	1	2		80
10								2		6			8
11					3	1		2			16		22
12					1	1				1	2	8	13
Total	4	18	36	8	100	27	66	126	66	11	27	11	500

Note: Class 1 – water bodies; 2 – mid-rice; 3 – late-rice; 4 – settlement; 5 – Bengal gram; 6 – tobacco; 7 – waste land; 8 – other crops; 9 – early rice; 10 – chilies; 11 – cotton; and 12 – red gram.

TABLE 6.14 LULC Classification Accuracy

Class Name	Producer's accuracy, %	User's accuracy, %
Early-rice	96.97	80.00
Mid-rice	83.33	68.18
Late-rice	97.22	87.50
Bengal gram	81.00	91.01
Chillies	54.55	75.00
Cotton	59.26	72.73
Other crops	79.37	80.00
Red gram	72.73	61.54
Settlement	62.50	83.33
Tobacco	77.78	70.00
Waste land	80.30	84.13
Water bodies	50.00	100.00

The classification accuracy estimated from image classification is in the range of very good to excellent category [34].

6.6.5 NDVI BASED CROP CONDITION ASSESSMENT

The Normalized difference vegetation index (NDVI) for the command area is estimated using digital number (DN) data in visible red (R) and near-infrared (NIR) wavelengths of IRS P6 AWiFS sensor. The NDVI is observed to vary spatially in the entire command area in relation to cropping intensity/crop vigor. The NDVI generated from eight images of AWiFS data are shown in Figure 6.11, which shows temporal variability of NDVI in different locations of the command area.

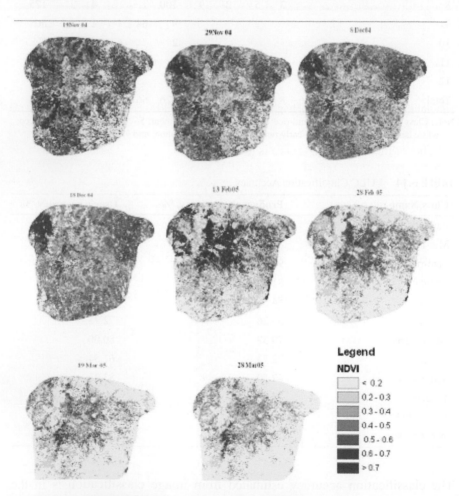

FIGURE 6.11   NDVI distribution within the ABC command area.

6.6.6   PERFORMANCE INDICATORS

The performance indicators for assessment of irrigation project under study are productivity, equity, and adequacy.

6.6.6.1   Productivity

The rice productivity retrieved from the Kriged surface has been recorded as 6277, 4874, 4469, 4065 and 3386 kg/ha, respectively, for the five distributary commands of Rajupalem, Paidipadu, Kotapadu, Dronadula and Nutalapadu (Figure 6.12). The continuous decline in rice productivity has been observed from the upper to lower distributaries, which is due to non-uniform supply of canal water in different reaches of the command with scanty rainfall during the cropping season.

6.6.6.2   Equity (NDVI Based)

The crop vigor that is explained by remote sensing based NDVI has shown considerable variations in the command area. To assess equity, NDVI values of distributary commands are analyzed. The maximum NDVI has varied from 0.72 to 0.35 and average NDVI from 0.43 to 0.32 from upper to lower distributary commands, respectively (Table 6.15). The maximum,

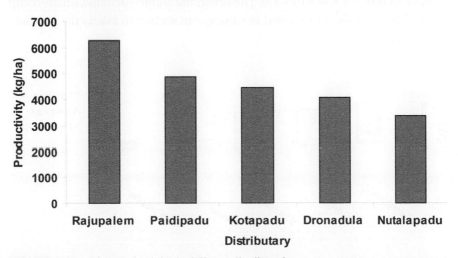

FIGURE 6.12   Rice productivity in different distributaries.

TABLE 6.15   Spatial Distribution of NDVI of Mid-Rice Among Distributaries

NDVI	Rajupalem	Paidipadu	Kotapadu	Dronadula	Nutalapadu
Max.	0.72	0.65	0.63	0.43	0.35
Ave.	0.43	0.40	0.38	0.36	0.32
Min.	0.32	0.30	0.25	0.26	0.27

average, and minimum NDVI of mid-rice indicate declining trend from upper to lower distributary commands (Figure 6.13). This implies that the crop canopy growth has considerably reduced from upper to lower distributary commands.

6.6.6.3   Adequacy (Relative Water Supply)

Seasonal irrigation water supply and demand for each distributary commands are compared for assessing surplus or deficit water supply situations (Figure 6.14). During the entire cropping season, the relative water supply (RWS) of 1.13, 0.47, 0.89, 0.61, and 0.64 has been experienced in Rajupalem, Paidipadu, Kotapadu, Dronadula, and Nutalapadu distributary commands, respectively. The RWS of 1.13 received at Rajupalem made it to come under 'adequately water surplus distributary' [47] category and remaining four distributaries classified as 'water deficit distributaries' as all of them recorded a RWS of <1.0. However, the supply-demand relationship at monthly intervals is discussed in subsequent section to assess the gap during critical crop growth stage.

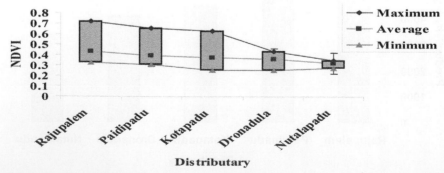

FIGURE 6.13   NDVI variability among distributaries.

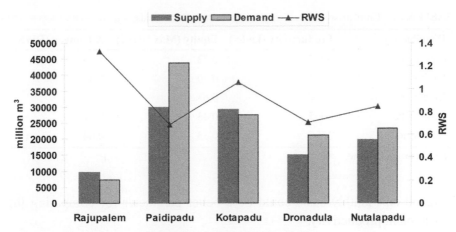

FIGURE 6.14 Relative water supply in the canal command.

6.6.6.4 Performances of Different Distributaries

The performance indicators estimated in different distributary commands of the irrigation project are compared in Table 6.16.

The Rajupalem distributary recorded higher productivity (6277 kg/ha) than that of average productivity (3529 kg/ha) of Prakasam district atlas of Andhra Pradesh [19]. Moreover, Rajupalem distributary (upper reach) shows greater irrigation performance indicators (productivity, equity, and adequacy) in comparison to Nutalapadu distributary (lower reach), which has exhibited poor performance with low productivity and inadequate water supply.

6.6.6.5 Impact of Supply–Demand Relationship on Irrigation Performance

Crop productivity is the most prominent indicator in any irrigation performance assessment. In the command area, the entire farmers community are habituated of applying higher quantity of inorganic fertilizers (nitrogen: phosphorus; potassium) than the recommended dosages of 240–300 kg/ha, whereas the release of canal water at the head regulator is controlled by the State I & CAD authority. Therefore, the variation in rice productivity at various reaches of the irrigation command is influenced by another major input such as irrigation water. The ratio of irrigation water supply to its crop water

TABLE 6.16   Comparison of Performance Indicators Commands

Distributary	Productivity (kg/ha)	Equity (Max NDVI)	Adequacy (RWS)
Rajupalem	6277	0.72	1.13
Paidipadu	4874	0.70	0.47
Kotapadu	4469	0.63	0.89
Dronadula	4065	0.43	0.61
Nutalapadu	3386	0.35	0.64

Note. M m^3 – million m^3.

demand on monthly-basis could be a better parameter for comparing the variation in productivity.

Rice of 120 days duration is cultivated in the command area. The critical crop growth period of rice is falling within 35 to 65 days from the date of transplanting, which coincides within November 2004 to February 2005. Hence, assessment of irrigation water supply and crop demand gap in the entire crop growth period and particularly during critical growth stage is a driving parameter to regulate crop productivity.

6.6.6.6   Water Supply-Demand Ratio

Using CRIWAR model, monthly net irrigation requirement (NIR) of cropping pattern, derived by RS is calculated for each distributary commands. The crop water demand (gross irrigation requirement) is estimated by dividing NIR by irrigation system efficiency of 60% [22]. Temporal variability in supply-demand ratio during November 2004 to February 2005 for two distributary commands is shown in Figure 6.15. The supply-demand ratios for Nov-04, Dec-04, Jan-05, and Feb-05 are found to be 1.65, 1.00, 0.87, and 0.79 in Rajupalem distributary command (upper-reach), and 0.50, 0.61, 0.68, and 0.49 in Nutalapadu distributary (lower-reach), respectively (Table 6.17). The upper reach distributary recorded an average irrigation supply-demand ratio of 1.07 against 0.57 in the lower reach. It is observed that Rajupalem received dominant canal supply in comparison to Nutalapadu distributary (Figure 6.15).

Hence, crop water demand simulation based on integrated use of geo-statistical and RS image classification, enabled to quantify water supply-demand relationship in the command area during the cropping season. Timely and required quantity of irrigation water supply to upper distributary has resulted in doubling crop productivity in comparison to lower distributary.

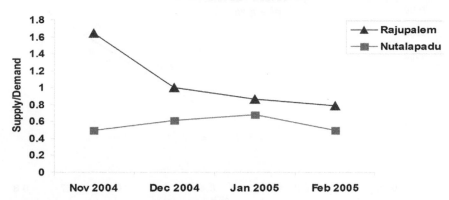

FIGURE 6.15 Monthly supply-demand ratio.

TABLE 6.17 Supply–Demand Ratio Between Upper and Lower Distributaries

Month	Rajupalem	Nutalapadu
October 2004	1.20	1.43
November 2004	1.65	0.50
December 2004	1.00	0.61
January 2005	0.87	0.68
February 2005	0.79	0.49
March 2005	2.49	2.40
Average during Nov. 04–Feb. 05	1.07	0.57

Quantity of irrigation water supply to any distributary command depends on its shape and size. Hence, lower distributary with larger command area of 8800 ha and canal length of 20 km has recorded lower performance in terms of crop productivity as compared to upper distributary with smaller command area of 2648 ha and 8 km length of irrigation canal.

6.6.7 PREDICTION OF RICE PRODUCTIVITY

6.6.7.1 Simple Linear Regression Model

Two models have been developed for prediction of rice productivity against $NDVI_{max}$ (denoted as TCVI) (Figure 6.16) and $NDVI_{ave}$ (Figure 6.17) of different pixels.

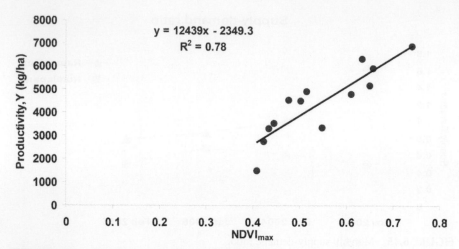

FIGURE 6.16   Rice productivity based on NDVI$_{max}$.

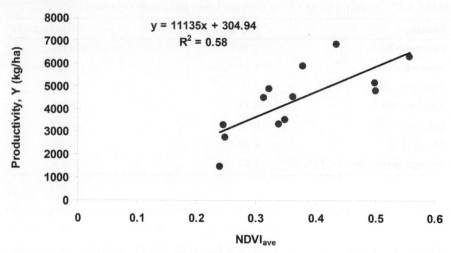

FIGURE 6.17   Rice productivity based on NDVI$_{Ave}$.

The crop productivity models are plotted by regression in scatter diagrams between crop productivity (y) and the NDVI (x). Since AWiFS images acquisition period covers the growing period of mid-rice (date of sowing = 10th Nov 2004) and late-rice (date of sowing = 5th Dec 2004) completely, the models have been developed for productivity data of mid-rice and late-rice and represented in the following expressions. In both the models, the productivity increased linearly with NDVI.

$$Y = 12439 \ NDVI_{max} - 2349.3, R^2 = 0.78 \qquad (17)$$

$$Y - 11135\ NDVI_{ave} + 304.94,\ R^2 = 0.58 \tag{18}$$

The model based on $NDVI_{max}$ yielded higher R^2 (coefficient of determination) of 0.78 as compared to model developed on $NDVI_{ave}$ with R^2 of 0.58. Hence, model that shows higher statistical significance (R^2) is selected and validated for productivity of early-rice at disaggregated village level CCE data in the command area (Table 6.18).

The maximum deviation of predicted yields from observed yields was less than 10 percent (Table 6.18), indicating the acceptability of the model. Using this model, further spatial rice yield map of the study area is developed by extrapolating the predicted rice yield for the entire command area (Figure 6.18). It shows that more rice intensity is seen at upper reaches and near to the canal in comparison to tail reach areas.

6.6.7.2   Non-Linear Multiple Regression Model for Crop Productivity

Based on the CCD experiments, which have been carried out in triplicates, the crop productivity was expressed as a non-linear function of input parameters as follows:

$$Y = -3586.35 + 10850.7\,WD + 8.8\,NPK - 3243.5$$
$$WD^2 - 0.004\ NPK^2 - 2.51\ WD \times NPK \tag{19}$$

where, Y = rice productivity (kg/ha), WD = water depth (m); and NPK = total chemical fertilizer applied (kg/ha).

Based on the ANOVA (analysis of variance) results (Table 6.19), it is observed that linear (ANOVA, p=0.000) and square terms (ANOVA, p = 0.004) have significant effect on crop productivity. But contribution of interaction term is found to have insignificant effect on productivity. Further surface plot indicates non-linear relationship between interaction and productivity (Figure 6.19). The predicted values are in close agreement with the experimental results (Table 6.20) and range of residuals lie within the limits, which indicates the good prediction accuracy and generalization ability of the predicted model. The coefficient of multiple regression equation, adjusted R^2, was found to be 96.8%, which indicate the fitness and adequacy of the model. The coefficient of the regression model indicates

TABLE 6.18   Validation of Linear Regression Model

Village in the command area	NDVI$_{max}$	Rice productivity (kg/ha)		
		Predicted	Observed	% Deviation from observed yield
Ballikurava	0.64	5587	6048	7.63
Cherukuru	0.64	5664	5958	4.94
Darsi	0.57	4704	5208	9.68
Guntupalli	0.68	6153	6479	5.04
Kopperapalem	0.61	5239	5616	6.72
Nagulapalem	0.59	4935	5447	9.40
Santhamagullur	0.69	6234	6115	−1.94
Veerannapalem	0.67	5978	5916	−1.04
Vinjanampadu	0.55	4475	4858	7.88
Yanamadala	0.43	3018	3306	8.73

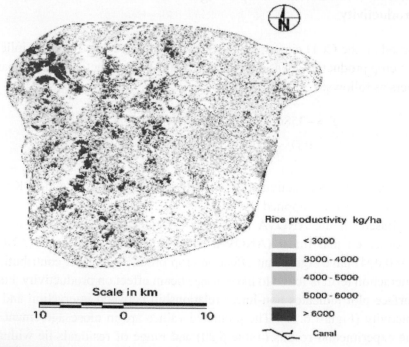

Rice productivity kg/ha

< 3000

3000 - 4000

4000 - 5000

5000 - 6000

> 6000

Canal

FIGURE 6.18   Spatial rice map (NDVI based).

that depth of water applied influence crop productivity to a large extent as compared to nutrients application in the command area.

TABLE 6.19   Analysis of Variance for Rice Productivity (kg/ha)

Source	DF	Seq SS	Adj SS	Adj MS	F	p
Regression	5	12154854	12154854	2430971	62.12	0.000
Linear	2	10393083	10393083	5196542	132.80	0.000
Square	2	1566407	1566407	783203	20.01	0.004
Interaction	1	195364	195364	195364	4.99	0.076
Residual error	5	195656	195656	39131		
Lack-of-fit	3	185508	185508	61836	12.19	0.077
Pure error	2	10149	10149	5074		
Total	**10**	**12350511**				

SS = 197.8; R^2 = 98.4%; R^2 (adj) = 96.8%

Note: Seq SS = Sequential sum of squares; Adj SS = Adjusted sum of squares; Adj MS = Adjusted mean squares.

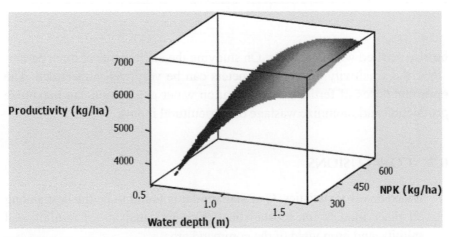

FIGURE 6.19   Rice productivity with application of water depth and nutrients (NPK).

The Response surface plot of rice productivity with water depth and NPK (Figure 6.19) indicates sensitivity of the use of these parameters on crop productivity. The productivity is linearly increased up to 5956 kg/ha against a depth of 1.08 m of applied water and use of 284 kg/ha of NPK. Thereafter, the effect of input seemed to be ineffective as rice productivity is declining with the application of higher depth of water and doses of fertilizer. Hence, the proposed model facilitates the better understanding of the effect of water management and nutrients usage on rice productivity and can

TABLE 6.20   Central Composite Design (CCD) with Experimental and Predicted Responses

Water depth (m)	Nutrients NPK (kg/ha)	Crop productivity (experimental)	Crop productivity (predicted)	Residual
0.612	284	3386	3579	−6.89388
1.546	660	7040	6928	1.35170
1.079	472	6277	6324	−0.24920
1.079	660	6568	6612	−0.28839
0.612	660	4814	4882	−0.72222
1.546	284	6496	6508	−0.12183
1.079	472	6338	6324	0.07407
1.079	284	5956	5751	1.76428
1.546	472	6760	6860	−0.68342
0.612	472	4634	4373	3.12366
1.079	472	6196	6324	−0.71581

be also verified with field trails. On studying the coefficients of input parameters, the sensitivity of crop parameters can be very well understood. The optimum doses of fertilizer and irrigation water application can maximize production and minimize wastage on agricultural inputs.

6.7   CONCLUSIONS

a.   Based on error statistics, Gaussian model is found to be the best among all other models for geostatistical (Kriging) analysis of rainfall, soil salinity, and crop yield of the command area.

b.   Using unsupervised classification procedure, the image processing of LISS III satellite imageries is carried out to develop accurate land use/land cover of the irrigation command with Kappa statistics of 0.7787.

c.   The upper reach of the irrigation system (Rajupalem distributary) recorded nearly double in rice productivity in comparison to the lower reach (Nutalapadu distributary).

d.   The maximum and average NDVI values from upper to lower reach distributary commands varied considerably from 0.72 to 0.34 and from 0.43 to 0.32, respectively, which indicates a great variation in canal water availability from upper to lower reaches of the canal network.

e. During four months of critical growth period of rice (November 2004 to February 2005), Rajupalem distributary recorded higher performance in terms of water supply-demand ratio (1.65, 1.00, 0.87, and 0.79) as compared to Nutalapadu distributary (0.5, 0.61, 0.68, and 0.49).

f. On relating the rice productivity with other parameters (total depth of water, NPK, NDVI), the following findings have been observed:

g. The linear regression model developed correlating rice productivity with TCVI ($R^2 = 0.77$) could be able to predict the productivity spatially for the entire command area.

h. The development of non-linear multiple regression model for rice productivity against seasonal depth of water and nutrients applications ($R^2 = 0.96$) enabled to identify the sensitivity of input parameter. The model also identified the insignificant role of applied nutrients on rice yield.

i. Hence it is inferred that remote sensing based performance study not only helpful in estimating the irrigation demand of a large irrigated command quickly but also useful in predicting the crop productivity from different locations based on NDVI. It also helps to study the crop condition at various points in the command area and similarly, application of crop production functions is a useful technique in estimating the significant role of input parameters on crop production.

6.8   SUMMARY

Greater challenges in irrigation projects are to bridge the gap between the water demand and supply as it largely influences spatial variability in crop productivity. An integrated approach is needed to understand spatial and temporal variability in water supply, water demand, and crop productivity for performance assessment of an irrigation project. The aim of performance assessment is to identify the parameters causing variability in crop productivity so as to improve the functioning of the irrigation system and there will be further scope to validate the criterion to other irrigation projects. The Addanki branch canal command, located in the right main canal of the Nagarjuna Sagar irrigation project, is selected for performance assessment in the present study.

As the study area is located in semiarid conditions, variations in rainfall, salinity, and nutrients status of soil play an important role on crop productivity. Using different models of geostatistical tool of ArcGIS software, ordinary Kriging is performed for interpolating parameters at unmeasured locations. The best-fit model is confirmed only after performing cross validation and verifying the error statistics. Cumulative rainfall of Kriged values during crop period from October 2004 to March 2005, has been recorded as 178.6 mm at upper distributary (Rajupalem) and 183.7 mm at lower distributary (Nutalapadu) commands. Similarly, depth of canal water applied during cropping period in case of Rajupalem and Nutalapadu distributaries was 904 and 570 mm, respectively.

Kriging surface of rice productivity has shown a significant variation from 6277 to 3386 kg/ha from upper to lower reach distributary commands. High (> 150 mm/m), medium (100–150 mm/m), and low (50–100 mm/m) soil moisture content was observed in upper, middle, and lower reaches of the command area, respectively, as noticed in Prakasam district atlas. It is observed that soil moisture content is in accordance with the canal water release pattern at upper, middle, and lower reaches.

The required satellite data sets are identified based on existing crop calendar and water release schedule. The LISS-III and AWiFS data of Resourcesat-1 satellite (IRS P6), which is especially designed for land/water management and agricultural applications, have been procured from NRSA. LISS III images are selected for LULC generation. Similarly, AWiFS images are selected for assessment of crop condition spatially in the entire crop growth period. The ERDAS Imagine 8.5 version software is used for processing of images.

Unsupervised classification was carried out with three images of LISS III of November 19, 2004; December 13, 2004; and March 19, 2005. Eight possible land cover dynamics are generated in a composite image by aggregating the vegetation and no-vegetation layers. These layers are labeled as individual LULC based on ground truth, crop calendar information, and feed back of technical personnel from the DOA and the I & CAD authority of Government of Andhra Pradesh. Classification of LISS III data performed in ERDAS Imagine has resulted in twelve LULC classes, containing rice staggered in three sowing dates (early-rice, mid-rice, and late-rice) and dry crops (chilies, cotton, red gram, Bengal gram, and tobacco), other crops, water bodies, waste land, and settlement. The classification accuracy has been estimated, considering 500 random points, in ERDAS imagine software, using random points method. The accuracy assessment of LULC classification has been estimated in terms of overall classification accuracy of 81.2% and Kappa statistics of 0.7787.

The satellite derived NDVI has been used as an index to assess the condition of crops across the command area. NDVI is calculated using reflected radiation in infrared (0.7 µm to 1.1 µm) and red (0.6 µm to 0.7 µm) bands. The observation of NDVI has clearly depicted variability in rice crop conditions during its entire growth period. The higher value of NDVI has been recorded at upper distributary command as compared to the lower one. The peak value of NDVI recorded around February 13, 2005 has clearly represented heading stage of the mid rice.

The performance of irrigation project is assessed in terms of three indicators namely, productivity, equity, and adequacy. The distributary-wise rice productivity has been predicted as 6277, 4874, 4469, 4065 and 3386 kg/ha, respectively, in the canal commands of Rajupalem, Paidipadu, Kotapadu, Dronadula, and Nutalapadu. The continuous decline in rice productivity has been observed from upper to lower distributary commands, synchronizing with the proportional decline in canal water supply. The equity is estimated in terms of NDVI values among different distributary commands. Maximum NDVI has varied from 0.72 to 0.34 and average NDVI from 0.43 to 0.32 from upper to lower distributary commands, respectively. This implies that crop canopy growth has considerably reduced from upper to lower distributary commands in proportion to irrigation water supply.

During the entire cropping season, a relative water supply (RWS) of 1.13, 0.47, 0.89, 0.61, and 0.64 has been experienced in Rajupalem, Paidipadu, Kotapadu, Dronadula, and Nutalapadu distributary commands, respectively. The RWS of 1.13 in Rajupalem distributary is categorized as 'adequately water surplus distributary.' Comparison of performance indicators among the distributaries shows that Rajupalem (upper) performed best as compared to Nutalapadu (lower).

Irrigation water supply and demand is analyzed for each distributary commands for correlating the variability in crop productivity in different canal reaches. Hence, the CRIWAR model is used to estimate crop water demand to assess monthly supply-demand ratio during the cropping season. Supply-demand ratio for Nov-04, Dec-04, Jan-05, and Feb-05 are found to be 1.65, 1.00, 0.87, and 0.79 in Rajupalem distributary (upper reach); and 0.50, 0.61, 0.68, and 0.49 in Nutalapadu distributary (lower reach) command, respectively. The upper reach distributary recorded an average irrigation supply-demand ratio of 1.07 against 0.57 in the lower reach. The trend shows that Rajupalem received major share of canal supply in comparison to Nutalapadu distributary. Integrated use of geostatistical and RS image classification is enabled to quantify water supply-demand relationship in the command area during the cropping season. Timely supply of required quantity of irrigation water to upper distributary has resulted in higher crop productivity in comparison to the lower one.

Linear regression models have been developed for the prediction of rice productivity relating to $NDVI_{max}$ (TCVI) and $NDVI_{ave}$. The regression model based on $NDVI_{max}$ has been selected based on higher R^2 (0.78) and validated with crop productivity recorded at scattered points of the command area. Using the developed model, rice productivity can be predicted at any locations of the command area.

Another nonlinear multiple regression model has been developed for the prediction of rice productivity against seasonal depth of water and nutrients applied. It is concluded that depth of water has dominant role on crop productivity as compared to nutrients. The model has been validated with data points at other locations of the command area. The model also indicates that higher dosages of nutrients application have insignificant impact on crop yield beyond certain level.

KEYWORDS

- **Binary images**
- **Branch canal**
- **Chemical fertilizer**
- **Command area**
- **Crop cutting experiment**
- **Crop development stage**
- **Crop duration**
- **Crop production function**
- **Crop productivity**
- **Distributary**
- **Effective precipitation**
- **Equity**
- **Evapotranspiration**
- **Geographical information system**
- **Geostatistics**

- **Gross irrigation**
- **Image processing**
- **Initial stage**
- **Irrigation demand**
- **Irrigation supply**
- **Kriging technique**
- **Land use/land cover**
- **LISS III**
- **Lower reach**
- **Major irrigation project**
- **Penman model**
- **Reference evapotranspiration**
- **Reflected radiation**
- **Regression model**
- **Relative water supply**
- **Remote sensing**
- **Reservoir**
- **Rice dry crop**
- **Satellite data**
- **Satellite imageries**
- **Saturation vapor pressure**
- **Soil moisture content**
- **Soil salinity**
- **Upper reach**
- **Waste land**
- **Water requirement**
- **Water supply-demand**
- **Wind velocity**

REFERENCES

1. Abernethy, C. L. (1986). Performance measurements in canal water management: a Disc-ussion. *ODI-IIMI Irrigation Management Network Paper No. 86/2d,* Overseas Development Institute, London.

2. Abernethy, C. L. and Pearce, G. R. (1987). Research needs in third world irrigation. *Hydraulics Research Limited*, Wallingford, U. K.

3. Allen, R .G., Pereira, L. S., Raes, D. and Smith, M. (1998). *Crop evapotranspiration, guidelines for computing crop water requirements*. Irrigation and Drainage, Paper No.56, FAO, Rome, Italy.

4. Ambast, S. K., Singh, O. P., Tyagi, N. K., Menenti, M., Roerink, G. J. and Bastiaanssen, W. G. M. (1999). *Appraisal of irrigation system performance in saline irrigated command using SRS and GIS*. Balkema, Rotterdam, 457–462.

5. Avalos, J. M. M., Gonzalez, A. P., Vazquez, E. V. and Fouz, P. S. (2007). Mapping monthly rainfall data in Galicia (NW Spain) using inverse distance and geostatistical methods. *Advances in Geosciences*, 10, 51–57.

6. Bacchi, B. and Kottegoda, N. T. (1995). Identification and calibration of spatial correlation patterns of rainfall. *J. Hydro,* 165, 311–348.

7. Bastiaanssen, W. G. M. (1998). Remote sensing in water resources management: the state of the art. *International Water Management Institute (IWMI)*, Colombo, Sri Lanka, pp.118.

8. Bastiaanssen, W. G. M., Molden, D. J. and Makin, I. W. (2000). Remote sensing for irrigated agriculture: examples from research and possible applications. *Agric. Water Manage.,* 46, 137–155.

9. Bhutta, M. N. and Van der Velde, E. J. (1992). Equity of water distribution along secondary canals in Punjab, Pakistan. *Irrig. Drain. Syst.*, 6(2), 161–177.

10. Biggs, T. W., Thenkabail, P. T., Gumma, M. K., Scott, C. A., Parthasaradhi, G. R. and Turral, H. (2006). Irrigated area mapping in heterogeneous landscapes with MODIS time series, ground truth and census data, Krishna Basin, India. *Int. J. Rem. Sens.*, 27, 4245 4266.

11. Bos, M. G., Vos, J. and Feddes, R. A. (1996). *CRIWAR 2.0 A Simulation model on Crop Irrigation Water Requirements*. ILRI Publication 46, Wageningen, The Netherlands.

12. Campling, P., Gobin, A. and Feyen, J. (2001). Temporal and spatial rainfall analysis across a humid tropical catchment. *Hydrol. Proc.,* 15(3), 359–375.

13. Chakraborti, A. K., Rao, V. V., Shanker, M. and Sureshbabu, V. A. (2002). Performance evaluation of an irrigation project using satellite remote sensing, GIS & GPS *(http://www.gisdevelopment.net/application/agriculture/irrigation/agriir001.htm)*.

14. Chief Planning Officer. (2006). Agriculture. In: *Hand Book of Prakasam District Statistics (2004–05),*Ongole, Andhra Pradesh, India.

15. Christel, P. and Reed, D. W. (1999). Mapping extreme rainfall in a mountainous region using geostatistical techniques: A case study in Scotland. *Int. J. of Climatology*, 19(12), 1337–1356.

16. Chung, S. and Sung, J. (2006). Spatial variability of yield, chlorophyll content, and soil properties in a Korean rice field. Precision farming machinery laboratory, bio- production division, *National Agricultural Mechanization Research Institute*, RDA Suwon, Republic of Korea.

17. Cohen, J. (1960). A coefficient of agreement of nominal scales. *Educational Psychological Measurements,* 20, 37–46.

18. Da Silva, J. R. M. and Alexandre, C. (2005). Spatial variability of irrigated corn yield in relation to field topography and soil chemical characteristics. *Precision Agriculture, 6,* 453–466.

19. Directorate of Agriculture. (2004). *Prakasam District Atlas*. Department of Agriculture, Hyderabad.

20. Dooorenbos, J. and Pruitt, W. O. (1977). *Guidelines for Predicting Cop Water Requirements*. FAO Irri. and Drain. Paper No. 24, FAO, Rome, Italy.

21. Gaur, A. (2006).Squeezed dry: Implications of drought and water regulation in the Krishna basin, India. *World Water Week*, SIWI, Stockholm.

22. Gaur, A., Biggs, T. W., Gumma, M. K., Parthasaradhi, G. and Turral, H. (2008). Water scarcity effects on equitable water distribution and land use in a major irrigation project-Case study in India. *J. Irrig. Drain. Eng.*, ASCE, 134(1), 26–34.

23. Gontia, N. K. (2005). *Crop water stress indices and evapotranspiration estimation for Irrigation scheduling and yield modeling of wheat crop using Remote sensing and GIS*. Unpublished PhD thesis submitted to the Department of Agricultural and Food Engineering, Indian Institute of Technology, Kharagpur.

24. Goovaerts, P. (1997). *Geostatistics for Natural Resources Evaluation*. Oxford University Press, New York, 60:201.

25. Goovaerts, P. (2000). Geostatistical approaches for incorporating elevation into the spatial interpolation of rainfall. *J. of Hydro.*, 228,113–129.

26. Gundogdu, K. S. and Guney, I. (2007). Spatial analyzes of groundwater levels using universal Kriging. *Journal of Earth System Science*, 116, 49–55.

27. Hatfield, J. L. (1983). Remote sensing estimators of potential and actual crop yield. *Remote Sensing of Environment*, 13(4), 301–311.

28. Heady, E. O. and Dillon, D. L. (1960). *Agricultural Production Functions*. Kalyani Publishers. Ludhiana, pp. 86–91.

29. Isaaks, E. H. and Srivastava, R. M. (1989). *An Introduction to Applied Geostatistics*. Oxford University Press, New York.

30. Jahromi, S. S. and Feyen, J. (2001). Spatial and temporal variability performance of the water delivery in irrigation schemes. *Irrig. Drain. Syst.*, 15, 215–233.

31. Johnson, K., Hoef, J. M. V., Krivoruchko, K. and Lucas N. (1996). *Using ArcGIS Geostatistical Analysis: GIS User Manual*, ESRI, New York, 120–187.

32. Kitanidis, P. K. (1997). *Introduction to Geostatistics: Application to Hydrology*, Cambridge University Press, Cambridge, U K, pp. 249.

33. Levine, G. (1982). Relative water supply: An explanatory variable for irrigation systems. *Technical Report No. 6.* Cornell University, Ithaca, New York, USA.

34. Manserud, R. A. and Leemans, R. (1992). Comparing global vegetation maps with the Kappa statistics. *Ecological Modeling*, 62, 275–279.

35. Menenti, M., Visser, T. N. M., Morabito, J. A. and Draovandi, A. (1989). Appraisal of irrigation performance with satellite data and georeferenced information. In: *Rydzewsky, J. R., Ward, K. (Eds.), Irrigation Theory and Practice*. Pentech Press, London, pp. 785 – 801.

36. Ministry of Water Resources. (2003). *Vision for integrated water resources development and management*. Ministry of Water Resources, Govt. of India, New Delhi.

37. Mishra, P. K., Tiwari, K. N., Chowdary, V. M. and Gontia, N. K. (2005). Irrigation water demand and supply analysis in the command area using remote sensing and GIS. Hydro. J., 28(1–2), 59–69.

38. Murthy, C. S., Thiruvengadachari, S., Raju, P. V. and Jonna, S. (1996). Improved ground sampling and crop yield estimation using satellite data. *Int. J. Rem. Sens.*, 17(5), 945–956.

39. National Land and Water Resources Audit. (2000). *Australian water resources assessment of 2000. Surface water and groundwater – Availability and quality*, Braddon, Australian Capital Territory, Australia.

40. National Remote Sensing Agency. (2006). *Irrigation performance assessment of Krishna western delta using Remote Sensing.* NRSA-ANGRAU Project Report. pp. 37–63.

41. Oza, M. P., Bhagia, N., Patel, J. H., Dutta, S. and Dadhwal, V. K. (1996). National wheat acreage estimation for 1995–1996 using multidate IRS-IC WiFS data. *J. Ind. Soc. Rem. Sens.,* 24(4), 243–254.

42. Pinter, P. J., Jackson, R. D., Idso, S. B. and Reginato, R. J. (1981). Multidate spectral reflectance as predictors of yield in water stressed wheat and barley. *Int. J. Rem. Sens.,* 2, 43–48.

43. Rao, D. P. (2000). Geographic data needed in the interpretation of Indian satellite-based remote sensing data: Opportunities and realities. *Current Science,* 79(4), 468–473.

44. Rao, N. H., Brownee, S. M. and Sarma, P. B. S. (2004). GIS-based decision support system for real time water demand estimation in canal irrigation systems. *Current Science,* 87(5), 628–636.

45. Raul, S. K., Panda, S. N., Hollaender, H. and Billib, M. (2008). Sustainability of rice dominated cropping system in the Hirakud canal command, Orissa, India. *Irrigation and Drainage,* 57, 93–104.

46. Ray, S. S. and Dadhwal, V. K. (2001). Estimation of crop evapotranspiration of irrigation command area using remote sensing and GIS. *Agric. Water Manage.,* 49, 239–249.

47. Ray, S. S., Dadhwal, V. K. and Navalgund, R. R. (2002). Performance evaluation of an irrigation command area using remote sensing: A case study of Mahi command, Gujarat, *India. Agric. Water Manage.,* 56, 81–91.

48. Sadeghi Pour Marvi, M. (2008). A comparison of three mathematical models of response to applied nitrogen using spinach. *American-Eurasian J. Agric. & Environ. Sci.,* 4 (5), 611– 616.

49. Sakthivadivel, R. S., Thiruvengadachari, A. U., Bastiaanssen, W. G. M. and Molden, D. (1999). *Performance evaluation of the Bhakra irrigation system, India, using remote sensing and GIS Techniques.* Research report 28, International Water Management Institute. Colombo, Srilanka.

50. Sarma, P. B. S. and Rao, V. V. (1997). Evaluation of an irrigation water management scheme. A case study. *Agric. Water Manage.,* 32, 181–195.

51. Sarangi, A., Madramootoo C. A. and Enright, P. (2006). Comparison of spatial variability techniques for runoff estimation from a Canadian watershed. *Biosystems Engineering,* 95(2), 295–308.

52. Shakoor, A., Shehzad, A. and Asghar, M. N. (2006). Application of Remote sensing techniques for Water resources planning and management. (*http://ieeexplore.ieee.org/ xpls /abs_all.jsp.arnumber=410642,* Accessed on January 5, 2009).

53. Srinivasaprasad, A., Umamahesh, N. V. and Viswanath, G. K. (2006). Optimal irrigation planning under water scarcity. *J. Irrig. Drain. Eng.,* ASCE, 132(3), 228–237.

54. Takeshi, Y., M. Man, L. H., and Khang, V. T. (2006). Good Soil Care Practice in the Tropics: Toward a New Challenge. *Japan International Research Center for Agricultural Sciences.* Japan.

55. Thiruvengadachari, S. and Sakthivadivel, R. (1997). *Satellite remote sensing for assessment of irrigation system performance: A. Case study in India.* Research Report No. 9, International Irrigation Management Institute, Colombo, Sri Lanka.

56. Tucker, C. J., (1979). Red and photographic infrared linear combination for monitoring vegetation. *Rem. Sens. Environ.,* 8,127–150.

57. United Nations, (2005). World population prospectus. *United Nations bulletin*, New York, USA.
58. Vijay Kumar, A., and Remadevi, D., (2006). Kriging of groundwater levels – A case study. *Journal of Spatial Hydrology*, 6: 81–92.
59. Wackernagel, H. (2003). *Geostatistics. In Multivariate Geostatistics: An Introduction with Applications,* Springer-Verlag, New York, 145–169.
60. Wang, R. Y. (1986). An approach to tree-classifier design based on hierarchical clustering. *Int. J. Rem. Sens.,* 7, 75–88.
61. Webster, R. and Oliver, M. A. (2001). Cross-correlation, coregionalization, and co-Kriging. In: *Geostatistics for Environmental Scientists*, John Willey and Sons, Chichester, UK.
62. Wolff, P., and Stein, T. M., (1999). Efficient and economic use of water in agriculture: Possibilities and limits. *Natural Resources and Development.* 49/50, 151–159.
63. Yurtseven, E., Kesmez, G. D., and Unlukara, A. (2005). The effects of water salinity and potassium levels on yield, fruit quality and water consumption of a native central Anatolian tomato species (Lycopersiconesculantum). *Agric. Water Manage.,* 78, 128–135.

42. United Nations (2007), World population prospects, United Nations population database.

43. Vörösmarty, C.J., and Sahagian, D. (2000), Anthropogenic disturbance of the terrestrial water cycle, Bioscience, 50(9), 753-65.

44. Wackernagel, H. (2003), Geostatistics for natural resources characterization, Springer, New York, 130-190.

45. Wang, F.Y. (1996), Approaches to feasibility allocation based to watershed management, J. Water Res., 7, 73-88.

46. Watkins, R., and Dhir, V.K. (1996), Crop coefficient, crop configuration, and co-Kharif, in: Development and ground management science, John Wiley and Sons, Chichester, UK.

47. Weld, V., and Scott, C.M. (1994), Efficient and equitable use of water in agriculture, Economics and ethics, Natural Resources and Development, 10/98, 123-139.

48. Zimmerman, B., Kramer, K.D., and Douthit, A. (2005), The effects of water scarcity and population growth on yield, field design, and water consumption of a staple crop and associated human ecosystem services (Agroecosystems), degree water, Methuen, 24(2), 1-12.

CHAPTER 7

EVALUATION OF PERFORMANCE INDICES FOR WATER DELIVERY SYSTEMS: CANAL IRRIGATION

S. K. RAUL and M. L. GAUR

Department of Soil and Water Engineering, College of Agricultural Engineering and Technology, Anand Agricultural University, Godhra–389001, Gujarat, India; Mobile: +91 9428152757, E-mail: sanjaykraul@gmail.com, mlgaur@yahoo.com

CONTENTS

7.1 INTRODUCTION

The rising demand for water in irrigated agriculture, which consumes more than 80% of the developed surface water resources of India, has been a great challenge for the water resources planners and managers for its efficient management. Irrigated agriculture of India, which covers around 40% area, contributes 56% of all food grain production in the country [14]. Agriculture has been the primary source of livelihood for 75% of the population of India, and it contributes to 30% of GDP and 60% employment. Provision of assured water supply to the agriculture is, therefore, the topmost objective of the water resources program of the country. Canal irrigation is one of the principal methods of irrigation used for the improvement of crops. This has been extensively used in many developing countries such as India, Pakistan, and Egypt. Besides providing water for irrigation, canals also act as a source of intensive seepage below the ground.

The seepage contribution from a canal in the Indus basin of Pakistan, to the groundwater recharge was 15.1% of the inflow at the head of the canal [2]. Similar study by [3] yielded 26% seepage loss. Further study concluded that if the canal system is not managed properly, then seepage loss can increase even upto 35 to 40% of the diversion into the canal resulting in regional groundwater table build-up, increased soil salinity and reduced water use efficiency (WUE) [17]. On the other hand, there exist large imbalances between the irrigation water supply and demand due to faulty irrigation scheduling.

In a field study conducted in the command area of Guvvalagudem major distributary of the Nagarjunasagar Left Canal, Andhra Pradesh it was found that most of the times canal supplies were lesser than the design capacity and there were wide gaps between weekly canal water demands and supplies [15]. The mismatch between the annual supplies and the annual demands were lesser in magnitude (upto 20%) when compared to the weekly mismatches, which went up to 100% in some cases.

It has also been the most common practice in many of the canal irrigation projects that when the canal runs at its full capacity, the head and middle reach farmers draw excess water by keeping their outlet open continuously. Hence, over the years excessive seepage from the canal system has resulted in waterlogging at the head and middle reaches of canal command areas, whereas farmers in the tail reach suffer from severe water shortage. This happens mostly due to lack of adequate knowledge among the beneficiaries regarding the right quantum of crop-specific water allocation. On a study involving monitoring and quantification of water flows at four control points and six minor canals under the Rahad irrigated scheme in Sudan, it was observed that there was considerable seasonal variability in the performance of the indicators used and mal operation of canal structures [6]. The water delivery performance in the minor canal was also found to decrease towards the tail end. MIKE-11 hydraulic model was used in the Right Bank Main Canal system of the Kangsabati project in West Bengal – India to compute a performance ratio, which was used as an indicator for assessing the degree of uniformity in flow deliveries along the length of the canal [12]. A sharp decline was seen in the performance ratio along the length of the canal because most of the distributaries of the head and middle reaches have drawn more than their desired shares.

It is, therefore, necessary to study various performance measures of irrigation water delivery systems so that proper care can be undertaken to improve the system efficiency. The canal performance can be best evaluated by analyzing the four most important measures such as adequacy, efficiency, dependability and equity [13]. These measures provide a quantitative assessment of the overall system performance as well as the contributions to performance from the structural and management components of the system. A framework for the performance assessment of irrigation water management in heterogeneous irrigation schemes has been proposed by [7]. The irrigation water management was divided into three phases, like planning, operation and evaluation. Two types of allocation measures (productivity and equity) and five types of scheduling measures (adequacy, reliability, flexibility, sustainability and efficiency) together with the methodologies for their estimation for an irrigation scheme during different phases of irrigation water management were identified. Multi-temporal RS study was undertaken to compute distributary-wise performance indices, namely, adequacy, equity and water use efficiency of the Mahi right bank canal command in Gujarat, India [16]. The water availability was found to be more along main canals and branch canals and it was less in cropped areas and crop condition was poor towards the tail end areas.

Water delivery performance of the Menemen Left Bank Irrigation System in Lower Gediz irrigation system in the west of Turkey was studied by [24]. Performance at tertiary canal level was evaluated by using the adequacy, efficiency, dependability and equity indicators. Water delivery performance of the canals in each irrigation season rated worse for adequacy, dependability and equity than for efficiency. The results of the spatial and temporal dimensions of these indicators showed that factors causing this problem derive in part from physical structure, and in part from management. Key among these are inadequate water measurement and control at the head of the canals, canal capacity limitations, non-compliance with the rotation plan, and mismatch between the reservoir release plan and irrigation demand. Water delivery performance of a minor under the left main canal of Som-kagdar irrigation project was examined by [20]. Outlet-wise performance was evaluated using the adequacy, equity, dependability and relative water supply indicators, which showed poor performance of the system. The analysis of results of the spatial and temporal dimensions of these indicators showed that factors causing the problem are derived partly due to physical state of system and partly due to improper operation and management.

Hayrabolu Irrigation Scheme of the Thrace district in Turkey was evaluated by using the indicators, namely, agricultural, economic, water-use, physical and environmental performance [18]. Analyses of water-use performance showed that the water distribution was not tightly related to crop water demand. Economic performance indicators showed that the scheme had a serious problem about the collection of water fees. Physical performance, evaluated in terms of irrigation ratio and sustainability of irrigated land, were poor. Irrigation water delivery performance on the independent branch of Wushantou reservoir in Taiwan Chianan irrigation association was evaluated by [22] in which the lateral serves four tertiary canals. Adequacy, efficiency, equity and reliability of irrigation water delivery were assessed for four-rice crop growing seasons in 2005 and 2006. The results showed a very high irrigation efficiency of 94.5% over the two years while the values of adequacy indicators were 0.83 and 0.77 resulting to a fair and poor performance in 2005 and 2006, respectively. The values of reliability index of 0.22 and 0.24 were regarded as a very poor performance over the two years.

Water delivery performance of the Menemen Left Bank Irrigation system in Turkey at secondary and tertiary canal level for the irrigation seasons 2005–2007 was determined by [9]. At secondary canal level, water supply ratio was used, and at tertiary level, the indicators of adequacy, efficiency, dependability, and equity were used. The water supply ratios at the secondary canal level was found to be more than one, whereas the performance indicators at the tertiary level were found to be poor for each of the three years of the study, with efficiency rising to "fair" level only in 2005.

Similar study was conducted to evaluate the water distribution systems of Igomelo irrigation scheme in Tanzania at three levels such as head, middle and tail end of the scheme [11]. Irrigation performance indicators such as dependability, equity and adequacy of water supply, conveyance efficiency and structure condition index were used to evaluate the system. The system was found to be performing well. Hydraulic performance of secondary canal based on delivery performance ratio, adequacy based on relative water supply and variability in discharges using coefficient of variation was evaluated to study the effect of variability in discharges on equity of water distribution of outlets [21]. The delivery performance ratio of minor and head outlets indicated that head outlets draw more than their design discharge, whereas the middle and tail reach of the canal receive less than allocated discharge.

In reality because of poor operation and technical deficiencies most of water is ended at the head outlets and inequity exist at system level. The relative water supply of the first head outlet indicated that supply is more than demand, whereas at remaining outlets supply was less than demand. The variability in discharges along the outlet had increasing trend.

Distributary-wise adequacy and equity on Mehasana district command of Sabarmati Right bank main canal in North Gujarat of India was computed using multi temporal, multidate remote sensing (RS) data, which was processed using Erdas imagine software [19]. One of the notable findings of this study was that water availability was in excess along main canals and branch canals. In cropped area, it was less. It was also noted that the crop condition was poor towards the tail ends of the command area. Numerous techniques addressing the performance assessment of irrigation systems was studied in the past [10]. Most of them give the performance index, which is highly complicated. A system is said to be performing acceptable only if the supplier (irrigation system manager) as well as beneficiary (farmer) are satisfied. From the point of view of farmer, he is satisfied when the irrigation water available is dependable and also to meet his requirement and the supplier is satisfied when he is able to supply water in a socially acceptable manner. To have a comprehensive view from both the sides different parameters are used. From the farmer's side adequacy, dependability and deficiency can be taken whereas the indicators like efficiency, equity and wastage can be taken from the point of view of an irrigation system manager.

The Vadhavana tank-based medium irrigation project in Vadodara district and Panam multipurpose major irrigation project in Panchmahal district of Gujarat are faced with large spatiotemporal variation in irrigation water supply and demand. The canal supplies in both the command areas are mainly controlled in accordance with the reservoir storage availability, which in turn depends mainly on the monsoon rainfall in the catchments. The CCA of the Vadhavana command is only 4,612 ha whereas the CCA of the Panam canal command is around 411.16 km^2. Though the Panam irrigation project has an irrigation potential of 493.70 km^2, but so far a maximum of 32% of CCA and 37% of CCA has been brought under irrigation in *Kharif* and *Rabi* season, respectively.

This chapter presents the research results to evaluate the performance of irrigation water-delivery systems of the canal irrigation projects in Vadodara district and Panchmahal district of Gujarat.

7.2 METHODOLOGY

7.2.1 STUDY AREA

The study was undertaken within the periphery of Vadhavana tank-based medium irrigation project and Panam multipurpose major irrigation project in central Gujarat. The Vadhavana canal command is located near the Vadhvana village, which comes under the periphery of Vadodara district covering an area of 7,380 ha. It is situated in and around 22.17° N latitude and 73.48° E longitude. The command area gets its irrigation supply from the Vadhavana tank, which is fed by the Jojva-Vadhvana reservoir.

Similarly, the Panam canal command, which comes under the periphery of Panchmahal district, lies between latitude 22°27' to 22°48' N and longitude 73°24' to 73°32' E. The command area gets its irrigation supply from the Panam reservoir, impounded in the upstream of Panam dam, through a huge network of canals.

The major crops grown in the study area are paddy, bajara, jowar/ bajra, cotton, castor and wheat. *Kharif* and *Rabi* are the two major crop growing seasons in the command area. Area under different crops during two major crop-growing seasons is given in Table 7.1. The selected study area has a CCA of 4,612 ha under Vadhvana canal command and a CCA of 3,489 ha under the Panam canal command. Paddy is the major crop cultivated during *Kharif* season covering around 60 to 65% of CCA, whereas wheat is the major crop practiced during *Rabi* season covering about 40 to 45% of CCA.

7.2.2 DATA COLLECTION

Canal network map including canal geometry, canal flow, seasonal cropping pattern, geohydrological and climatological data of both the command areas were collected from respective irrigation authorities/State Department of Agriculture. Field data related to canal flow at certain locations in the said command areas were monitored by using current meter and flow probe. Rainfall data on fortnight basis during the period 2009–2011 was collected from the Irrigation Department, Vadhavana Tank site, Dabhoi, Vadodara; and daily rainfall data for the period 2010–2012 was collected from the Veganpur sub-divisional office, Panam irrigation project circle, Panchmahal, which

TABLE 7.1   Area (ha) Under Major Crops in the Study Region

Crop	Vadhvana canal command				Panam canal command			
	2009–10		2010–11		2010–11		2011–12	
	Kharif	*Rabi*	*Kharif*	*Rabi*	*Kharif*	*Rabi*	*Kharif*	*Rabi*
Bajra/ Jowar	37	–	30	–	–	220	–	101
Castor	–	–	–	–	–	97	–	297
Cotton	80	–	70	–	–	–	–	–
Maize	–	–	–	–	–	329	–	210
Paddy	2,920	–	2,798	–	442	–	632	–
Wheat	–	2,041	–	1,820	–	426	–	340
Other crops	257	897	323	2,189	14	378	73	333

was used to estimate effective rainfall by using USDA-SCS (US Department of Agriculture Soil Conservation Service) method.

Daily canal flow data of five main canals (Table 7.2) under the Vadhavana canal command for crop years 2009–10 and 2010–11 was collected from the Office of the Executive Engineer, Vadodara irrigation circle, Vadodara. Daily canal flow data of the three distributaries (26/R, 29/R and 31/R) under Panam irrigation project for the crop period 2010–11 and 2011–2012 were collected from the Veganpur sub-divisional office, Panam irrigation project circle, Panchmahal. These data were analyzed. Daily canal flow data and seasonal crop data were used to estimate the seasonal water supply and demand of both the command areas by using standard methodology.

7.2.3   COMPUTATION OF IRRIGATION WATER REQUIREMENT

In the present context, irrigation water is applied to crops based on some preset thumb rules of the Government or by adopting some adhoc norms on farmers' experience. Sometimes more water is applied to the crops with a view to combat the risk on crop failure, which result in inequitable distribution among different reaches of the canal. Hence, the adoption of any crop water management program mandatorily requires accurate prediction of crop-wise irrigation water requirement. The irrigation requirement of a crop can be determined as follows [23].

TABLE 7.2 Canal Data Under the Study Region

Study area	Canal name	Code	Length (km)	CCA (ha)
Vadhvana canal command	Bhimpura main canal	C1	3.24	431
	Vadhvana Boriad main canal	C2	5.61	824
	Dabhoi main canal	C3	7.06	2398
	Vasai Dangiwada main canal	C4	8.59	847
	Simliya canal	C5	1.21	112
Panam canal command	26/R distributary		5.9	728
	29/R distributary		10.7	1620
	31/R distributary		2.8	941

$$GIR = \frac{(ET_c - RF_{ef})}{\eta_a} + SPR - SMC - GWC \tag{1}$$

where, GIR = gross irrigation requirement in field (mm); ET_c = crop evapotranspiration (mm) which is equal to $(ET_o \times K_c)$; ET_0 = reference evapotranspiration (mm); K_c = crop coefficient; SPR = special purpose requirements (mm); SMC = soil moisture contribution; GWC = capillary contribution from groundwater; RF_{ef} = effective rainfall (mm), and η_a = field application efficiency (fraction).

The two components, SMC and GWC are not considered in this study due to unavailability of realistic information. Various methods of estimation of ET_0 are available in literature [1, 5]. Although the FAO Penman-Monteith method [5] has been recommended as the sole standard method, yet the Hargreaves method [8] is followed in this study due to unavailability of all necessary meteorological information needed for the FAO Penman-Monteith method. This method requires the extraterrestrial radiation and mean monthly maximum and minimum temperature data for ET_0 estimation, which is mathematically expressed as:

$$ET_0 = 0.0023 \times R_a \times (T_{av} + 17.8) \times (T_{max} - T_{min})^{0.5} \tag{2}$$

where, R_a = extraterrestrial solar radiation (mm/month), which is calculated from the information on latitude and day of the year (Allen et al., 1998); T_{av} = mean monthly air temperature $(T_{max} + T_{min})/2$ (°C); T_{max} = mean monthly maximum air temperature (°C); and T_{min} = mean monthly minimum air temperature (°C).

Crop coefficient (K_c) values for each crop were taken from the literature [1, 5]. The U.S. Department of Agriculture Soil Conservation Service (USDA-SCS) method [4] was used to determine the effective rainfall. The value of field application efficiency was taken as 32% for rice and 58% for non-rice crops [5]. Water requirements for special purpose consist of land preparation, nursery raising and transplanting of rice. Land preparation usually requires pre-sowing irrigation for easy plowing, disking and land smoothing operations. Water requirement for land preparation is considered as 70 mm for all crops other than rice, whereas rice requires 200 to 250 mm of water for nursery raising and transplanting [23]. The lower limit of 200 mm has been assumed as special purpose water requirements of rice.

7.2.4 COMPUTATION OF CANAL PERFORMANCE INDICES

Canal performance measures in terms of water delivery such as adequacy (PI_A), efficiency (PI_{EF}), dependability (PI_D) and equity (PI_E) were estimated by using the methodology adopted by Molden et al. [13]. These performance measures can provide the framework for assessing the system improvement alternatives. The performance indicators expressed in terms of measurable quantities are called state variables. The major state variables that determine water-delivery-system performance may be defined in terms of an amount of water Q, which may refer to rate, volume, frequency, or duration of water delivery. In the present study, focus has been given on volumes, which was estimated from the known values of rates and the duration of water delivery. At a point 'x' in the system and at time 't',

The $Q_D(x, t)$ is the actual amount of water delivered to the system at a point 'x' in time 't' and $Q_R(x, t)$ is the actual amount of water required for consumptive use downstream of the delivery point 'x'.

7.2.4.1 Adequacy (PI$_A$)

Adequacy is defined as the ability of the irrigation system to meet the crop water requirement. It is dependent on the water supply, the delivery schedules, capacity of hydraulic structures to deliver water as per the schedules, and the operation and maintenance of the hydraulic structures. Point performance function relative to adequacy (PI_A) is given by:

$$PI_A = \frac{1}{T}\sum_T\left(\frac{1}{R}\sum_R P_A\right), \text{ where:} \tag{3}$$

$$P_A = \frac{Q_D}{Q_R}, \text{ if } Q_D \leq Q_R \tag{4}$$

$$P_A = 1, \text{ otherwise} \tag{5}$$

where, P_A = point performance function relative to adequacy.

7.2.4.2 Efficiency (PI_{EF})

The relative amount of water lost in a reach due to canal seepage and overflow, termed as conveyance efficiency, is typically used to address the objective of efficiency in irrigation and water-delivery systems. The overuse, or loss of water not directly reflected in the concept of conveyance efficiency, is the delivery of a more than adequate supply of water to diversion points within the system. The excess water deliveries to farms promote conditions of waterlogging and salinity. Point Performance function relative to efficiency (PI_{EF}) is given by:

$$PI_{EF} = \frac{1}{T}\sum_T\left(\frac{1}{R}\sum_R P_{EF}\right), \text{ where:} \tag{6}$$

$$P_{EF} = \frac{Q_R}{Q_D}, \text{ for } Q_R \leq Q_D \tag{7}$$

$$P_{EF} = 1, \text{ otherwise} \tag{8}$$

where, P_{EF} = point performance function relative to efficiency.

7.2.4.3 Dependability (PI_D)

Dependability is a temporal uniformity of the ratio of the delivered amount of water to the required or scheduled amount. A system that performs in a consistent manner may be considered dependable. A system that dependably delivers an inadequate amount of water may be more desirable than one that delivers an adequate but unpredictable supply. A farmer can plan for a dependable delivery of an inadequate supply of water by planting less or growing different crops or adjusting other farming inputs, but he cannot plan easily, when the supply of water is unpredictable. Dependability (PI_D) is expressed by:

$$PI_D = \frac{1}{R}\sum\nolimits_R CV_T\left(\frac{Q_D}{Q_R}\right) \tag{9}$$

where, $CV_T\left(\dfrac{Q_D}{Q_R}\right)$ = temporal coefficient of variation (ratio of standard deviation to mean) of the ratio $\left(\dfrac{Q_D}{Q_R}\right)$ over the time period T.

7.2.4.4 Equity (PI_E)

Equity is defined as spatial uniformity of the ratio of the delivered amount of water to the required or scheduled amount, which indicates the delivery of a fair share of water to users throughout a system. A share of water represents a right to use a specified amount. The fair share of water may be based on a legal right for water, or may be set as a fixed proportion of a water supply, as is done in many rotational delivery schemes. Equity of water delivery is a very difficult objective to measure. Still, it is important to define measures relating to equity so that systems can be designed to deliver water in a judicial manner to the users served by the system. Performance measure relative to equity is given by,

$$PI_E = \frac{1}{T}\sum\nolimits_T CV_S\left(\frac{Q_D}{Q_R}\right) \tag{10}$$

where, $CV_S\left(\dfrac{Q_D}{Q_R}\right)$ – spatial coefficient of variation (ratio of standard devia-

tion to mean) of the ratio $\left(\dfrac{Q_D}{Q_R}\right)$ over the region S.

The standards for these indicators are furnished in Table 7.3.

7.3 RESULTS AND DISCUSSION

This chapter includes the presentation of results obtained from the analysis of secondary data such as canal flow, rainfall and cropping pattern and estimation of irrigation water requirement for deriving the canal performance measures of the selected canals under the Vadhavana irrigation project and Panam irrigation project, located in central Gujarat.

7.3.1 VADHAVANA IRRIGATION PROJECT

7.3.1.1 Computation of Gross Irrigation Requirements

Seasonal gross irrigation requirement (*GIR*) of crops in *Kharif* and *Rabi* seasons during 2009–10 and 2010–11 were estimated as shown in Figure 7.1. *Kharif* bajra and jowar, cotton and other crops required 152, 796 and 463 mm of irrigation water, respectively, whereas *Kharif* paddy needed 1,068 mm of irrigation water during 2009–10. Wheat and other *Rabi* crops required 866 mm of irrigation water during 2009–10. On the other hand, *Kharif* bajra and jowar, cotton and other crops needed 48, 679 and 343 mm irrigation water, respectively, whereas *Kharif* paddy demanded 790 mm of irrigation

TABLE 7.3 Performance Standards for Irrigation Systems

Performance indices	Performance class		
	Good	**Fair**	**Poor**
PI_A	0.90–1.0	0.80–0.89	< 0.80
PI_{EF}	0.85–1.0	0.70–0.84	< 0.70
PI_E	0–0.10	0.11–0.25	> 0.25
PI_D	0–0.10	0.11–0.20	> 0.20

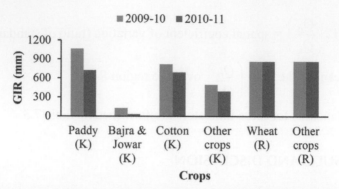

FIGURE 7.1 Gross irrigation requirement of crops under Vadhavana canal command: K = *Kharif*, R = *Rabi*.

water during 2010–11. Wheat and other *Rabi* crops needed 866 mm of irrigation water during 2010–11. Low values of irrigation requirements during 2010–11 were due to comparatively much rainfall during that year.

Total volume of irrigation water required by paddy, bajra and jowar, cotton and other crops during the *Kharif* season and wheat and other crops during the *Rabi* season of 2009–10 were 15.1, 0.37, 0.42, 0.71, 12.1 and 5.32 M.m³, respectively. Similarly, total volume of irrigation water required by paddy, bajra and jowar, cotton and other *Kharif* crops and wheat and other *Rabi* crops during 2010–11 were 11.82, 0.074, 0.27, 0.67, 10.8 and 11.93 M.m³, respectively.

7.3.1.2   Performance Indices of Vadhavana Canals

The total operation periods during the crop year 2009–10 for the canals C1, C2, C3, C4 and C5 (Table 7.2) were 132, 138, 143, 120 and 55 days, respectively. Similarly, the total operation periods during the crop year 2010–11 for the canals C1, C2, C3, C4 and C5 were 130, 141, 151, 132 and 68 days, respectively. This variation was mainly due to supply position in the Vadhavana tank.

7.3.1.2.1   Adequacy (PI$_A$)

The adequacy during the months September, December and March of the crop year 2009–10 was greater than 0.8, whereas other months possessed values from 0.45 to 0.75 (Figure 7.2). The average value of PI_A in Vadhavana canal was 0.69 indicating a 'poor' performance as far as the adequacy is concerned. The low values of PI_A were mainly due to uneven distribution of rainfall in the season. Similarly, the adequacy during the months December, January and February for the crop year 2010–11 was greater than 0.9 whereas other months possessed values between 0.4 and 0.8. The average value of PI_A in Vadhavana canal was 0.74 indicating a 'poor' performance as far as the adequacy is concerned. The low values of PI_A were mainly due to uneven distribution of rainfall in the season.

7.3.1.2.2 Dependability (PI_D)

The performance index relative to dependability was done based on their respective turns for year 2009–10 in order to assess the dependency of the farmers on canal water over the year within the Vadhavana canal command. The spatial variations in terms of PI_D in the Vadhavana canal for the year 2009–10 were worked out and are presented in Figure 7.3. During the year 2009–10, the calculated values of PI_D were above 0.2, indicating 'poor' performance in terms of dependability of the system. It indicates that the water deliveries were not uniform over time in accordance to demand, thus poor timeliness. Also, sometimes closure of irrigation canal in response to high rainfall during the months of July and August might have resulted in high PI_D values. The average value of PI_D for the year 2009–10 was 0.3. The average value of PI_D for the year 2010–11 was 0.49, which fall above the upper

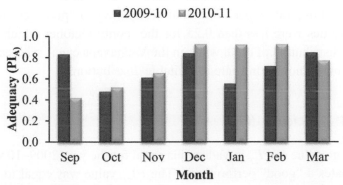

FIGURE 7.2 Adequacy index (PI_A) of Vadhavana canals.

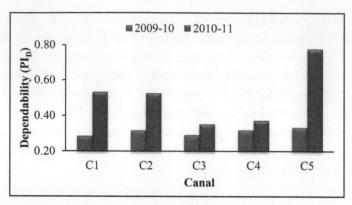

FIGURE 7.3   Dependability index (PI_D) of Vadhavana canals.

limits accounting to 'poor' performance. This indicates that the water deliveries were not in accordance with the demand over the area. Also, sometimes, closure of irrigation canal in response to high rainfall during the months of July, August and September might have resulted in high PI_D values.

7.3.1.2.3   Equity (PI_E)

The average value PI_E in Vadhavana canal for the year 2009–10 was 0.31, which falls above the upper limits accounting to "poor" performance. The PI_E values were less than 0.25 for the months October, November and December that indicates equitable distribution of canal water in the Vadhavana canal, whereas higher values in other months indicates inequitable distribution. The monthly PI_E values for the crop year 2009–10 are presented in Figure 7.4. The average values PI_E in Vadhavana canal for the year 2010–11 were 0.44, which fall above the upper limits that accounting to "poor" performance. The PI_E values were less than 0.25 for the month October which indicate equitable distribution of canal water in the Vadhavana canal, whereas higher values in other months indicate inequitable distribution.

7.3.1.2.4   Efficiency (PI_EF)

The average value of PI_{EF} in Vadhavana canal for the year 2009–10 was 0.95 that indicates a "good" performance. The PI_{EF} value was equal to 1.0 during October and January. It indicates that the system was efficient to meet

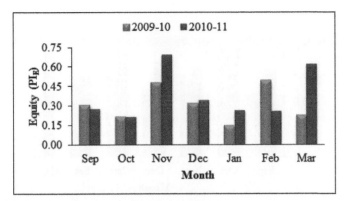

FIGURE 7.4 Equity index (PI_E) of Vadhavana canals.

the requirements of the region. The monthly PI_{EF} values during the crop year 2009–10 are presented in Figure 7.5. The average values efficiency in Vadhavana canal for the year 2010–11 was 0.81 that indicates a "fair" performance. The lower value of PI_{EF} indicates that the system was not efficient to meet the requirements of the region. The PI_{EF} value was 1.0 during the month of October.

7.3.2 PANAM IRRIGATION PROJECT

7.3.2.1 Computation of Gross Irrigation Requirement

Seasonal gross irrigation requirement (GIR) of different crops grown in the command area during 2010–11 and 2011–12 were estimated by Hargreaves equation [8] considering mean monthly maximum and minimum air temperature and stage-wise crop coefficients. Seasonal GIR of crops was found to vary widely ranging between 137.9 and 996.5 mm over years (Figure 7.6). The GIR of crops during the crop year 2011–12 was comparatively low due to relatively more rainfall during the year that continued up to the end of September.

7.3.2.2 Performance Indices of Panam Distributaries

The performance measures of the three distributaries under the study area for the crop years 2010–11 and 2011–12 were assessed as follows.

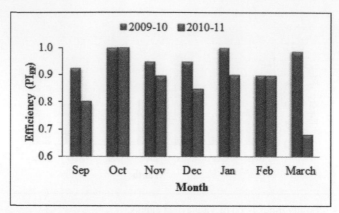

FIGURE 7.5   Efficiency index (PI_{EF}) of Vadhavana canals.

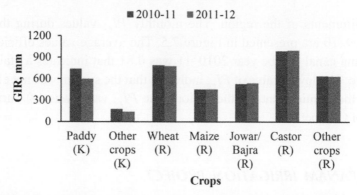

FIGURE 7.6   Gross irrigation requirement of crops under Panam canal command: K = *Kharif*, R = *Rabi*.

7.3.2.2.1   *Adequacy (PI$_A$)*

The value of PI_A during 2010–11 was found to vary between 0.57 and 1.0, average being 0.75 that indicates 'poor' performance of the system. Similarly, the value of PI_A was found to vary between 0.61 and 1.00 during the crop year 2011–12 (Figure 7.7). The average value for all the distributaries was 0.81 indicating a 'fair' performance.

7.3.2.2.2   *Dependability (PI$_D$)*

The performance index relative to dependability was done based on their respective turns for year 2010–11 in order to assess the dependency of the

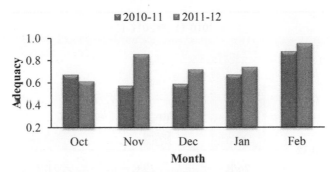

FIGURE 7.7 Adequacy index (PIA) of Panam distributaries

farmers on canal water over the year within the study area. During the year 2010–11, the calculated values of PI_D were above 0.2, indicating 'poor' performance of the system. The average value of PI_D for the year 2010–11 was 0.37. Similarly, the average value of PI_D for the year 2011–12 was 0.28 (Figure 7.8), which fall above the upper limits accounting to 'poor' performance. This indicates that the water deliveries were not in accordance with the demand over the area, thus poor timeliness. Also, sometimes, closure of irrigation canal in response to high rainfall during the months of July, August and September might have resulted in high PI_D values.

7.3.2.2.3 Equity (PI$_E$)

The average value of PI_E in the distributaries for the year 2010–11 was 0.36, which falls above the upper limits accounting to "poor" performance. The values were less than 0.25 for the months September, February and March, that indicates equitable distribution of canal water in the distributaries, whereas higher values in other months indicates inequitable distribution (Figure 7.9). Similarly, the average values of PI_E in the distributaries for the year 2011–12 were 0.29, which fall above the upper limits accounting to "poor" performance. The PI_E values were less than 0.25 for the months November, February and March which indicate equitable distribution of canal water in the distributaries, whereas higher values in other months indicate inequitable distribution.

7.3.2.2.4 Efficiency (PI$_{EF}$)

The average value of PI_{EF} in the distributaries for the year 2010–11 was 0.79 that indicates a "poor" performance. The value of PI_{EF} was equal to

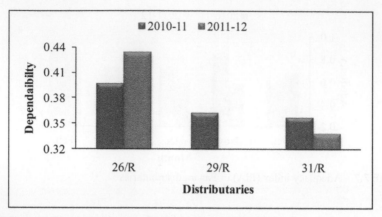

FIGURE 7.8   Dependability index (PI_D) of Panam distributaries.

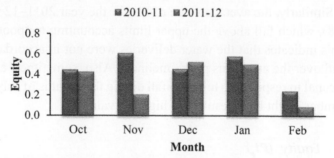

FIGURE 7.9   Equity index (PI_E) of Panam distributaries.

1.0 during November and December, which indicates that the system was efficient to meet the requirements of the region (Figure 7.10). Similarly, the average value of efficiency in the distributaries for the year 2011–12 was 0.80 that indicates a "fair" performance.

The average values of the performance indices like adequacy, dependability, equity and efficiency of the Vadhavana canals and Panam distributaries under consideration during the study period 2009–10 to 2011–12 are summarized in Table 7.4. It can be concluded that the overall performance of the selected command areas are 'poor', which needs to be improved by adopting suitable remedial measures. Judicial management of irrigation water needs to be done through participatory irrigation management, with the involvement of water users in various aspects of irrigation water management. Proper irrigation scheduling also needs to be followed for

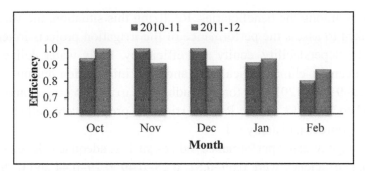

FIGURE 7.10   Efficiency index (PI_{EF}) of Panam distributaries.

TABLE 7.4   Performance Indices of Irrigation Water Delivery Systems in Gujarat

Irrigation project	Year	Adequacy (PI_A)		Dependability (PI_D)		Equity (PI_E)		Efficiency (PI_{EF})	
Vadhavana	2009–10	0.69	poor	0.30	poor	0.31	poor	0.95	good
	2010–11	0.74	poor	0.49	poor	0.44	poor	0.81	fair
Panam	2010–11	0.75	poor	0.37	poor	0.36	poor	0.79	poor
	2011–12	0.81	fair	0.28	poor	0.29	poor	0.80	fair

achieving this goal. Improving water delivery system requires upgrading framers' confidence to operate the system in addition to the good physical efficiency and accurate crop water requirement estimation. Since farmers are the real stakeholders, they have to step forward to look after their interest so that they get water from the system according to the predetermined time and space for proper crop planning.

7.4   SUMMARY

Rising demand of water for irrigation has brought new challenges to the water resources planners and managers for sustainable livelihood security. Spatial temporal variation in rainfall as well as short duration rainy season has resulted in severe agricultural water shortage in many parts of India. Panam irrigation project in Panchmahal district and Vadhavana irrigation project in Vadodara district of central Gujarat, India, are such irrigation projects, where large mismatch in irrigation water supply and demand prevails mainly due to improper irrigation scheduling and lack of awareness among the beneficiaries. Realizing this situation, the study was undertaken to assess the performance of the irrigation projects in terms of adequacy, dependability, equity and efficiency. Daily water deliveries of different canals, climatological data and crop information during the crop period 2009–10 to 2010–11 for the Vadhavana irrigation project and during 2010–11 to 2011–12 for the Panam irrigation project were collected from potential sources and analyzed.

The range of canal performance indices such as adequacy, dependability, equity and efficiency were 0.69–0.81, 0.28–0.49, 0.29–0.44 and 0.79–0.95, respectively, which indicates that the canals are performing under "poor" category. It is, therefore, highly essential to adopt suitable remedial measures for improving the performance of the system, which will be possible by judicial management of irrigation water through participatory irrigation management, with the involvement of water users in various aspects of irrigation water management.

7.5 CONCLUSIONS

Temporal variation in rainfall and lack of adequate management of the canal system has resulted in large gap in supply and demand of irrigation water over years that call for proper irrigation scheduling. The overall performance of the canals under study was found to be under 'poor' category. It is therefore, necessary to enhance the performance of the system by adopting suitable remedial measures. The adequacy, dependability and equity indices are required to be improved to more than 0.80, less than 0.20 and less than 0.25, respectively for achieving better performance of the canal system.

ACKNOWLEDGMENTS

The authors express their sincere thanks to: the Panam sub-divisional office, Veganpur; Irrigation department, Vadhavana tank site, Dabhoi, Vadodara; Office of the Executive Engineer, Vadodara irrigation circle; Vadodara and Panam irrigation project circle, Godhra for providing all necessary data and information to carry out this study. The necessary support received from the Anand Agricultural University to execute the study is also gratefully acknowledged.

KEYWORDS

- **Adequacy**
- **Canal system**
- **Crop planning**
- **Dependability**
- **Effective rainfall**
- **Efficiency**
- **Equity**
- **Gross irrigation requirement**
- **Irrigation scheduling**
- **Irrigation water**
- **Kharif**
- **Panam irrigation project**
- **Participatory irrigation management**
- **Performance class**
- **Performance evaluation**
- **Performance indices**
- **Rabi**
- **Reference evapotranspiration**
- **Reliability**
- **Salinity**
- **Sustainability**
- **Vadhavana irrigation project**
- **Water delivery system**
- **Water measurement**
- **Water use efficiency**
- **Water users**
- **Waterlogging**

REFERENCES

1. Allen, R. G., Pereira, L. S., Raes, D., and Smith, M. (1998). *Guidelines for Computing Crop Water Requirements*. Irrig. Drain. Paper 56, FAO: Rome, Italy.

2. Arshad, M., Ahmad, N., and Usman, M. (2009). Simulating seepage from branch canal under crop, land and water relationships. *Intl. J. Agri. Biol.*, 11(5), 529–534.

3. Bashir, A. C. and Aftab, R. (1997). Water: an economical good for sustainable irrigated agriculture of Pakistan. In: *Proc. Int. Symposium: Water for 21st Century – Demand, Supply, Development and Socio-environment Issues.* Center of Excellence in Water Resour. Eng. (CEWRE), Lahore, Pakistan.

4. Dastane, N. G. (1977). *Effective Rainfall in Irrigated Agriculture.* Irrig. Drain. Paper 25, FAO: Rome, Italy.

5. Doorenbos, J. and Pruitt, W. O. (1977). *Guidelines for Predicting Crop Water Requirements.* Irrig. Drain. Paper 24, FAO: Rome, Italy.

6. Gidoen, C., Saeed, A. B., and Mohamed, H. I. (2007). Evaluation of hydraulic performance of major canals in the Rahad agricultural scheme. *J. Sci. Tech. Sudan Univ. Sci. Tech.*

7. Gorantiwar, S. D. and Smout, I. K. (2005). Performance assessment of irrigation water management of heterogeneous irrigation schemes: 1. A framework for evaluation. *Irrig. Drain. Sys.,* 19(1), 1–36.

8. Hargreaves, G. H. and Samani, Z. A. (1985). Reference crop evapotranspiration from temperature. *Appl. Eng. Agri.*, *ASABE*, 1, 96–99.

9. Korkmaz, N., Avci, M., Unal, H. B., Asik, S., and Gunduz. M. (2009). Evaluation of the water delivery performance of the Menemen left bank irrigation system using variables measured on-site. *J. Irrig. Drain. Eng., ASCE,* 135(5), 633–642.

10. Mattamana, B. M., Varghese, D., and Paul, K. (2013). Irrigation system assessment-farmer's and manager's view. *Intl. J. Eng. Sci. Innov. Tech.*, 2(2), 145–159.

11. Mchelle, A. R., Tarimo, A. R., and Nganga, I. (2010). Evaluation of water distribution systems at Igomelo farmer-managed irrigation scheme in Tanzania. In: *2nd RUFORUM Biennial Meeting* 20–24 September 2010, Entebbe, Uganda, pp. 1405–1407.

12. Mishra, A., Anand, A., Singh, R., and Raghuvanshi, N. S. (2001). Hydraulic modeling of Kangsabati main canal for performance assessment. *J. Irrig. Drain. Eng., ASCE,* 127(1), 27–34.

13. Molden, D. J. and Gates, T. K. (1990). Performance measures for evaluation of irrigation-water-delivery systems. *J. Irrig. Drain. Eng., ASCE,* 116(6), 804–823.

14. Planning Commission (2002). *Report of the steering committee on irrigation for the 10th 5 year plan (2002–2007),* Govt. of India, New Delhi (India).

15. Rao, B. K. and Rajput, T. B. S. (2006). Mismatch between supplies and demands of canal water in a major distributary command area of the Nagarjunasagar left canal. *J. Agr. Eng. ISAE,* 43(3), 47–51.

16. Ray, S. S., Dadhwal, V. K., and Navalgund, R. R. (2002). Performance evaluation of an irrigation command area using remote sensing: a case study of Mahi command, Gujarat, India. *Agr. Water Manage.,* 56(2), 81–91.

17. Sarki, A., Memon, S. Q., and Leghari, M. (2008). Comparison of different methods for computing seepage losses in an earthen watercourse. *Agricultura Tropica Et Subtropica*, 41(4), 197–205.

18. Şener, M., Yüksel, A. N., and Konukcu, F. (2007). Evaluation of Hayrabolu irrigation scheme in Turkey using comparative performance indicators. *J. Tekirdag Agr. Fac.,* 4(1), 43–54.

19. Shah, S. and Dalwadi, H. J. (2012). Diagnostic analysis of an irrigation command by satellite remote sensing: Case study of Mehasana district command of Sabarmati right bank main canal, North Gujarat, India. *Intl. J. Civil Str. Eng.,* 3(1), 178–185.

20. Singh, S., Kothari, M., and Purohit, R. C. (2006). Performance evaluation of water delivery system for command area of Amarpura minor of Som-Kagdar irrigation project, Rajasthan. *J. Agr. Eng. ISAE,* 43(1), 13–21.
21. Tariq, J. A. and Kakar, M. J. (2010). Effect of variability of discharges on equity of water distribution among outlets. *Sarhad J. Agri.,* 26(1), 51–59.
22. Tongongar, B., Kan, C.-E., and Chen, J. H. (2008). Can efficiency reliability in irrigation system? *Ame.-Eur. J. Agr. Env. Sci.,* 3(2), 269–278.
23. Tyagi, N. K. (1980). *Crop planning and water resources management in salt affected soils – A systems approach.* PhD dissertation, Jawaharlal Nehru Technological University, Hyderabad (India).
24. Unal, H. B., Asik, S., Avci, M., Yasar, S., and Akkuzu, E. (2004). Performance of water delivery system at tertiary canal level: A case study of the Menemen left bank irrigation system, Gediz basin, Turkey. *Agr. Water Manage.,* 65(3), 155–171.

CHAPTER 8

EFFECTIVE IRRIGATION MANAGEMENT USING DECISION SUPPORT SYSTEM: WHEAT, PEANUT AND MAIZE

S. K. BEHERA[1] and R. K. PANDA[2]

[1]Senior Scientist (SWCE), All India Coordinated Research Project on Dry Land Agriculture (OUAT), AT/PO: Phulbani, Dist.: Kandhamal, Odisha-762001, India, Mobile: +91-9437619398, E-mail: subrat_behera@rediffmail.com

[2]Dean (R&D), Indian Institute of Technology Bhubaneswar, Samantapuri, Bhubaneswar, Odisha-751013, India, E-mail: rkpanda@iitbbs.ac.in

CONTENTS

8.1   INTRODUCTION

Dwindling water resources and increasing food requirements require greater water use efficiency (WUE), both in rainfed and in irrigated agriculture. Regulated deficit irrigation (RDI) provides a means of reducing water consumption while minimizing adverse effects on yield. Models can play an important role in developing practical recommendations for optimizing crop production under conditions of scarce water supply. Scarce water resources and growing competition for water will reduce its availability for irrigation.

At the same time, the need to meet the growing demand for food will require increased crop production from less water [4, 11, 14].

Achieving greater WUE will be a primary challenge for the near future and will include the employment of techniques and practices that deliver more accurate supply of water to crops [1, 2, 5, 12]. In this context, deficit irrigation can play an important role in increasing WUE. In some cases, periods of reduced growth may trigger physiological processes that actually increase yield and/or income. Such processes include flower-induction in the case of cotton, increased root development exploring deeper soil layers, early ripening of grains, and improved quality and flavor of fruits. However, stress applied during reproductive growth can affect fruit or grain set, resulting in decreased yields. The effects of stress on yields are complex and may differ with species, cultivar, and growth stage; and they have been the subject of many studies. Extensive field research is required to better understand the physical and biological processes that control crop responses to moisture stress [18, 20–22, 26].

Models that simulate crop growth and water flow in the rootzone can be a powerful tool for extrapolating findings and conclusions from field studies to conditions not tested, allowing predictions for deficit irrigation scheduling under various conditions of water supply and of soil and crop management. Furthermore, the use of models may be important to standardize research procedures in such coordinated research programs and thus facilitate meaningful comparisons between studies carried out in different locations and countries [3, 16, 18, 19, 23].

Simulation provides a tool to evaluate the water management options over long climatic records to address issues such as shifting to alternate crops, reducing water application and evaluation of adequacy of irrigation water supply for specific crops. Crop growth simulation models with resource and management inputs are being used particularly by irrigation engineers. These models have permitted the analysis of resource constraints. The relationship between crop production and the amount of irrigation water applied to the crop is important to engineers, agronomists, economists and water resources planners. This importance is currently accentuated due to competition among the users and declining ground water reserves [6, 9, 24].

In the present economic environment, there is a need for computerized tools to help farmers, consultants, planners and policy makers in their decision making process. With restricted access to economical and biophysical resources and due to the likelihood of environmental degradation, it is critical to make economically and environmentally correct decisions. Using computerized decision support systems, various scenarios can be analyzed and decisions can be made as to which scenario would be the best for a desired outcome. Today's fast computing system gives us the opportunity to examine many different scenarios without actually having to apply these scenarios to 'real life' conditions, until the best one has been identified [7, 8, 10, 25, 27].

This chapter discusses use of *Decision Support System for Agrotechnology Transfer* (DSSAT) for simulation modeling of effective irrigation management in wheat, maize and peanut under Indian conditions.

8.2 DECISION SUPPORT SYSTEM FOR AGROTECHNOLOGY TRANSFER: CROP GROWTH SIMULATION MODELS

The *Decision Support System for Agrotechnology Transfer* (DSSAT) software has been developed under the *International Benchmark Sites Network for Agrotechnology Transfer* (IBSNAT) project. IBSNAT developed the decision support software, DSSAT v3.5, which enables its users to match the biological requirements of crops to the physical characteristics of land so that objectives specified by the user, may be satisfied. The decision support software consists of:

   a. a Data Base Management System (DBMS) to enter, store and retrieve the data needed for validation and use by the crop models for solving problems;
   b. a set of validated crop models for simulating processes and outcomes of genotype by environment interactions; and
   c. an application program for analyzing and displaying outcomes of long-term simulated agronomic experiments.

DSSAT is a collection of computer programs integrated into a single software package in order to facilitate the application of crop simulation models in research and decision making. The resulting system is much more flexible than any other existing crop growth simulation models and has more functionality for data base manipulation and model application. The DSSAT was developed to allow users to interactively select any of the functions without knowing where the programs are or how they are communicating.

8.2.1   INPUT FILES

Input files of DSSAT and crop growth simulation models include experimental data file, weather data file, soil data file and genotype data file.

8.2.1.1   Experimental Data File

The experimental file was developed to allow great flexibility in retrieving data needed to simulate various experiments from different locations and different years. The file contains the experiment code and name, the treatment combinations, and details of the experimental conditions (field characteristics, soil analysis data, initial soil water and inorganic nitrogen conditions, seedbed preparation and planting geometries, irrigation and water management, fertilizer management, organic residue applications, chemical applications, tillage operations, environmental modifications, harvest management), and simulation controls. The experiment code uses the same convention as the file naming system to provide information on institute, site, planting year, experiment number, and crop. The file can also contain the names of the people supplying the data set and information on the plot sizes, etc., used in the experiment. It may also contain any incidents that occurred during the course of the experiment that may affect the interpretation of the data.

8.2.1.2   Weather Data File

Weather file contains all the available weather data. Daily weather data are required and must be available for the duration of the growing season, beginning with the day of planting and ending at crop maturity. Ideally, the weather file should contain data collected before planting to post-maturity. This would allow a simulation to be started before planting, thus providing an estimate of soil conditions at planting time. Weather data much prior to planting date would also allow users to select alternate planting dates, and simulate planting decisions based on weather and soil conditions. It is not necessary to have data for all variables, but the minimum data required for DSSAT v3.5 crop models are: solar radiation, minimum and maximum air temperature and rainfall. The standard format for variables should be followed. In DSSAT, this file is independent of crop type.

8.2.1.3   Soil Data File

The soil data file contains data on the soil profile properties. The file generally contains information that is available for the soil at a particular experimental site. This file contains soil identifier, information on soil texture and depth and the country, geographic data together with taxonomic information, information on soil properties that don't vary with depth, data on the first layer and so on. The number of layers in this file and the thickness of each layer must be consistent with the initial conditions. The file may contain properties for several soils of the same classification, provided each soil has its own code number.

8.2.1.4   Genotype Data File

Genotype data file contains genetic coefficient data namely: variety number and the different genetic coefficient, which describe specific cultivar characteristics of each crop. Three files are suggested for dealing with the morphological and physiological characteristics of a particular genotype: File for specific species (crop) characteristics, file for the "ecotype" characteristics within a species and file for the specific cultivar characteristics within an ecotype grouping. These files would contain all genotype specific inputs required for simulation. The use of at least one genotype file is highly recommended. For such a file, a standard format is recommended with each line beginning with a cultivar identification code.

8.2.2   OUTPUT FILES

Output files contains the overview of input conditions and crop performance, summery of soil characteristics and cultivar coefficient, crop and soil status at the main development stages, temporal distribution of simulated crop variables with time, simulated soil water with time, and selected harvest components and development duration for management strategy analysis. The output files are temporary information transfer files, created during simulation, and they are overwritten when a new simulation session is started.

8.2.3   GOVERNING EQUATIONS IN DSSAT GROUP OF MODELS

DSSAT, crop growth simulation models, was used for three major crops. Wheat and maize crops are under CERES model and groundnut crop under PNUTGRO model in DSSAT models. So the governing equations, which are used in DSSAT model for these three major crops, are discussed in this section.

8.2.3.1   Soil Water Balance

The soil water balance is calculated in DSSAT models in order to evaluate the possible yield reduction caused by soil and plant water deficits. The model evaluates the soil water balance of a cropped field or a fallow land using the following equation:

$$S = P + I - ET - R - D \tag{1}$$

where, S = the quantity of resultant soil water (storage); P = precipitation; I = irrigation; ET = evapotranspiration from soil and plants; R = runoff; and D = drainage from the profile.

Water content in any soil layer can decrease by soil evaporation, plant transpiration, root absorption or flow to an adjacent layer. The values of drained upper limit (field capacity) and drained lower limit (wilting point) are quite important in situations where the water input supply is marginal. The values of field capacity and wilting point should be estimated in the field as the traditional laboratory measured wilting point and field capacity water contents have frequently proved inaccurate for establishing field limits of water availability.

8.2.3.2   Infiltration and Runoff

Infiltration of water into the soil is calculated as the difference between precipitation or irrigation and runoff. Runoff is calculated using USDA-SCS procedure but with a small modification. The SCS technique considers the wetness of the soil, calculated from the previous rainfall amount, as an additional variable in determining runoff amount. The modified technique for layered soils replaces the wetness of the soil in the layers near the surface for the antecedent rainfall condition.

8.2.3.3 Drainage

Since the plants can take up water while drainage is occurring, therefore, the drained upper limit soil water content is not always the appropriate limit of soil water availability. Many productive agriculture soils drain quite slowly and may thus provide an appreciable quantity of water to plants before drainage practically stops. In DSSAT v3.5, drainage rates are calculated using an empirical relation that evaluates field drainage reasonably well. The drainage formula assumes fixed saturated volumetric water content, θ_0 and fixed upper limit water content, θ_u. Thus drainage takes place when the water content, θ_t, at any time, t, after field saturation is between θ_0 and θ_u. The equation used is:

$$\theta_t = (\theta_0 - \theta_u)\exp(-K_d t) + \theta_u \qquad (2)$$

where, θ_t = Water content, at any time, t; θ_0 = Saturated volumetric water content; θ_u = Upper limit water content; K_d = Fraction of excess water drained per day; and t = time.

The value of K_d is assumed to be constant for the whole soil profile because, in many soils, the most limiting layer to water flow dominates the drainage rate from all parts of the soil profile. A problem with in Eq. (2) for drainage evaluation in the field is that soils seldom reach saturation and it becomes difficult to determine an initial value for t. Ritchie presented a method for drainage of water above the drained upper limit, $\theta_u(L)$. The method was modified to account for restricting layers within the soil profile. The original soil water model assumed that water in access of $\theta_u(L)$ would drain out of the profile at an exponential rate defined by:

$$\frac{d\theta_t(L)}{dt} = -K_d(\theta_t(L) - \theta_u(L))\exp(-K_d t) \qquad (3)$$

where, $\theta_u(L)$ = volumetric water content in layer L; $\theta_t(L)$ = volumetric water content at time t in layer L.

Thus, drainage D_t, was computed by:

$$D_t = \sum_{i=1}^{N} -K_d(\theta_t(L) - \theta_u(L))z(L) \qquad (4)$$

where, N = number of soil layers; z (L) = depth of layer, mm.

In the model, constant drainage throughout a day is assumed and the value of K_d represents the fraction of water between θ_u and θ_t that drains in one day. For Eq. 4 to be used, θ_t must be greater than θ_u. The value of θ_t at any depth is updated daily to account for any infiltration, water flow or root absorption.

8.2.3.4 Root Water Absorption

The model calculates root water absorption using an approach in which the soil or root resistance determines the flow rate of water into roots. The soil limited water absorption rate, q_r considers radial flow to single roots and is expressed as:

$$q_r = \frac{4\pi K(\theta)(\psi_r - \psi_s)}{\ln(c^2/r^2)} \tag{5}$$

where, q_r = soil limited root water absorption rate, mm³ water/mm of root/d; $K(\theta)$ = soil hydraulic conductivity, mm/d; ψ_r= water potential at the root surface, mm; ψ_s= bulk soil water potential, mm; r = root radius, mm; and c = radius of the cylinder of the soil through which water is moving, mm.

The hydraulic conductivity $K(\theta)$ is approximated based on the assumption that all soils have a constant conductivity of 5×10^{-5} mm per day at the lower limit of water available to the plant (θ_{LL}). The relationship is given below:

$$K(\theta) = 5*10^{-5} \exp\left[C_T\left(\theta_t - \theta_{LL}\right)\right] \tag{5a}$$

where, C_T is texture dependent coefficient that is approximated from θ_{LL} and can be expressed as:

$$C_T = 100\left(1.2 - 2.5\theta_{LL}\right) \tag{6}$$

The sum of the maximum root absorption from each soil depth gives the maximum possible uptake from the profile. If the maximum uptake exceeds the maximum calculated transpiration rate, the maximum root water absorption rates calculated for each depth are reduced proportionally so that the uptake becomes equal to transpiration rate. If the maximum uptake is less than the maximum transpiration, transpiration rate is set equal to the maximum absorption rate.

8.2.4 DRY MATTER PRODUCTION

In DSSAT models, the potential dry matter production is a linear function of intercepted photo synthetically active radiation (PAR). The percentage of incoming PAR intercepted by the canopy is an exponential function of leaf area index (LAI). The actual rate of dry matter production is usually less than the potential rate due to the effects of non-optimal temperature or water stress. A weighted daytime temperature is calculated for the minimum and maximum temperature for the use in biomass evaluation. The optimum daytime temperature is considered as 18°C to 20°C. Water stress reduces dry matter production rates below the potential, whenever crop extraction of soil water falls below the potential transpiration rate calculated for the crop.

8.2.5 LEAF AREA INDEX (LAI)

Plant leaf area has an important influence on light interception and dry matter production. The rate of leaf area expansion is a component of plant growth that is quite sensitive to environmental stresses. For example, leaf growth is more sensitive to plant water deficits than photosynthesis. Cool temperature or moderate drought stresses reduce the expansion growth more than photosynthesis is reduced, causing increase in specific leaf weight and increasing the proportion of assimilate partitioned to the roots. DSSAT models accounts for these plant responses by using separate relationships to calculate the influence of temperatures water deficits on photosynthesis and leaf growth.

8.3 CASE STUDY: WHEAT

Data availability and adequacy determined the selection of the case studies from India as being appropriate for analyzing the suitability of the DSSAT model in deficit irrigation scheduling.

8.3.1　MATERIALS AND METHODS

8.3.1.1　Experimental Site

Field experiments were conducted at the experimental farm of the Agricultural and Food Engineering Department, Indian Institute of Technology, Kharagpur, India (22°19' N latitude and 87°19' E latitude). The local climate is sub-humid and sub-tropical with an average rainfall of 1200 mm concentrated over the months of June to September. The soil at the experimental site is an acid lateritic sandy loam. Wheat is usually a 100–110 days cereal crop in this region and suits the prevailing climate in the winter season (December–March). Experiments were conducted during three consecutive years between 1995 and 1998.

8.3.1.2　Field Layout and Experimental Details

The experimental area was divided into 20 plots of 5 m × 4 m size maintaining a buffer of 1 m between adjacent plots. A seed rate of 100 kg/ha was used. The seed was sown at a row spacing of 20 cm and a plant spacing of 5 cm and the depth of sowing was 5 cm during all the three experiments. The three experiments were in fact three years of trial such as: experiment 1 (1995–96), experiment 2 (1996–97) and experiment 3 (1997–98).

8.3.1.3　Irrigation Treatments

The irrigation treatments consisted of irrigation scheduling based on maximum allowable depletion (MAD) of available soil water (ASW) criteria. These were:

- T_1 = 10% maximum allowable depletion (MAD) of available soil water (ASW)
- T_2 = 30% MAD of ASW
- T_3 = 45% MAD of ASW
- T_4 = 60% MAD of ASW
- T_5 = 75% MAD of ASW

8.3.1.4 Irrigation Scheduling

Irrigation scheduling was based on the percentage depletion of available soil water in the root zone. The available soil water was taken as the difference between root zone water storage at field capacity and permanent wilting point. The maximum allowable depletion of available soil water was fixed at 10, 30, 45, 60 and 75%. Using the data of soil moisture measured by neutron probe and gravimetric measurements, the percentage depletion of available soil water in the effective root zone was estimated [16] as below:

$$Depletion\ (\%)\ =\ 100*\frac{1}{n}\sum_1^n \frac{FC_i - \theta_i}{FC_i - WP} \tag{7}$$

where, n is the number of sub-divisions of the effective rooting depth used in the soil moisture sampling, FC_i is the soil moisture at field capacity for i^{th} layer, θ_i is the soil moisture in i^{th} layer and WP is the soil moisture at permanent wilting point.

The amount of water applied after the attainment of predefined MAD was calculated as:

$$V_d = \frac{MAD\ (\%)*(FC - WP)*Rz*A}{100} \tag{8}$$

where, V_d is the volume of irrigation water, R_z is the effective rooting depth and A is the surface area of the plot. The surface area of each plot was 20 m². Each plot was made into a small basin, furrowed and watered individually. Measured amounts of water were applied to the furrows using a hosepipe and water meters.

8.3.2 DATA COLLECTION

For modeling the water balance and crop response to deficit irrigation, it was necessary to collect data relating to weather variables, profile soil moisture content and the growth attributes of the crop.

8.3.2.1   Weather Data

Daily values of the weather variables (solar radiation, maximum and minimum temperature, maximum and minimum relative humidity, wind speed and precipitation) for the experimental period were obtained from an automatic weather station installed near the experimental crop field.

8.3.2.2   Soil Profile Moisture Data

In order to assess the change in soil water status, soil moisture was measured in 0–15, 15–30, 30–45, 45–60, 60–90 and 90–120 cm soil profiles. The moisture content of topsoil layer (0–15 cm) was measured gravimetrically and that of the lower layers were measured using a neutron probe. Moisture measurements were done at 2–3 days interval.

8.3.2.3   Crop Data

Crop parameters were measured during different stages of growth. The crop data included: planting date, date of emergence, 20% cover date, full cover date, maturity date, harvest date, maximum rooting date, crop coefficient at full cover, planting depth and maximum root depth. The data on grain yield, above ground dry matter yield and leaf area index were recorded at different stages of crop growth during each crop experiment. The field water use efficiency (expressed as grain yield per unit cropped area per unit water applied to the field) was estimated for each treatment in each experiment.

8.3.3   *PERFORMANCE OF CERES WHEAT MODEL*

8.3.3.1   Calibration

The model was calibrated using the experimental data on grain yield, above ground dry matter and maximum leaf area index. The well-watered treatment (10% MAD or T_1) of each experiment was selected for calibration. The values of genetic coefficients were estimated using the best-fit method. Model calibration was performed for each experiment separately and an average value of each genetic coefficient was considered for further use.

All the genetic coefficients have a pre-assigned range and calibrated values needs to fall within that range. A fairly good agreement was found between simulated and measured grain yield of wheat. It was also found that the simulated and measured above ground dry matter and leaf area index matched reasonably well.

8.3.3.2   Validation

The model was validated for the four treatments of irrigated crops, i.e., 30%, 45%, 60% and 75% depletion of available soil water (ASW). The genetic coefficients determined by the process of calibration were used for validation. For validation, the model was run independently for the T_2, T_3, T_4 and T_5 treatments of each experiment.

8.3.3.3   Simulation of Wheat Grain Yield

Comparison of the simulated and measured grain yield at harvest for different treatments in all the experiments is presented in Figure 8.1. The simulated grain yields for T_1, T_2 and T_3 were found to be at par in all the three-crop experiments because the plants were not under any soil water stress particularly under these treatments. The grain yield was reduced for T_4 and T_5 because the plants experienced some stress during the growth cycle.

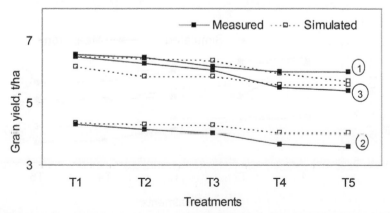

FIGURE 8.1   Measured and simulated grain yield (t/ha) of wheat under different scheduling of irrigation during the three experiments: 1(95–96), 2(96–97) and 3(97–98).

Similar trend of variation was observed in all experiments. A reasonably good agreement (Figure 8.4a) was found between simulated and measured values of grain yield of wheat during all experiments ($R^2 = 0.97$).

8.3.3.3.1 Simulation of Above Ground Dry Matter

The simulated above ground dry matter (ADM) was found to be in good agreement with the measured values. Comparison of the simulated and measured above ground dry matter at harvest under different treatments during all experiments is presented in Figure 8.2.

In general, a good agreement was found between simulated and measured values of above ground dry matter during all experiments (Figure 8.4b) except the second experiment, where the model over estimated above ground dry matter for the treatments ($R^2 = 0.94$).

8.3.3.3.2 Simulation of Leaf Area Index

Comparison of the simulated maximum and measured maximum leaf area index under different treatments during all experiments of wheat is presented in Figure 8.3. In general, a good agreement (Figure 8.4c) was found between simulated and measured values of leaf area index during all experiments ($R^2 = 0.92$).

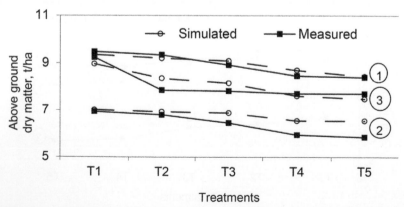

FIGURE 8.2 Measured and simulated above ground dry matter (t/ha) of wheat under different scheduling of irrigation during the three experiments.

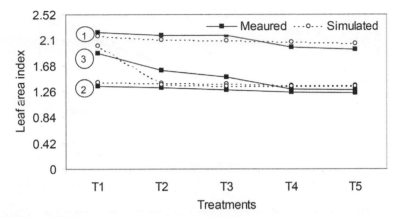

FIGURE 8.3 Measured and simulated leaf area index of wheat under different scheduling of irrigation during the three experiments numbered as 1 (95–96), 2 (96–97) and 3 (97–98).

8.3.3.3.3 Simulation of Temporal Variation of Soil Water

CERES-wheat model was calibrated for the soil water variation in different soil layers under the irrigation schedule T_1, which is based on 10% maximum allowable depletion (MAD) of available soil water (ASW). Comparison of the measured and simulated temporal variation of soil water was made for different soil layers such as: 0–15, 15–30, 30–45, 45–60 and 60–90 cm for this irrigation scheduling in experiment 1 (1995–96), as shown in Figure 8.5. A very good agreement was noted between measured and simulated profile soil moisture content for each soil layer. Regression analysis of measured and simulated profile soil moisture content gave the values of coefficient of determination (R^2) as 0.91, 0.89, 0.88, 0.80 and 0.73 respectively for the soil layers of 0–15, 15–30, 30–45, 45–60 and 60–90 cm (Table 8.1).

8.4   CASE STUDY: MAIZE

8.4.1   MATERIAL AND METHODS

8.4.1.1   Experimental Site

Field experiments were conducted at the experimental farm of the Agricultural and Food Engineering Department, Indian Institute of Technology, Kharagpur, India (22° 19' N latitude and 87°19' E latitude). The local climate is sub-humid and sub-tropical with an average rainfall of 1200 mm concentrated

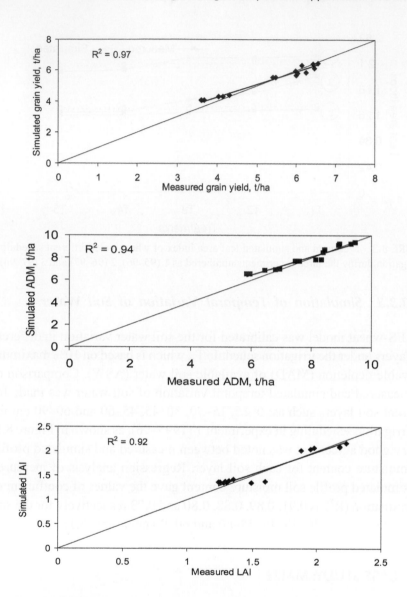

FIGURE 8.4 Comparison of simulated and measured results of wheat crop under different scheduling of irrigation during experiments 1 through 3: (a) grain yield; (b) above ground dry matter (ADM); (c) LAI.

over the months of June to September. The soil at the experimental site is an acid lateritic sandy loam. The soil is partly eroded due to high intensity rainfall in the area during the monsoon season. The field capacity and permanent

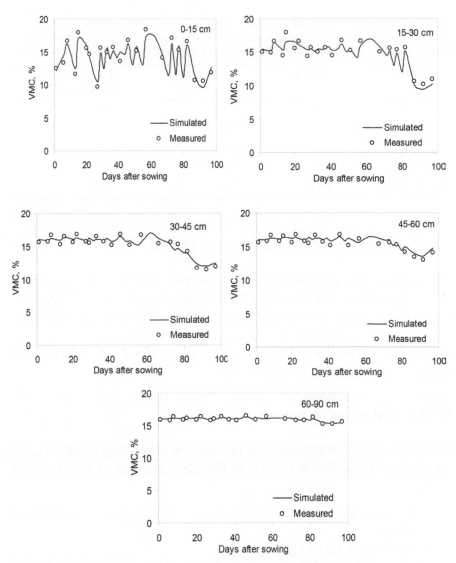

FIGURE 8.5 Calibration results of CERES-wheat model showing the comparison of simulated and observed soil water variation (VMC, %) in different soil layers for the irrigation scheduling based on 10% MAD (T_1) of available soil water during experiment 1.

wilting point of the soil at the crop field are 16% and 7% respectively. Maize is usually a 95–100 days cereal crop in this region and suits the prevailing climate in the summer season (March to June). Experiments were conducted during three consecutive years such as: 1995–98.

TABLE 8.1 Regression Analysis (Y = mX + C) of Measured and Calibrated Soil Profile Moisture Content Within and Below the Root Zone of Wheat for the Irrigation Schedule Based on 10% MAD (T$_1$) of ASW During the First Experiment (1995–96)

Soil profile	R^2 value	Slope of the equation, m	Intercept of the equation, C
0–15 cm	0.91	1.10	−1.51
15–30 cm	0.89	0.91	1.46
30–45 cm	0.88	1.08	−1.28
45–60 cm	0.80	1.19	−3.19
60–90 cm	0.73	1.12	−1.93

8.4.1.2 Field Layout and Experimental Details

The experimental area was divided into 20 plots of 5 m × 4 m size maintaining a buffer of 1 m between adjacent plots. A seed rate of 17–20 kg/ha was used. The seeds were sown at a row spacing of 60 cm and a plant spacing of 25 cm during all the three experiments. The three experiments were in fact three years of trial such as: experiment 1 (1995–96), experiment 2 (1996–97) and experiment 3 (1997–98).

8.4.1.3 Irrigation Treatments

The irrigation treatments consisted of irrigation scheduling based on maximum allowable depletion (MAD) of available soil water (ASW) criteria, defined as:

- T$_1$ = 10% maximum allowable depletion (MAD) of available soil water (ASW)
- T$_2$ = 30% MAD of ASW
- T$_3$ = 45% MAD of ASW
- T$_4$ = 60% MAD of ASW
- T$_5$ = 75% MAD of ASW

8.4.2 *DATA COLLECTION*

For modeling the water balance and crop response to deficit irrigation, it was necessary to collect data relating to weather variables, profile soil moisture content and the growth attributes of the crop.

8.4.2.1 Weather Data

Daily values of the weather variables such as: solar radiation, maximum and minimum temperature, maximum and minimum relative humidity, wind speed and precipitation for the experimental period were obtained from an automatic weather station installed near the experimental crop field.

8.4.2.2 Soil Profile Moisture Data

In order to assess the change in soil water status, soil moisture was measured in 0–15, 15–30, 30–45, 45–60, 60–90 and 90–120 cm soil profiles. The moisture content of topsoil layer (0–15 cm) was measured gravimetrically. The lower layers were measured using a neutron probe. Moisture measurements were done on 2–3 day intervals.

8.4.2.3 Crop Data

Crop parameters were measured during different stages of growth. The crop data included planting date, date of emergence, 20% cover date, full cover date, maturity date, harvest date, maximum rooting date, crop coefficient at full cover, planting depth and maximum root depth. The data on grain yield, above ground dry matter yield and leaf area index were recorded at different stages of crop growth during each crop experiment. The field water use efficiency, expressed as grain yield per unit cropped area per unit water applied to the field was estimated for each treatment in each experiment.

8.4.3 PERFORMANCE OF CERES MAIZE MODEL

8.4.3.1 Calibration

The model was calibrated using the experimental data on grain yield, above ground dry matter and maximum leaf area index. The well-watered treatment (10% MAD or T_1) of each experiment was selected for calibration. The values of genetic coefficients were estimated using the best-fit method. Model calibration was performed for each experiment separately and an average value of each genetic coefficient was considered for further use.

All the genetic coefficients have a pre-assigned range and calibrated values needs to fall within that range. A fairly good agreement was found between simulated and measured grain yield of maize. It was also found that the simulated and measured above ground dry matter and leaf area index was matching reasonably well.

8.4.3.2　Validation

The model was validated for the four treatments of irrigated crops, i.e., 30%, 45%, 60% and 75% depletion of available soil water (ASW). The genetic coefficients determined by the process of calibration were used for validation. For validation, the model was run independently for the T_2, T_3, T_4 and T_5 treatments of each experiment.

8.4.3.3　Simulation of Grain Yield

Comparison of the simulated and measured grain yield at harvest pertaining to different treatments for all the experiments is presented in Figure 8.6. The simulated grain yields for T_1, T_2 and T_3 were found to be at par during all the three-crop experiments because the plants were not under any soil water stress particularly under these treatments. The grain yield reduced for T_4 and T_5 because the plants experienced some stress during their growth cycle.

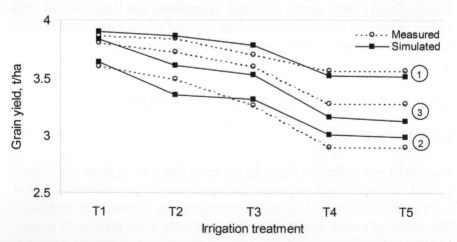

FIGURE 8.6　Measured and simulated grain yield (t/ha) of maize under different levels of irrigation during all the three experiments numbered as 1, 2 and 3.

Similar trend of variation was observed in all experiments. A reasonably good agreement (Figure 8.9a) was found between simulated and measured values of grain yield of maize crop during all experiments ($R^2 = 0.91$).

8.4.3.4 Simulation of Above Ground Dry Matter

The simulated above ground dry matter (ADM) was found to be in good agreement with the measured above ground dry matter. Comparison of the simulated and measured above ground dry matter at harvest under different treatments during all experiments is presented in Figure 8.7. In general, a good agreement was found between simulated and measured values of above ground dry matter during all experiments (Figure 8.9b) except the second experiment where the model over estimated above ground dry matter for the treatments ($R^2 = 0.93$).

8.4.3.5 Simulation of Leaf Area Index

Comparison of the simulated maximum and measured maximum leaf area index under different treatments during all experiments of maize is presented in Figure 8.8. In general, a good agreement (Figure 8.9c) was found between simulated and measured values of leaf area index during all experiments ($R^2 = 0.98$).

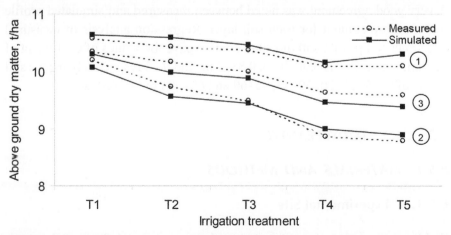

FIGURE 8.7 Measured and simulated above ground dry matter (t/ha) of maize under different levels of irrigation during all the three experiments numbered as 1, 2 and 3.

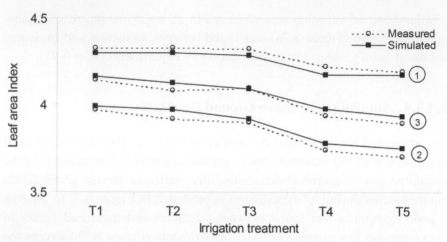

FIGURE 8.8 Measured and simulated leaf area index of maize under different levels of irrigation during all the three experiments numbered as 1, 2 and 3.

8.4.3.6   Simulation of Temporal Variation of Soil Water

CERES-maize model was calibrated for the soil water variation in different layers under the irrigation schedule T_1, which is based on 10% maximum allowable depletion (MAD) of available soil water (ASW). Comparison of the measured and simulated temporal variation of soil water was made for different soil layers such as: 0–15, 15–30, 30–45, 45–60 and 60–90 cm for this irrigation schedule in experiment 1 (1995–96), as shown in Figure 8.10. A very good agreement was noted between measured and simulated profile soil moisture content for each soil layer. Regression analysis of measured and simulated profile soil moisture content gave the values of coefficient of determination (R^2) as 0.90, 0.80, 0.83, 0.82 and 0.73 respectively for the soil layers of 0–15, 15–30, 30–45, 45–60 and 60–90 cm (Table 8.2).

8.5   CASE STUDY: PEANUT

8.5.1   MATERIALS AND METHODS

8.5.1.1   Experimental Site

Field experiments were conducted at the experimental farm of the Agricultural and Food Engineering Department, Indian Institute of Technology,

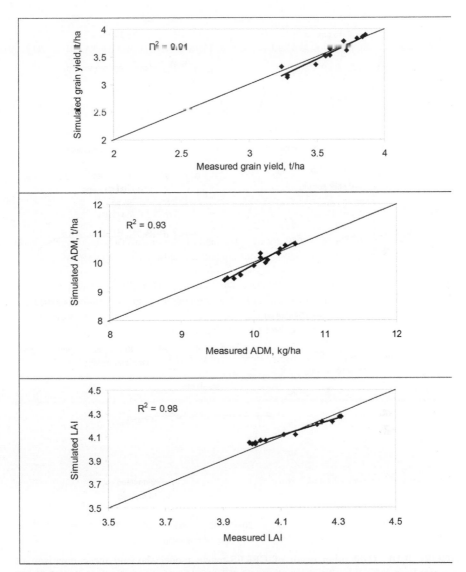

FIGURE 8.9 Comparison of simulated and measured results of maize crop under different level of irrigation during experiments 1 through 3: (a) grain yield (b) above ground dry matter (ADM) (c) LAI.

Kharagpur, India (22°19' N latitude and 87°19' E latitude). The local climate is sub-humid and sub-tropical with an average rainfall of 1200 mm concentrated over the months of June–September. The soil at the experimental site is an acid lateritic sandy loam and taxonomically grouped under the order

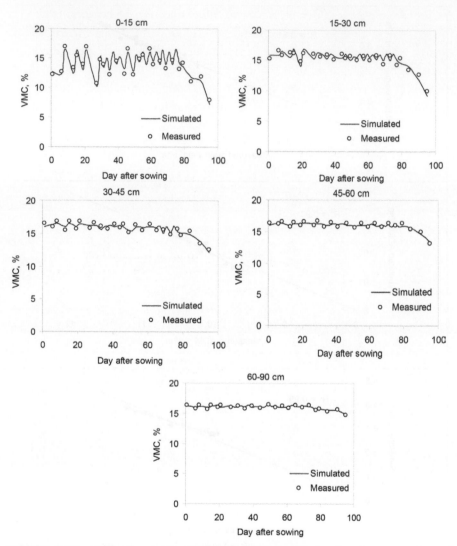

FIGURE 8.10 Calibration result of CERES-maize model showing the comparison of simulated and observed soil water variation in different layer for the irrigation schedule based on 10% MAD (T_1) of Available soil water during experiment 1.

ALFISOL. The soil is partly eroded due to high intensity rainfall in the area during the monsoon season. The average field capacity and permanent wilting point of the root zone soil in the crop field are 16% and 7%, respectively. The field capacity and wilting point of each soil layer in the root zone is given in Table 8.1. Peanut, cultivar AK-12–24, a short duration (105–115 days)

TABLE 8.2   Regression Analysis (Y = mX + C) of Measured and Calibrated Soil Profile Moisture Content Within and Below the Root Zone of Maize for the Irrigation Schedule Based on 10% MAD (T$_1$) of ASW During the First Experiment (1995–96)

Soil profile	R^2 value	Slope of the equation, m	Intercept of the equation, C
0–15 cm	0.90	0.911	0.836
15–30 cm	0.80	0.799	3.128
30–45 cm	0.83	0.955	0.786
45–60 cm	0.82	1.059	–0.87
60–90 cm	0.73	1.093	–1.5

crop with wide adaptability and stable performance was selected for the present study. Peanut is a popular summer (February–May) irrigated crop of this region, which suits to the prevailing climate. Experiments were conducted during three consecutive years such as: 1996, 1997 and 1998. The dates of sowing of peanut crop during the three experiments were 25th February (1996), 28th February (1997) and 20th February (1998).

8.5.1.2   Field Layout and Experimental Details

The experimental area was divided into 20 plots of 5 m × 4 m size maintaining a buffer of 1 m between adjacent plots. The field experiments were designed as per Randomized Block Design (RBD) with irrigation schedules or treatments as the factors. There were five irrigation treatments in all the three experiments. There were four replications for each treatment, out of which results of best three replications in each case were considered for further analysis. A seed rate of 120 kg/ha was used. The seeds were sown at a row spacing of 30 cm and a plant spacing of 20 cm during all the three experiments. This gave a plant density of 166,666 plants per hectare. The three experiments were in fact three years of trial such as: experiment 1 (1996), experiment 2 (1997) and experiment 3 (1998).

8.5.1.3   Irrigation Treatments

The irrigation treatments consisted of irrigation scheduling based on maximum allowable depletion (MAD) of the total available soil water (ASW) criteria. Each irrigation treatment was based on a predefined level of MAD, which was a fixed percent of the total ASW. Irrigation water was applied whenever the threshold value of MAD for the particular irrigation treatment was attained. The irrigation treatments considered in the study were:

- T_1 = 10% maximum allowable depletion (MAD) of available soil water (ASW)
- T_2 = 30% MAD of ASW
- T_3 = 45% MAD of ASW
- T_4 = 60% MAD of ASW
- T_5 = 75% MAD of ASW

8.5.2 DATA COLLECTION

For modeling the water balance and crop response to deficit irrigation, it was necessary to collect data relating to weather variables, profile soil moisture content and the growth attributes of the crop.

8.5.2.1 Weather Data

Daily values of the weather variables such as: solar radiation, maximum and minimum temperature, maximum and minimum relative humidity, wind speed and rainfall for the experimental period were obtained from an automatic weather station installed close to the experimental crop field.

8.5.2.2 Soil Profile Moisture Data

In order to assess the change in soil water status, soil moisture was measured in 0–15, 15–30, 30–45, 45–60, 60–90 and 90–120 cm soil profiles. The moisture content of topsoil layer (0–15 cm) was measured gravimetrically. The lower layers were measured using a neutron probe. Moisture measurements were done at every 2–3 days interval. Neutron probe was calibrated using the measured volumetric moisture content of different soil layers such as: 15–30, 30–45, 45–60 and 60–90 cm and then a calibration curve was developed between count ratio of the neutron probe on Y-axis and volumetric moisture content on X-axis. The field count was divided by the standard count to obtain the count ratio. The standard count was measured using a tank of 1 m × 1 m filled with water.

8.5.2.3   Crop Data

Crop parameters were measured during different stages of growth. The crop data included planting date, date of emergence, 20% cover date, full cover date, maturity date, harvest date, maximum rooting date, crop coefficient at full cover, planting depth and maximum root depth. The crop coefficient at full cover was determined by using the measured daily evapotranspiration (ET) of peanut crop and daily reference crop evapotranspiration (ET_0) computed by Penman-Monteith method. The reason for selecting Penmen-Monteith method to estimate the reference crop evapotranspiration is that the estimated values by this method were found to be very close to the lysimeter measured reference ET_0 using grass as a reference crop (Kashyap and Panda, 2001). The crop parameter such as maximum root depth was determined by measuring the root depth through destructive plant sampling at different stages of the growth. Maximum root depth was obtained for treatment T_1 for all the three experiments due to the least soil water stress experienced by the crop at this treatment. The data on grain yield, above ground dry matter and leaf area index were recorded at different stages of crop growth during each crop experiment. The field water use efficiency, expressed as grain yield per unit cropped area per unit water applied to the field was estimated for each treatment in each experiment.

8.5.3   PERFORMANCE OF CROPGRO PEANUT MODEL

8.5.3.1   Calibration

The model was calibrated using the experimental data on grain yield, above ground dry matter and maximum leaf area index. The well-watered treatment (10% MAD or T_1) of each experiment was selected for calibration. The values of genetic coefficients were estimated using the best-fit method. Model calibration was performed for each experiment separately and an average value of each genetic coefficient was considered for further use. All the genetic coefficients have a pre-assigned range and calibrated values needs to fall within that range [12].

A fairly good agreement was found between simulated and measured grain yield of peanut. It was also found that the simulated above ground dry matter and leaf area index matched reasonably well with their measured counterparts.

8.5.3.2 Simulation of Crop Growth Parameters

The model was validated for the four irrigation treatments that is, 30%, 45%, 60% and 75% maximum allowable depletion of available soil water (ASW) for the crop growth parameters such as grain yield, above ground dry matter and maximum leaf area index. Comparison of the simulated and measured grain yield at harvest pertaining to different treatments for all the experiments is presented in Figure 8.11.

The simulated grain yields for T_1, T_2 and T_3 were found to be at par during all the three-crop experiments because the plants were not under any soil water stress particularly under these treatments. The grain yield was lower for T_4 and T_5 because the plants experienced some stress during their growth cycle. Similar trend of variation was observed in all experiments. A reasonably good agreement (Figure 8.14a) was found between simulated and measured values of grain yield of peanut crop during all the experiments ($R^2 = 0.95$).

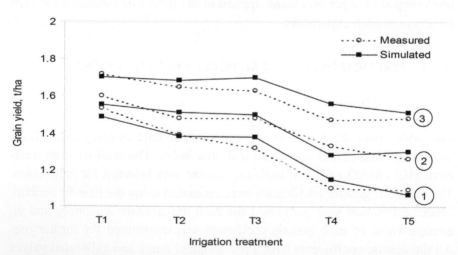

FIGURE 8.11 Measured and simulated grain yield (t/ha) of peanut under different levels of irrigation during all the three experiments numbered as 1, 2 and 3.

The simulated above ground dry matter (ADM) was found to be in good agreement with the measured above ground dry matter. Comparison of the simulated and measured above ground dry matter at harvest under different treatments during all experiments is presented in Figure 8.12. In general, a good agreement was found between simulated and measured values of above ground

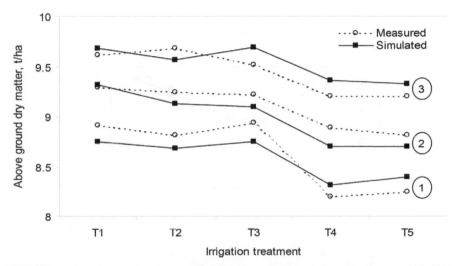

FIGURE 8.12   Measured and simulated above ground dry matter (t/ha) of peanut under different levels of irrigation during all the three experiments numbered as 1, 2 and 3.

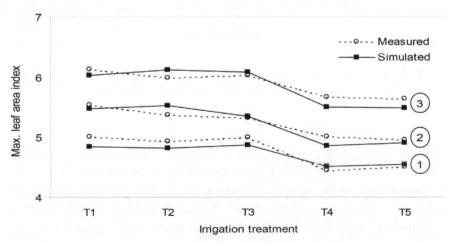

FIGURE 8.13   Measured and simulated maximum leaf area index of peanut under different levels of irrigation during all the three experiments numbered as 1, 2 and 3.

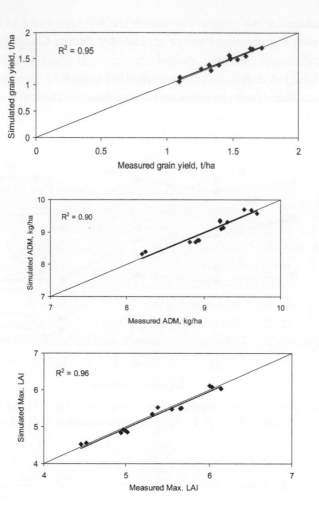

FIGURE 8.14 Comparison of simulated and measured results of peanut crop under different level of irrigation during experiments 1 through 3: (a) grain yield (b) above ground dry matter (ADM) (c) maximum LAI.

dry matter during all experiments (Figure 8.14b) except the second experiment where the model over estimated above ground dry matter ($R^2 = 0.90$).

Comparison of the simulated maximum and measured maximum leaf area index under different treatments during all experiments of peanut is presented in Figure 8.13. In general, a good agreement (Figure 8.14c) was found between simulated and measured values of leaf area index during all experiments ($R^2 = 0.96$).

8.5.3.3 Simulation of Temporal Variation of Soil Water

CROPGRO peanut model was calibrated for the soil water variation in different soil layers under the irrigation schedule T_1, which is based on 10% maximum allowable depletion (MAD) of available soil water (ASW). Comparison of the measured and simulated temporal variation of soil water was made for different soil layers such as: 0–15, 15–30, 30–45,

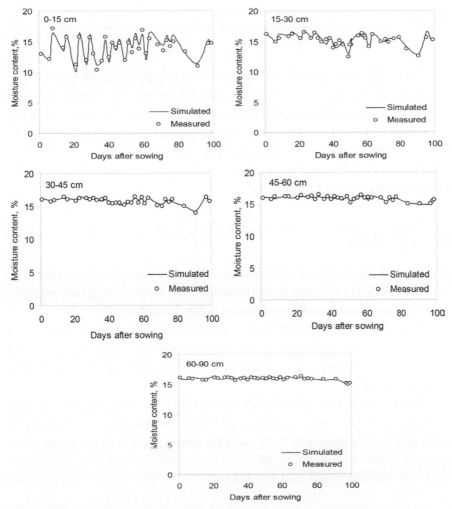

FIGURE 8.15 Calibration result of CROPGRO-peanut model showing the comparison of simulated and observed soil water variation in different layer for the irrigation schedule based on 10% MAD (T_1) of Available soil water during experiment 1.

TABLE 8.3 Regression Analysis (Y = mX + C) of Measured and Calibrated Soil Profile Moisture Content Within and Below the Root Zone of Peanut for the Irrigation Schedule Based on 10% MAD (T_1) of ASW During the First Experiment (1995–96)

Soil profile	R^2 value	Slope of the equation, m	Intercept of the equation, C
0–15 cm	0.89	0.83	2.17
15–30 cm	0.82	0.90	1.32
30–45 cm	0.78	0.99	0.08
45–60 cm	0.74	0.96	0.70
60–90 cm	0.73	0.73	4.26

45–60 and 60–90 cm for this irrigation schedule in experiment 1 (1996), as shown in Figure 8.15. A very good agreement was noted between measured and simulated profile soil moisture content for each soil layer. Regression analysis of measured and simulated profile soil water content gave the values of coefficient of determination (R^2) as 0.90, 0.80, 0.83, 0.82 and 0.73 respectively for the soil layers of 0–15, 15–30, 30–45, 45–60 and 60–90 cm (Table 8.3).

8.6 SUMMARY

This chapter discusses use of *Decision Support System for Agrotechnology Transfer* (DSSAT) for simulation modeling of effective irrigation management in wheat, maize and peanut under Indian conditions.

The CERES-Wheat, CERES-Maize and CROPGRO-peanut models can effectively be used to simulate yield, above ground dry matter and maximum leaf area index of wheat, maize and peanut respectively in sub-humid subtropical regions.

The CERES-Wheat, CERES-Maize and CROPGRO-peanut models simulate the profile soil water variation with considerable accuracy. Therefore, these models can successfully be used for determination of irrigation management depth and scheduling of irrigation without going through the rigorous experimentation and data monitoring.

KEYWORDS

- **CERES-maize**
- **CERES-wheat**
- **Crop growth model**
- **Crop yield**
- **CROPGRO-peanut**
- **Decision support system**

- **Deficit irrigation**
- **Dry matter**
- **DSSAT**
- **Irrigation management**
- **Irrigation scheduling**
- **Leaf area index**
- **Maize**
- **Moisture content**
- **Peanut**
- **Root water uptake**
- **Simulation modeling**
- **Soil water balance**
- **Water use efficiency**
- **Wheat**

REFERENCES

1. Bandyopadhyay, P. K. and Mallick, S. (2003). Actual evapotranspiration and crop coefficients of wheat (*Triticum aestivum*) under varying moisture levels of humid tropical canal command area. *Agricultural Water Management*, 59(1), 33–47.
2. Collino, D. J., Dardanelli, J. L., Sereno, R. and Racca, R. W. (2002). Physiological responses of argentine peanut varieties to water stress: water uptake and water use efficiency. *Field Crops Research*, 68, 133–142.
3. Craciun, I. and Craciun, M. (1999). Water and nitrogen use efficiency under limited water supply for maize to increase land productivity. In: *Crop yield response to deficit*

irrigation. Kirda, C., Moutont, P., Hera, C., Nielsen, D. R. (eds.). Kluwer Academic Publishers, The Netherlands, 87–94.

4. Dhanpal, A. H. (1992). Simulation of soil moisture regime: Application of the SWATRE model to a maize crop on the reddish brown earths in the dry zone of Sri Lanka. *Agriculture Systems*, 38, 61–73.

5. English, M. J. and Nakamura, B. (1989). Effects of deficit irrigation and irrigation frequency on wheat yields. *Journal of Irrigation and Drainage Engineering*, ASCE 115, 172–184.

6. FAO (1993). CLIMWAT for CROPWAT, a climatic database for irrigation planning and management by M. Smith. *FAO Irrigation and Drainage Paper No. 49*. Rome.

7. FAO (1977). Guidelines for predicting crop water requirements by J. Doorenbos and W.O. Pruitt. *FAO Irrigation and Drainage Paper No. 24*. Rome.

8. FAO (1979). Yield response to water by J. Doorenbos and A. Kassam. *FAO Irrigation and Drainage Paper No. 33*. Rome.

9. FAO (1992). CROPWAT, a computer program for irrigation planning and management by M. Smith. *FAO Irrigation and Drainage Paper No. 26*. Rome.

10. FAO (1998). Crop evapotranspiration by R. Allen, LA. Pereira, D. Raes & M. Smith. *FAO Irrigation and Drainage Paper No. 56*. FAO, Rome.

11. Hartkamp, A. D., White, J. W., and Hoogenboom, G. (2003). Comparison of three weather generators for crop modeling: a case study for subtropical environments, *Agriculture Systems*, 76, 539–560.

12. Hoogenboom, G., Jones, J. W., Wilkens, P. W., Batchelor, W. D., Bowen, W. T., Hunt, L. A., Pickering, N. B., Singh, U., Godwin, D. C., Baer, B., Boote, K. J., Ritchie, J. T. and White, J. W. (1994). Crop models. In: G. Y. Tsuji, G. Uehara, S. Balas (eds.) DSSAT v3. Vol 2–2. Univ. of Hawaii, Honolulu, Hawaii.

13. Iqbal, M. M., Shah, S. M., Mohammad, W. and Nawaz, H. (1999). Field response of potato subjected to water stress at different growth stages. *In: Crop yield response to deficit irrigation*. Kirda, C., Moutont, P., Hera, C., Nielsen, D. R. (eds.). Kluwer Academic Publishers, The Netherlands.

14. Jagtap, S. S. and Jones, J. W. (2002). Adaptation of evaluation of the CROPGRO-soybean model to predict regional yield and production. Agriculture Ecosystem Environ-ronment, 93, 73–85.

15. Jefferies, E. C., Hook, J. E., and Blair, S. L. (1994). Using crop models to plan water withdrawal for irrigation in drought years. *Agriculture Systems,* 45(3), 271–289.

16. Ma, L., Nielson, D.C, Ahuja, L. R., Kiniry, J. R., Hanson, J. D. and Hoogenboom, G. (2002). An evaluation of RZWQM, CROPGRO and CERES-Maize for responses to water stress in the central great plains of the US. p. 119–148. In: L. R. Ahuja, L. Ma and T. A. Howell (eds.) *Agricultural system models in field research & technology transfer*. Lewis Publ. Boca Raton, FL.

17. Martin, D. L., Stegman, E. C., and Freres, E. (1990). Irrigation scheduling principles. p. 155–372 In: G. L. Hoffman, T. A. Howell and K. H. Solomon (eds.) *Management of farm irrigation systems*. ASAE Monograph.

18. Musick, J. T. (1994). General guidelines for deficit irrigation management. Paper presented at Central Plains Irrigation Short Course, February, 7–8, 1994. Garden City, Kansas, USA.

19. Oosterom, E. J. van., O'Leary, G. J., Carberry, P. S., and Craufurd, P. Q. (2002). Simulating growth, development and yield of tillering pearl millet. III. Biomass accumulation and partitioning. *Field Crop Research,* 79(2–3), 85–106.

20. Panda, R. K. and Behera, S. K. (2005). Irrigation water management strategy for peanut under deficit conditions. *Journal of Applied Irrigation Science*, 40 (1), 91–115.

21. Panda, R. K., Behera, S. K. and Kashyap P. S. (2003). Effective management of irrigation water for wheat under stressed conditions, *Agricultural Water management*, 63(1), 37–56.

22. Panda, R. K., Behera, S. K., and Kashyap P. S. (2004). Effective management of irrigation water for maize under stressed conditions, *Agricultural Water management*, 66(3), 181–203.

23. Pandey R. K., Maranville J. W., and Admou A. (2000). Deficit irrigation and nitrogen effects on maize in a Sahelian environment: I. Grain yield and yield components, *Agricultural Water Management*, 46(1), 1–13.

24. Stastana, M., Trnka, M., Dubrovsky, M., and Zalud, Z. (2002). Evaluation of the CERES models in different production regions of the Czech Republic, *Rostlinka Vyroba*, 48, 125–132.

25. White, J. W., Hoogenboom, G., and Hunt, L. A., (2005). A Structured Procedure for Assessing How Crop Models Respond to Temperature. *Agronomy Journal*, 97, 426–439.

26. Zhang, X., You, M., and Wang, X. (1999). Effects of water deficits on winter wheat yield during its different development stages. *Acta Agricultural Boreali-Sinica*, 14, 79–83.

27. Ziaei, A. N. and Sepaskhah, A. R. (2003). Model for simulation of winter wheat under dry land and irrigated conditions. Agricultural Water Management, 58(1), 1–17.

PART III

RESEARCH ADVANCES IN SOIL AND WATER ENGINEERING

CHAPTER 9

INTEGRATED FARMING SYSTEM AND BIODRAINAGE: MANAGEMENT OF WATERLOGGED AREA

SUSANTA KUMAR JENA,[1] S. ROY CHOWDHURY,[1] R. K. MOHANTY,[1] N. SAHOO,[2] D. K. KUNDU,[3] and M. S. BEHERA[3]

[1]ICAR – Indian Institute of Water Management, Opposite Rail Vihar, Chandrasekharpur, Bhubaneswar-751023, India.
Mobile: +91-9437221616, E-mail: skjena_icar@yahoo.co.in, somnath_rc@yahoo.com, rajeebm@yahoo.com

[2]Retired Principal Scientist of ICAR-IIWM, Bhubaneswar-751023, India. E-mail: narayansahoo65@yahoo.in

[3]ICAR – Central Research Institute for Jute and Allied Fibers, Barrackpore, Kolkata-700120, India.
E-mail: kundu_crijaf@yahoo.com, behera_ms@rediffmail.com

CONTENTS

9.1   INTRODUCTION

The task of providing food security to India's burgeoning population is becoming increasingly difficult. Around 70% of the India's population is living in rural area with agriculture as their livelihood support system. The vast majority of Indian farmers are small and marginal. The farm size is decreasing further due to population growth. The quality of land is also degrading due to various reasons resulting decline in agricultural productivity leading to food insecurity. Land degradation can be defined as a temporary or a permanent lowering of land productivity through deterioration of land's physical, chemical and biological conditions. It represents a complex ensemble of water erosion, wind erosion, soil compaction, salinization and waterlogging. An area is said to be waterlogged when the water table rises to an extent that the soil pores in the root zone of a crop become saturated, resulting in restriction of the normal circulation of air, decline in the level of oxygen and increase in the level of carbon dioxide. The water table, which is considered harmful would depend upon the type of crop, type of soil and the quality of water [6].

In India the total degraded land due to waterlogging is 6.41 Mha out of which 1.66 Mha is mainly wasteland due to surface ponding and rest area of 4.75 Mha is under subsurface waterlogging [9]. High intensity of rainfall combined with saucer shaped physiography and flat land near the coastal area in deltaic alluvial region is the most important reason for waterlogging [7]. The problem of waterlogging is very severe in coastal and deltaic region of eastern India in which water stagnation and rise of water table above ground surface is more than 1 m in many places during monsoon. So no crop or paddy with an average yield of 0.5–0.75 t.ha^{-1}

is obtained during *kharif* season. After monsoon also due to rise in water table no other cash crop or remunerative crop is possible except paddy with very low return in terms of yield and pricing. The quality of water is very good for irrigation as well as aquaculture purpose. Many researchers have worked on on-farm reservoir design, etc. using water balance model in medium and upland, where the pond water was used for supplemental irrigation or life-saving irrigation [1, 7, 10, 11]. Since the land resources are finite, requisite measures are required to reclaim degraded and wastelands, so that areas going out of cultivation due to social and economic reasons are replenished by reclaiming these lands and by arresting further loss of production potential.

There are several measures to reclaim waterlogged area. Drainage is one of the measures to control waterlogging, which is defined as the natural or artificial removal of surface and subsurface water from a given area. Traditionally management strategies to address waterlogging problem have often focused on engineering approaches such as deep open ditches, vertical drainage (groundwater pumping) or horizontal sub-surface drainage which all require expensive capital investment and operation and maintenance.

Biodrainage is the use of vegetation to manage water fluxes in the landscape through evapotranspiration, and is an alternative technique that has recently attracted interest in drainage and environmental management circles. Biodrainage can be either remedial, i.e., lowering water table after they have risen; discharge control, or preventative, i.e., intercepting soil water before it reaches the water table; recharge control [12]. Heuperman [5] found that the lowering of water table due to biodrainage for 10 years through planting trees has high water requirement. Biodrainage presents itself as a feasible and environment friendly option that farmers could adopt to reclaim their land. It is based on the ability of plants and trees to transpire water and thus remove excess water and salinity.

9.1.1 PRINCIPLES OF BIODRAINAGE

Biodrainage is a combined drainage-cum-disposal system and is less costly and more environmentally friendly. It relies on vegetation, rather than mechanical means, to remove excess water. The driving force behind the biodrainage concept is the consumptive water use of plants. It is economically attractive because it requires only an initial investment for planting the vegetation, and when established, the system could produce economic returns by means of fodder, wood or fiber harvested. There is consensus that biodrainage, when properly implemented, can lower the water table. It could solve problems associated with waterlogged areas and canal seepage. Biological systems make use of the evapotranspirative power of plants, especially of trees, to lower groundwater tables. Low cost technology such as biodrainage could be an alternative providing several advantages as below:

- the negative side effects of conventional drainage systems are reduced,
- they require less investment,
- may find quicker application,
- they are environmentally friendly,
- provide fuel wood, timber, fruits, shade and shelter,
- function as windbreaks and yield organic matter for fertilizer,
- they contribute to the enhancement of biodiversity, as flora and fauna flourish, and
- air pollution is diminished and they contribute to carbon sequestration.

Australian researchers [2] have reported the ability of different trees in influencing water table. Thus a new approach is gaining momentum to use different types of plants to control shallow water tables. These plants draw their main water supply from groundwater or from the capillary fringe just above it. Such types of plants are called phreatophytes. Main physiological features of such plants are luxuriant transpiration in contact with groundwater. Examples are tree species like poplar, *eucalyptus,* tamarix, muskit, *Acacia,* sissoo, etc.

Annual rate of transpiration from (Eucalyptus) plantation area over 6 year period (1991–1997) was 3446 mm [8]. The plantations were visualized as wells 500 m apart with pumping capacity of 33 m^3/h. The observed draw down during a period of 6 years was between 7.8 and 8.0 m at various point of the plantation area with maximum draw down being 13–15 m [8]. However, the efficacy of biodrainage has been established through various reports [4] after surveying of 80 sites in western Australia concluded that extensive planting covering as much as 70–80% of catchment area is necessary to achieve significant water table reduction in deep water table (often recharge area) situation. In shallow water table (often discharge areas) zones, for every 10 % increase in planted area water table was lowered by about 0.4 m. A comparative study of Casurina glanca and Eucalyptus camaldulensis [3] showed that former had greater potential to discharge saline groundwater.

The study in this chapter was undertaken with the objectives of optimizing micro-level water resources design in waterlogged area; enhancing productivity of waterlogged area through integrated farming system; and reclamation study of waterlogged area through bio-drainage and cultivation of water loving co-existing crops.

9.2 MATERIALS AND METHODS

9.2.1 STUDY AREA

The study was conducted in the ICAR- Indian Institute of Water Management (IIWM) research farm, Mendhasal, Khurda, Bhubaneswar, India (20° 30′ 0″ N latitude and 84° 48′ 10″ E longitude). There was a patch of 3 ha area under severe waterlogging. Continuous waterlogging has converted that land to wasteland. No crop could be grown in these fields and it was remaining fallow in almost all years. The soil pH ranged from 3.5 to 6.5; soil texture is sandy clay loam; soil organic carbon was low (< 0.5%); soil available nitrogen was low (< 280 kg ha^{-1}); soil available potassium was medium (50–170 mg/kg of soil); soil available phosphorous was medium (5–10 mg/kg of soil); iron toxicity was present. Depth of groundwater table range was 20–40 cm as minimum and 50–150 cm as maximum from ground surface during December to June. During monsoon, it is above ground surface. The yield of shallow aquifer is low. The land was unsuitable for plowing except during the months of May and early June, and was left fallow in almost all years.

9.2.2 ANALYSIS AND INTERVENTIONS

For determining the design and dimensions of the ponds, collection and analysis of climatic data (rainfall, pan evaporation, etc.) for the period 1975–2003 for Bhubaneswar, Odisha, India was done. Different *probability distribution functions* (PDF) were fitted to the maximum one-day rainfall data. SMADA (Storm water Management and Design Aid) software was used for this analysis. From this analysis maximum one-day rainfall for different return periods was found out which was utilized for further design of different hydraulic structures. From the hydrologic data, analysis the dimensions of the ponds were decided. Integrated farming system was undertaken in those fields keeping some area under paddy, growing fish in the pond and taking vegetable on the bund. The production and productivity of the integrated farming system has been discussed in this write up. Physico-chemical properties of the study area were done using standard laboratory procedures at different years to find out whether there is any improvement in soil properties over years.

9.2.3   LAY-OUT OF BIODRAINAGE EXPERIMENTAL PLOT

An area of 2640 m² of the waterlogged wasteland was converted into four elevated platforms (P1, P2, P3 and P4) of 20 m x 20 m each with the excavated soils from the adjacent 20 m x 10 m area (D1, D2, and D3) and also from a strip of 110 m x 4 m (Figure 9.1). There was a net increase in elevation of platforms by 0.65 m in comparison to original ground level. After the modification of the land there were four elevated platforms and three depressions. Platform 1 (P1) and platform 2 (P2) was under acacia plantation with pineapple as intercrop, platform 3 (P3) was under casurina with pine apple, turmeric and arrowroot as intercrop, and platform 4 (P4) was under casurina. Depression 1 (D1) is between P1 and P2, depression 2 (D2) is between P2 and P3 and depression 3 (D3) is between P3 and P4 from where soil had been removed.

Acacia mangium and *Casuarina equisetifolia* planting material having average length of 45 cm was procured from College of Forestry, Orissa University of Agriculture and Technology, Bhubaneswar, Odisha, India. They were planted in two platforms each with a spacing of 2 m x 2 m in the fourth week of July 2004. The layout of the experimental plot is given in Figure 9.1. Normal procedure of agro-forestry planting was followed. Pits of 30 cm length, breadth and depth were excavated. Those pits were filled with well-decomposed compost and farmyard manure @ 4 kg/pit and

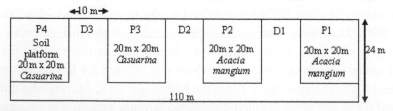

FIGURE 9.1   Layout of bio drainage experimental plot.

fertilizer (25 g DAP) was also applied per pit. One liter of water was applied per plant at the time of planting.

The different growth parameters observed were height (cm), collar diameter (mm), and diameter at breast height (DBH) (mm) at the time of planting and then at every 3 months after planting (MAP). Bottle gourd was taken as intercrop after the establishment of tree plantation for one season. After 12 months of planting pineapple was planted as intercrop in paired rows between two rows of acacia mangium as well as between two rows of casuarina. The spacing of pineapple between rows as well as plants was kept at 60 cm. In platform 1 and 2 the intercrop taken was pineapple and on platform 3 half of the area was covered with pineapple and rest half with turmeric and arrowroot as intercrop. The reason for taking pineapple, turmeric and arrowroot as intercrop was due to the fact that all these crops are shade loving and performs better in diffused light. All of them also perform relatively better in acid soils (where pH is low) in comparison to other type of crop.

Soil samples were taken from the experimental field prior to planting and then after one year of planting and after four years of planting. Standard procedure was followed for soil analysis in the laboratory of ICAR-IIWM. The different parameters used to find out the change of soil quality in this study are pH, electrical conductivity (EC), organic carbon (OC), N (nitrogen), P (phosphorous), K (potassium) availability and other micro-nutrients such as copper (Cu), manganese (Mn), iron (Fe) and zinc (Zn). In this paper the results and discussions are made on change of pH, EC and OC of the soil over time to know whether biodrainage has improved the soil quality and reclaimed the waterlogged wasteland or not.

Observation wells were installed one on the platform and other in depression to measure the depth of water table from ground level throughout the year. The observations were taken once in a week. The transpiration and stomatal conductance of the trees were also measured.

9.3   RESULTS AND DISCUSSIONS

9.3.1   RAINFALL ANALYSIS AND HYDROLOGY

For determining the design and dimensions of the ponds, collection and analysis of climatic data (rainfall, pan evaporation, etc.) for the period 1975–2003 for Bhubaneswar was done. Different probability distribution functions (PDF) were fitted to the maximum one-day rainfall data. SMADA software has been used for this analysis. From this analysis, maximum one-day rainfall for different return periods were found out, which was utilized for further design of different hydraulic structures.

The annual rainfall varied from 951.6 mm (1996) to 2218.7 mm (2001), with 55% of all the years have rainfall below normal. The 84.1% of the total rainfall occurs between June and October. Normal rainy days in a year are 105: maximum was 129 (1983), and minimum was 86 days (1979). Onset of effective monsoon is 15th June, earliest is 7th June, latest is 23rd June (based on both mean and median). Similarly cessation of effective monsoon is 8th October, earliest is 26th September, and latest is 20th October based on mean; and 10th October, 28th September, and 22nd October is normal, earliest and latest date for cessation of monsoon based on median. The weekly maximum, minimum and normal rainfall observed during 1975–2003 is given in Figure 9.2. The comparison of weekly rainfall and evaporation is given in Figure 9.3. From Figure 9.3, it is observed that the rainfall is higher than evaporation during 24th week to 43rd week causing water congestion and excess water is to be stored in ponds for aquaculture and for irrigating *rabi* crops including vegetable and other cash crops. Whereas evaporation is higher than rainfall during 44th week to 23rd week indicating irrigation is required if any crop is to be grown during this period.

The weekly rainfall at different probability level is given in Figure 9.4. Depending upon the requirement rainfall at different probability level would be considered for design of different structures such as field bunds, ponds,

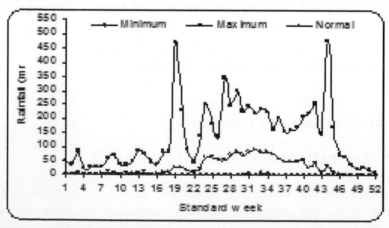

FIGURE 9.2   Weekly maximum, minimum and normal rainfall observed during 1975–2003.

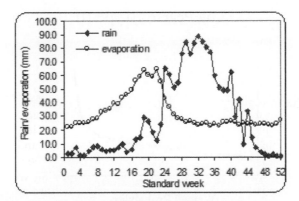

FIGURE 9.3 The comparison of weekly rainfall and evaporation.

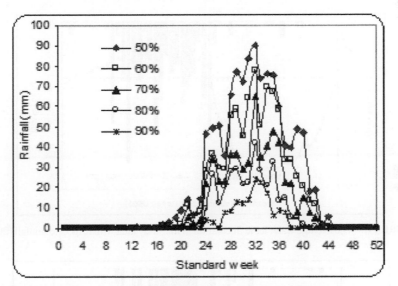

FIGURE 9.4 Weekly rainfall at different probability level at Bhubaneswar.

emergency spill way, drainage system, etc. It was found that Log Pearson type III PDF fits well (Figure 9.5) to the observed data. The average water table fluctuation in the study area with respect to rainfall (mm) in different standard meteorological week is given in Figures 9.6–9.8 for the years 2003, 2004, and 2005, respectively.

The water table fluctuation in the study area with respect to rainfall (mm) in different standard meteorological week is given in Figure 9.9. The different notations used in the figure are EP stands for experimental plot number and

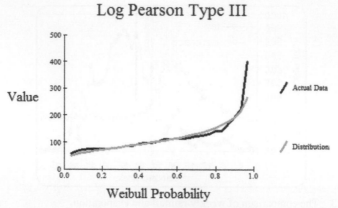

FIGURE 9.5   Fitting of maximum one day rainfall with Log Pearson type III distribution.

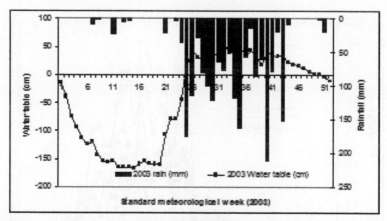

FIGURE 9.6   The average water table fluctuation in the study area with respect to rainfall (2003).

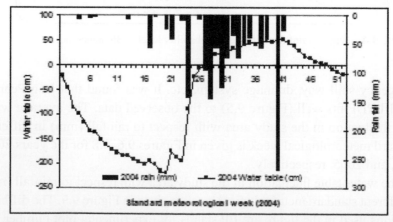

FIGURE 9.7   The average water table fluctuation in the study area with respect to rainfall (2004).

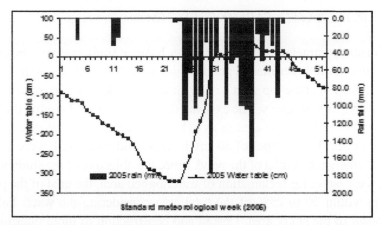

FIGURE 9.8   The average water table fluctuation in the study area with respect to rainfall (2005).

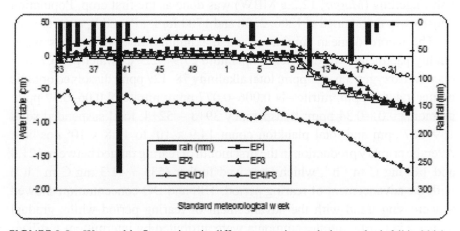

FIGURE 9.9   Water table fluctuations in different experimental plots and rainfall in 2004.

P denotes platform number (elevated portion where soil was deposited) and D denotes depressions from where soil was excavated to form the platform.

From Figure 9.9, it is observed that in Experimental plot 4 (EP 4), the water table below ground level is deeper in comparison to other experimental plots. The desirable condition for any crop to grow better is that the water table should be below root zone depth (i.e. 2 m in many cases), otherwise the area is called waterlogged. Hence after first year of work, where the bio drainage component is negligible, it is seen that the land modification alone has changed the water table regime making it better and suitable for crop growth in comparison to other plots.

9.3.2 INTEGRATED FARMING SYSTEM

The climatic parameter analysis and water balance study resulted the design dimensions of the experimental ponds which were 27 m x 27 m, 30 m x 30 m, and 34 m x 34 m at the top with 2 m depth and side slope 1:1 in experimental plot 1, 2 and 3 respectively. The excavated soils were spread around the pond to elevate the surrounding area so as to keep the water table below 2 m from ground surface. Hume pipes of 30 cm diameter and 4 m length were used as inlet and emergency outlet of the pond. Since the objective of the study was to store excess water for reclamation of waterlogged area, the area of the ponds are kept within 20 to 25% of the total area considering the water balance component of the study area. Design and construction of three micro water resources covering water surface area of 625 (P_1), 785 (P_2) and 1025 m^2 (P_3) was completed by March 2006. Treatment implementation and stocking of fish fingerling (*Magur*, 12.2 g MBW) was done as the first crop. Population density was maintained at 1200, 2100 and 1700 for P_1, P_2 and P_3 respectively.

The recorded mean minimum and maximum values of various water quality parameters were: water temperature 27.9–32.3°C; water pH 6.7–8.7; dissolved oxygen 3.6–9.1 ppm; total alkalinity 78–127 ppm; dissolved organic matter 1.4–6.4 ppm; nitrite –N 0.006–0.077 ppm; nitrate-N 0.06–0.57 ppm; ammonia 0.01–0.34 ppm; transparency 39+3 – 52+4; total suspended solid 169–367 ppm and total plankton count 14.9 x 10^3 to 19.8 x 10^4 nos/liter. Average primary production in the first month of rearing ranged between 121.4 and 149 mg C m^{-3} h^{-1}, which improved further (533 + 41.3 mg C m^{-3} h^{-1}) with the advancement of rearing period. TSS and DO concentration showed a decreasing trend with the advancement of rearing period while, gradual increase in nitrite, nitrate, ammonia were attributed by intermittent fertilization, increased level of metabolites and decomposition of unutilized feed. At any given point of time, other water quality parameters did not register any specific trend. In this experiment, average growth performance of *Magur* was highest in pond-1 (P_1) (163.5g) followed by pond-3, (P_3) (141.0 g) and pond-2 (P_2) (130.5 g). In this experiment, reductions in growth did not appear to be due to poor water quality, as water quality did not differ significantly among various treatments, may be due to behavioral interaction or physiological response to density itself. Relatively moderate survival rate (61–64.75%) was mainly due to cannibalism at the initial stage of rearing. In this, crop yield of fish ranged between 1632 and 1710 kg/ha/200 days, survival rate

(SR% – 61–64.75), feed conversion ratio (FCR) – 1.39–1.47, per day increment (PDI) was 0.595–0.623 g/day.

Indian major carps (IMC) were taken as subsequent crops in coming years and were released during 4th week of August. The recorded mean minimum and maximum values of various water quality parameters were presented in Table 9.1. All growth parameters were undertaken regularly. The results obtained are presented in Table 9.2. The catla has recorded a maximum growth in comparison to rohu and mrigal. As age of the pond increased the quality of water improved as the sides of the bunds have been stabilized, hence IMC was undertaken in place of magur to reduce the input cost and preference in market.

TABLE 9.1 Minimum and Maximum Average Values of Water and Soil Quality Parameters (*Based on Monthly Sampling*) in Fishponds Under Integrated Farming System Approach at IIWM Research Farm

Parameters	P_1 (625 m^2)	P_2 (785 m^2)	P_3 (1025 m^2)
Available-N in soil (mg 100 g^{-1})	8.1–11.1 (9.8)	11.9–14.6(12.3)	16.1–21.9 (17.3)
Available-P in soil (mg 100 g^{-1})	1.3–2.69 (2.21)	1.28–2.93(2.23)	1.63–2.89 (2.11)
Dissolved Organic Matter (ppm)	1.3–3.2 (2.2)	1.45–3.8(2.7)	0.55–3.6 (2.6)
Dissolved Oxygen (ppm)	3.7–5.3 (3.9)	3.3–6.4(4.6)	4.4–6.9 (4.3)
NH_4^+ water (ppm)	0.31–0.88 (0.65)	0.34–0.97 (0.68)	0.41–0.91 (0.59)
Nitrate – N (ppm)	0.06–0.53 (0.36)	0.05–0.47(0.34)	0.16–0.6 (0.33)
Nitrite – N (ppm)	0.009–0.06 (0.04)	0.013–0.075(0.037)	0.011–0.07(0.032)
Organic carbon in soil (%)	0.24–0.56 (0.41)	0.49–0.62(0.54)	0.57–0.7 (0.61)
pH	6.7–8.6 (6.83)	6.9–8.4(7.11)	6.7–8.1 (7.32)
Phosphate – P (ppm)	0.07–0.34 (0.21)	0.06–0.33(0.21)	0.13–0.54 (0.26)
Soil pH	6.6–7.1 (6.94)	6.8–7.1(7.01)	6.8–7.1 (6.97)
Temperature (°C)	27.8–31.2 (28.4)	27.7–31.3(28.4)	27.9–31.5(28.7)
Total alkalinity (ppm)	79–88 (82)	68–109(94)	73–107 (88)
Total plankton (units l^{-1})	1.4x10^3–2.3x10^3 (1.7x10^3)	2.9x10^3–3.7x10^3 (3.3x10^3)	9.4x10^2–2.8x10^3 (1.3x10^3)
TSS (ppm)	162–367 (211)	137–290 (220)	60–247 (178)

*Figures in parenthesis represent mean values.

TABLE 9.2 Species-Wise Growth Characteristics of IMCs (Fry to Advanced Fingerling Production)

Species	C. catla			L. rohita			C. mrigala			C. carpio		
	P_1	P_2	P_3	P_1	P_2	P_3	P_1	P_2	P_3	P_1	P_2	P_3
IMBW (g)	1.8	1.8	1.8	1.7	1.7	1.7	2.2	2.2	2.2	1.6	1.6	1.6
MBW-10/6	270.4	293.5	286.0	88	95	102	110	128	145	158	165	180
PDI (g)	0.93	1.01	0.98	0.3	0.32	0.35	0.37	0.43	0.49	0.54	0.56	0.62
K_n	1.02	1.1	1.08	0.89	0.96	0.96	0.99	1.03	1.0	0.98	1.16	1.1
GP rank		I			IV			II			III	
AFCR	P_1 (1.13)				P_2 (1.22)				P_3 (1.31)			P_3 (1.31)

Stocking density (SD)=30000 fry/ha (P_1 – 1900, P_2 – 2400, P_3 – 3100); stocking composition (SC)=30:30:40::SF:CF:BF

(P_1: pond-1, P_2: pond-2, P_3: pond-3; IMBW: Initial mean body weight; MBW: mean body weight; K_n: Ponderal index/condition factor=(weight/cube of length); SF: surface feeder; CF: column feeder; BF: bottom feeder; GP rank: growth parameter rank; AFCR: average feed conversion ratio)

The general observations on the aquaculture activities are listed below.

- Non-availability of fingerlings and stocking with fry, i.e., fry to advanced fingerling production option undertaken;
- Pre and post-stoking mortality due to size and transportation shock;
- Poor plankton population due to poor soil quality (newly excavated ponds) and thus the weak DO (dissolved oxygen) concentration;
- Recruitment of *A.mola* through stocking material, enhanced their population many fold and became competitor with IMCs for food and space – leading to poor growth performance.

Under on-dyke horticulture activities, there were 114 papaya, 89 banana, and 16 coconut plants around 1ˢᵗ pond, 69 banana, 9 papaya and 4 coconut plants around 2ⁿᵈ pond and 70 banana plants were planted around the 3ʳᵈ pond (Figure 9.10). Besides another 90 banana plants were planted in adjacent area. The different varieties of tissue culture banana planted are *G-9, Bantal, and Robosta*. Papaya variety was *"farm selection."*

In the first year under on-dyke horticulture activities vegetable such as bottle gourd in 386 m² area (7.8 t/ha), tomato in 252 m² area (2 t/ha) and brinjal on 66 m² (1.52 t/ha) were taken up. Different varieties of paddy such as *Khandagiri, Swarna, CR-1009 and Surendra* were grown in four different plots showed average yield of 2.72 t/ha.

In subsequent years on an average 220 bunches of banana were harvested. Different varieties of paddy such as *Khandagiri, Swarna, CR-1009 and Surendra* were grown in four different plots. During *kharif* the yield of Khandagiri was 2.1 t/ha, Surendra gave 3.2 t/ha and Swarna showed average yield of 2.7 t/ha. During *rabi* Khandagiri paddy gave a yield of 2.3 t/ha.

FIGURE 9.10   Integrated farming system in waterlogged area of ICAR – IIWM farm.

Different vegetables were taken as on-dyke horticultural activities as well as intercrops such as brinjal (6.25 t/ha), cowpea (1.5 t/ha), Bean (2 t/ha), ladies finger (4.9 t/ha) and 200 kg of bottle gourd was also obtained.

9.3.3   GROWTH PARAMETERS OF BIODRAINAGE PLANTATION

The different growth parameters observed were height (cm), collar diameter (mm), and diameter at breast height (DBH) (mm) at the time of planting and then at every 3 months after planting (MAP). In *Acacia mangium*, the net increment in plant height over initial was 128.8%, 270.8%, 632%, 803% after 3, 6, 9 and 12 months after planting (MAP), respectively. The net increment (NI) in collar diameter was 154%, 1057%, 2002%, and 2528% respectively during the same period. However, DBH (diameter at breast height, i.e., at 1.37 m from ground) was seen as 20.6 mm, 36.2 mm and 51.4 mm at 6, 9 and 12 MAP respectively. In *Casuarina equisetifolia* plant height increased to 105.3 cm (net increment of 56%), 209.6 cm (NI 210%), 342.4 cm (NI 407%), and 428.0 cm (NI 534%) after 3, 6, 9 and 12 months after planting, respectively. The net increment in collar diameter was 152%, 490%, 1157%, and 1420% after 3, 6, 9, and 12 MAP respectively. The diameter at breast height attained 11.3 mm, 20.7 mm and 31.1 mm after 6, 9, and 12 months after planting. Thus, *Acacia mangium* was faster both in height growth and collar diameter than *Casuarina*. However, *Casuarina* stem was less tapering than *Acacia mangium* at 12 months after planting. The average mortality of trees after one year for both the species were very less (< 6%).

After about four years of planting (during July 2008), for acacia the highest DBH reached up to 20.1 cm and in casuarinas it reached up to 12.5 cm. However, the average collar diameter at bottom, DBH, height and canopy area were 178 mm, 143 mm, 15.4 m and 3.7 m respectively by fourth year of planting. For casuarinas the average collar diameter at bottom, DBH, height and canopy area are 143.7 mm, 108 mm, 13.5 m and 3.85 m respectively by the same period. The growth characteristic curve of both the species are given in Figures 9.11 and 9.12.

The transpiration in acacia ranged between 1.95 and 2.32 m mol/m^2/s with stomatal conductance 69.4 to 84.5 m mol/m^2/s during March 2008 (after 44 months of planting). In case of casuarina the range of transpiration was between 2.34 and 2.75 m mol/m^2/s with stomatal conductance up to 183.1 m mol/m^2/s during the same time. The progressive shade has significantly reduced the intensity of incident radiation to up to 50%. This has

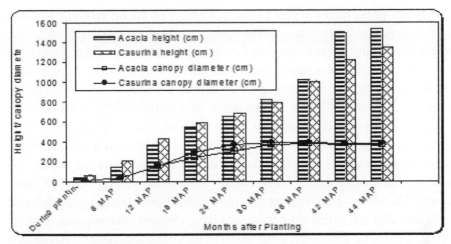

FIGURE 9.11 Height and canopy diameter growth of Acacia and casurina observed in modified waterlogged wasteland.

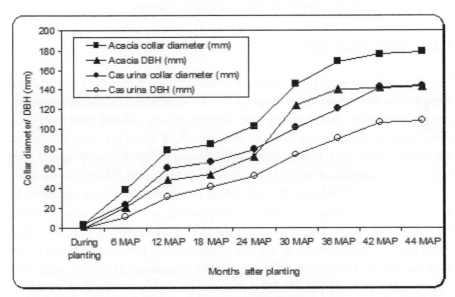

FIGURE 9.12 Progressive growth of collar diameter and diameter at breast height (DBH) of Acacia and casurina observed in modified waterlogged wasteland.

affected the transpiration efficiency and stomatal conductance of mainly bottom tier leaves of acacia and casuarina vegetation. However, up-tier leaves showed transpiration at normal range, i.e., up to 5.7 and 5.3 m/mol/m²/s in casuarina and acacia respectively.

Intercropping of bottle gourd among the trees on the raised platforms was undertaken after the monsoon season was over in the first year of planting. Due to earth work the top soil was less fertile and the soil was acidic. Therefore, bottle gourd which was planted in basins was chosen over other crops. One pit was made among four trees. So the plant to plant spacing in bottle gourd was also kept at 2 m. From one platform of dimension 20 m x 20 m on an average 360 kg of bottle gourd was harvested which is about 9 t/ha. After one year of planting as it is observed from Figure 9.11 that the average height and canopy of acacia was 3.61 m and 1.58 m and for casuarina the height was 4.28 m and canopy 1.62 m. The canopy cover restricted growing of other crops. Therefore, intercropping of crops which grow better under diffused light and also suit inside plantation area were chosen. Intercropping of pineapple, arrowroot, turmeric among the trees was done successfully. About 220 pieces of pine apple was harvested after 18 months of planting and continuing and 50 kg of turmeric seed and 40 kg of arrowroot seed was produced during each season as intercropping in bio-drainage plantation.

9.3.4   SOIL ANALYSIS

The soil analysis done prior to the plantation showed that the experimental plots had highly acidic soil with pH around 3.5, which might have happened due to continuous water logging and washing of top soil and base materials. The electrical conductivity (EC) of soil was 0.14 dS/m and organic carbon (OC) was 0.16% prior to the plantation. The soil analysis done after one year of plantation showed remarkable improvement in soil pH. The pH of the soils of the raised beds/platforms (P1 and P4) is comparatively better and close to neutral or slightly acidic after one year of planting, whereas EC is well within the permissible limit. The available organic carbon improved from very low status to low-medium but was not that remarkable in elevated platform under plantation or in depressions after one year of plantation. However, after four years of plantation the pH of all the elevated platforms as well as depressions have become near neutral (Figure 9.13.a). The EC of the soil was well within permissible limit. But there was no remarkable change even after four years of planting (Figure 9.13.b). There was improvement in organic carbon from very low status prior to plantation to little improvement after one year of plantation and marked improvement after four years of plantation (Figure 9.13.c).

Hence from Figure 9.13, it is observed that biodrainage plantation has improved the soil quality and enhanced the organic carbon status of the soil.

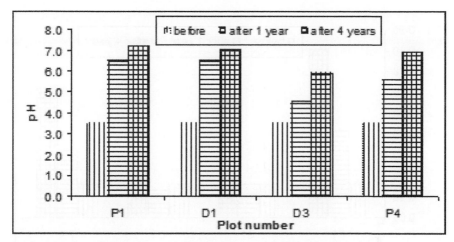

FIGURE 9.13a   Change of soil pH over time in bio-drainage plantation.

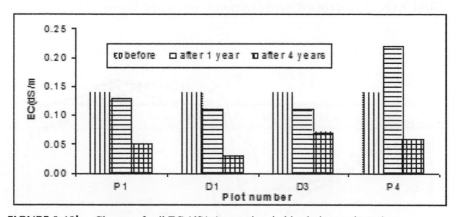

FIGURE 9.13b   Change of soil EC (dS/m) over time in bio-drainage plantation.

Improvement in soil organic carbon is due to incorporation of dry leaves and addition of organic manures in soils through intercropping. The soil from highly acidic has been improved to neutral due to well drained condition of the soils and restricting the washing out of base material.

9.3.5   IMPACT OF BIODRAINAGE PLANTATION ON WATER TABLE

The water table depth was observed every week. These observations were taken to find out whether biodrainage plantations have positive impact of lowering water table or not. The depth to water table in different standard meteorological week is presented in Figure 9.14. In the first year of

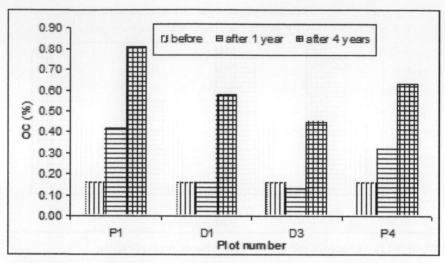

FIGURE 9.13c   Change of soil organic carbon (%) over time in bio-drainage plantation.

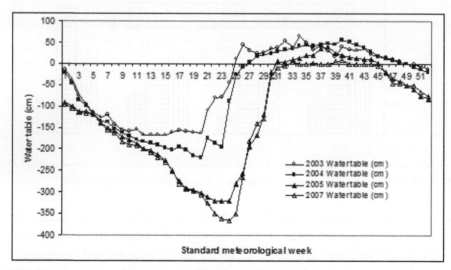

FIGURE 9.14   Water table fluctuation in biodrainage plantation.

experiment and prior to plantation, the water table remained above ground surface during 25th to 48th week whereas during the driest period it lowered up to 1.67 m. During the first year of planting due to land modification the drainage condition of the soil got improved as well as due to little consumption of water by the biodrainage plants the water table got lowered to 2.20 m during driest period and water table was above ground surface during 27th to 48th week. It further lowered to 3.20 m during 2005 summer. Water table observation after three years of planting showed the duration of water table above ground surface has minimized and during driest period it was lowered down to 3.65 m. However, the soil profile remains almost saturated during 33rd to 44th week.

Hence after initial years of work, where the bio drainage component is negligible, it is seen that the land modification has changed the water table regime making it better and suitable for crop growth and crop diversification in comparison to other plots. Once the tree grew it had prominent effect on water table and it is further expected that over time it would further lower and would help in reclaiming the waterlogged degraded land.

9.4   CONCLUSIONS

Rainfall is higher than evaporation during 24th week to 43rd week causing water congestion and excess water is to be stored in ponds for aquaculture and for irrigating *rabi* crops including vegetable and other cash crops. Evaporation is higher than rainfall during 44th week to 23rd week indicating irrigation is required if any crop is to be grown during this period. Land modification by excavating ponds for storing excess water is desirable.

The daily water balance study had resulted the design dimensions of the experimental ponds which were 27 m x 27 m, 30 m x 30 m, and 34 m x 34 m at the top with 2 m depth and side slope 1:1 in experimental plots 1, 2 and 3, respectively. The excavated soils were spread around the pond to elevate the surrounding area so as to keep the water table below 2 m from ground surface. Hume pipes of 30 cm diameter and 4 m length were used as inlet and emergency outlet of the pond. Since the objective of the study was to store excess water for reclamation of waterlogged area, the area of the ponds are kept within 20 to 25% of the total area considering the water balance component of the study area.

Integrated farming system with aquaculture in the pond such as rearing magur in the first year followed by Indian major carps in subsequent years was highly profitable and helps in improving the livelihood options of poor farmers. On-dyke horticulture such as banana, papaya and other vegetables as intercrop was possible in the system and helps in crop diversification and rural livelihood option.

Successful establishment of trees and intercrops and its vigorous growth revealed that bio drainage species of acacia and casuarinas can be adopted for reclamation of waterlogged wasteland. Bottle gourd can successfully be taken as an intercrop among biodrainage plantation during the first year of plantation where the canopy of the trees are limited and have not resulted complete shading. Pine apple, turmeric and arrowroot are suitable intercrops after one year of planting as they like diffused light for their better growth. The mortality of biodrainage plantation in highly acidic, non-fertile waterlogged wasteland is not much. Acacia mangium has better performance over casuarinas in the initial years of planting.

Biodrainage has improved the soil quality making the soil to near neutral from highly acidic condition. There is remarkable improvement of soil organic carbon from low to medium due to incorporation of dry leaves of biodrainage plantation as well as due to intercropping. The water table which was above ground surface during rainy season and within 2 m during summer season before biodrainage plantation was lowered down to more than 3.6 m depth from ground surface after three to four years of plantation. From all above observations it could be said that the reclamation of waterlogged degraded land through biodrainage gives encouraging results. Drainage engineers should no longer ignore the opportunities that biodrainage systems can offer. Biodrainage is an attempt for cheaper drainage technology that could replace some of or be complementary to the more expensive remedies for solving the waterlogging/ drainage problems.

9.5 SUMMARY

A study was under taken in a coastal, perennial waterlogged area in ICAR-IIWM research farm. Rainfall and other climatic factor analysis, soil analysis, water balance study, water table fluctuation study at shallow depth up to 2 m was done. A waterlogged wasteland was modified to elevated platforms and depression through land modification and plantations with high transpiration trees like *Acacia mangium* and *Casuarinas* were done. Its effect on groundwater table fluctuation was studied. Rainfall is higher than evaporation during 24th week to 43rd week causing water congestion and excess water is to be stored in ponds for aquaculture and for irrigating *rabi* crops including vegetable and other cash crops. Evaporation is higher than rainfall during 44th week to 23rd week indicating irrigation is required if any crop is to be grown during this period. Land modification by excavating ponds for storing excess water is desirable.

Integrated farming system approach enhances the production and productivity of the waterlogged land. *Acacia mangium* was faster both in height, growth and collar diameter than *Casuarina*. The average mortality of trees after one year for both the species were very less (< 6%). After four years of planting, the highest diameter at breast height (DBH) reached up to 20.1 cm for acacia and for casuarina it reached up to 12.5 cm. However, the average values of collar diameter, DBH, height and canopy area were 178 mm, 143 mm, 15.4 m and 3.7 m respectively. For casuarina the average values of collar diameter, DBH, height and canopy area were 143.7 mm, 108 mm, 13.5 m and 3.85 m respectively by the same four years period.

Intercropping of pineapple, arrowroot, turmeric among the trees was also done successfully. In a year on an average 220 pieces of pineapple was harvested and 50 kg of turmeric seed and 40 kg of arrowroot seed was produced as intercropping in bio-drainage plantation. The depth to pre-monsoon water table changed from 0.5 m to 1.67 m due to well drained condition as well as due to bio drainage after one year of plantation and even went down to 2.20 m in the next year and to 3.20 m below ground level in third year and continuously declining. The soils of the experimental plots were highly acidic (pH: 3.5–5.0), low organic carbon (0.13–0.67%), and low in available nutrients (N<280 kg/ha, K: 50–170 mg/kg of soil, P: 5–10 mg/kg of soil) and high iron contamination which was restricting the growth and yield of crop prior to intervention, but got improved by land modification and biodrainage plantations over time. Successful establishment of trees and intercrops and its vigorous growth revealed that bio drainage species of acacia and casuarinas can be adopted for reclamation of waterlogged wasteland.

KEYWORDS

- **Agro-forestry**
- **Aquaculture**
- **Biodrainage**
- **Canopy area**
- **Compost**

- **Crop diversification**
- **Drainage problems**
- **Evaporation**
- **Evapotranspiration**
- **Farm yard manure**
- **Integrated farming system**
- **Intercropping**
- **On-dyke horticulture**
- **On-farm-reservoir**
- **Organic carbon**
- **Rainfall**
- **Reclamation**
- **Tree plantation**
- **Wasteland reclamation**
- **Water congestion**
- **Water table fluctuation**
- **Waterlogging**

REFERENCES

1. Ambast, S. K., Sen, H. S., and Tyagi, N. K. (1998). *Rainwater Management for Multiple Cropping in Rainfed Humid Sunderbans Delta (W.B.).* Bulletin No. 2/98, Central Soil Salinity Research Institute, Karnal, India.
2. Bari, M. A., and Schofield, N. J. (1992). Lowering of shallow water table by extensive eucalypt reforestation. *Journal of Hydrology,* 133, 273–291.
3. Cramer, V. A., Thorburn, P. J., and Fraser, G. W. (1999). Transpiration and groundwater uptake from farm forest plots of Casuarina glauca and Eucalyptus camaldulensis in saline areas of southeast Queensland, *Australia. J. Ag. Water Man.,* 39, 187–204.
4. George, R. J., Nulsen, R. A., Ferdowsian, R., and Raper, G. P. (1999). Interactions between trees and groundwater in recharge and discharge areas – a survey of Western Australian sites. *J. Ag. Water Man.,* 39, 91–113.
5. Heuperman, A. (1999). Hydraulic gradient reversal by trees in shallow water table areas and repercussions for the sustainability of tree growing systems. *Agricultural water management.* 39, 153–167.
6. Jena, S. K., (2006). Waterlogged area management. *Journal of Indian Institute of Public Administration.* Vol. XIV, 357–368.

7. Jena, S. K., Sahoo, N., Roy Chowdhury, S., Mohanty, R. K., Kundu, D. K., and Mohanty, M. (2006). Optimizing micro water resources design and integrated farming system approach for enhancing productivity of waterlogged area. *J. of the Indian Society of Coastal Agricultural Research,* 24(1), 180–183.
8. Kapoor, A. S. (1999). Bio-drainage: to overcome waterlogging and salinization problems in irrigated lands. *17th Congress on irrigation and drainage.* Vol. 1-C, ICID, Granada, Spain, pages 9–22.
9. Maji, A. K., Reddy, G. P. O., and Sarkar, D. (2010). *Degraded and wastelands of India: status and spatial distribution.* ICAR, New Delhi, P. 158.
10. Panigrahi, B. (2001). *Water Balance Simulation for Optimum Design of On-Farm Reservoir in Rainfed Farming System.* PhD Thesis, Indian Institute of Technology, Kharagpur, India.
11. Sanchez-Cohen I., Lopes V. L., Slack, D. C., and Fagel, M. M. (1997). Water balance model for small-scale water harvesting systems. *Journal of Irrigation and Drainage Engineering,* ASCE, 123(2), 123–128.
12. Thiyagarajan, G., and Umadevi, R. (2006). Bio-drainage. *Science Tech Entrepreneur,* Nov, pages 1–10.

CHAPTER 10

ROLE OF SIMULATION MODELING OF AQUIFER SYSTEMS IN WATER RESOURCES PLANNING AND MANAGEMENT

S. MOHANTY[1] and MADAN KUMAR JHA[2]

[1]ICAR-Indian Institute of Water Management, Chandrasekharpur, Bhubaneswar-751 023, India, Mobile: +919438008253; E-mail: smohanty.wtcer@gmail.com

[2]Agricultural and Food Engineering Department, Indian Institute of Technology, Kharagpur–721302, West Bengal, India, E-mail: madan@agfe.iitkgp.ernet.in

CONTENTS

10.1 INTRODUCTION

Groundwater is an important and invaluable natural resource on earth. Because of its several inherent qualities (e.g., consistent temperature, widespread and continuous availability, excellent natural quality, low development cost, limited vulnerability, resilience against drought, etc.) as well as the relative ease and flexibility with which it can be tapped, it has been considered to be a reliable and safe source of water supplies in all climatic regions including both urban and rural areas of developed and developing countries [4, 7, 22, 25]. It can be drawn on demand, and in case of emergency it can be used as alternate source of water making it more attractive to many groups of users. It is estimated that groundwater provides about 50% of the current global domestic water supply, 40% of the industrial supply, and 20% of water use in irrigated agriculture [24]. However, the aquifer depletion due to over-exploitation and the growing pollution of groundwater are threatening our sustainable water supply and ecosystems.

In India, in spite of favorable national scenario on the availability of groundwater, there are several areas of the country that face water scarcity due to over-exploitation of groundwater. Excessive groundwater exploitation has led to alarming decrease in groundwater levels in several parts of the country: Tamil Nadu, Gujarat, Rajasthan, Punjab and Haryana [5]. In recent studies, the analysis of GRACE satellite data revealed that the groundwater reserves in the states of Rajasthan, Punjab and Haryana are being depleted at a rate of 17.7 ± 4.5 km^3/year [18]. The depletion of groundwater resources has increased cost of pumping, caused seawater intrusion in coastal areas and has raised questions about sustainable groundwater supply as well as environmental sustainability. Therefore, efficient and judicious utilization of groundwater resources is essential as part of sustainable land and water management strategies.

The groundwater simulation models have emerged as a preferred tool among water resources researchers and managers for studying the impacts of groundwater development on future scenario [3, 19]. These models are useful in simulating groundwater flow scenarios under different management options, and thereby taking corrective measures for the efficient utilization of water resources. The simulation approach attempts to replicate real world complexity by integrating components of the physical hydrogeologic system and providing insight into changes within the aquifer and their interaction with overlying surface water systems. Groundwater simulation models are currently in routine use for water supply management, pollution control, and environment protection.

Visual MODFLOW is a widely applied groundwater model used by various regulatory agencies, universities, consultants and industry both in developed and developing countries. It integrates the MODFLOW for simulating the flow, MODPATH for calculating advective flow path lines, MT3D/RT3D for simulating transport and SEAWAT for simulating coupled flow and transport processes. MODFLOW is a modular three-dimensional finite difference groundwater flow model, which simulates transient/steady groundwater flow in complex hydraulic conditions with various natural hydrological processes and/or artificial activities and can be used for large areal extent and for multi-aquifer modeling [9].

In the last four decades, groundwater simulation models have been widely used for developing optimal groundwater management strategies in different parts of the world [2, 8, 13, 14, 16, 20, 21, 27]. However, in developing countries like India, basin-scale groundwater modeling studies are scanty probably owing to limited or absence of good-quality spatial and temporal field data and the lack of technical expertise. As a result, so far very few studies on basin-wide groundwater-flow modeling [1, 6, 15, 17] have been carried out in India in general and eastern India in particular.

In this chapter, the concepts of groundwater modeling has been described so that these can be useful for other researchers to carry out such modeling works to develop management strategy for integrated water resources management in respective river basins. Finally, a case study of groundwater modeling in Kathajodi-Surua Inter-basin within Mahanadi deltaic system of Odisha has been presented.

10.2   BASIC CONCEPTS OF MODELING

A model is a tool designed to represent a simplified version of reality. It is a representation of a portion of the natural or human-constructed world. It is always simpler than the prototype system and can reproduce some but not all of its characteristics. Models can be used as a predictive tool, interpretive tool or generic tool. Different types of hydrologic/hydrogeologic models can be broadly classified into two major groups: material models and mathematical models. Material models can be either physical, scaled-down versions of a real system or analog, which use substances other than those in a real system.

Mathematical models are abstractions that represent processes as equations, physical properties as constants or coefficients in the equations, and measures of state in the system as variables. Mathematical models can be either empirical (black box) or theoretical and can be further classified as deterministic and stochastic. Theoretical models rely on physical laws and theoretical principles, whereas empirical models are based on observed input-output relationships only. Deterministic models mathematically characterize a system and give the same response or results for the same input data. Conversely, stochastic models use the statistical characteristics of hydrologic or hydrogeologic phenomena to predict possible outcomes.

Mathematical models integrate existing knowledge about processes occurring in a system into a logical framework of rules, equations and relationships to quantify how a system behaves. They range from a simple linear regression equation to highly complex partial differential equations such as water flow and solute transport in porous media. Mathematical models are extensively used in hydrological sciences and can be solved analytically after making several simplifying assumptions or numerically. Accordingly, they are classified as analytical models and numerical models, respectively.

10.3   PROTOCOL FOR GROUNDWATER MODELING

A groundwater model can be defined as simplified representation of real world groundwater systems. The major goal of groundwater modeling is to predict hydraulic head in an aquifer system and/or the concentration distribution of a particular chemical in the aquifer in time and space. Numerical modeling of groundwater systems has been evolving since the mid-1960s and today computer simulation models and interactive computer programs (called "modeling systems," i.e., generalized software packages) are commonplace. The tremendous advances in computer technology have made these the primary and standard tool for analysis and decision making in small-scale as well as large-scale groundwater problems related to quantity (groundwater flow) and quality (contaminant transport). The dominance of numerical models in groundwater studies has led to the use of phrase "groundwater models" as a synonym for 'numerical groundwater models'. These days, standard and robust computer programs for simulating flow and transport in aquifer systems are available and the model user can apply a computer program to the problem under study without writing computer codes. However, the protocol for numerical modeling (Figure 10.1) as suggested by Anderson and Woessner [3] can be followed in order to obtain reliable and useful modeling results. In addition, it is essential that the modeler has a thorough knowledge of the hydrologic/hydrogeologic processes being modeled and has a solid experience in modeling. The modelers should be aware of the details of the numerical method, including the derivative approximations, the scale of discretization, and the matrix solution techniques, otherwise significant errors can be introduced and remain undetected. The major steps in groundwater modeling are discussed in this section.

10.3.1   DEVELOPMENT OF CONCEPTUAL MODEL

A key step in groundwater modeling procedure is to develop a conceptual model of the system being modeled. A conceptual model is a pictorial representation of the groundwater flow system, frequently in the form of a cross-section. The nature of the conceptual model determines the dimensions of the numerical model and the design of the grid. The purpose of building a conceptual model is to simplify the complex field problem and organize the associated field data to make it more amenable to modeling [3, 19]. Simplification is necessary as a complete reconstruction of the field system is not feasible. The analysis of lithologic data collected across the study area will be useful for building the conceptual model. Based on the conceptual model, the governing equation of the model is decided.

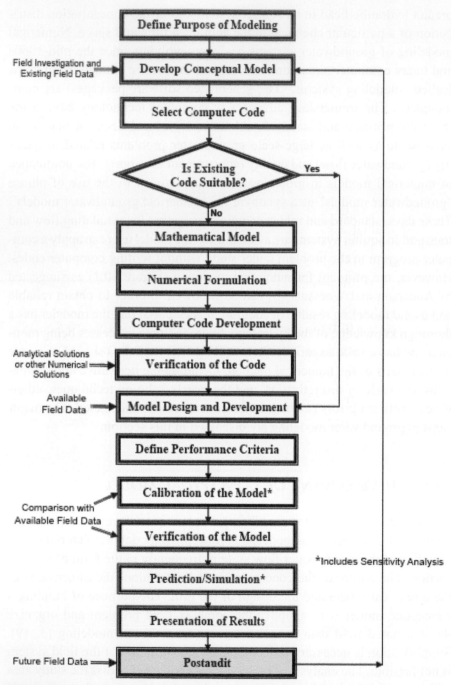

FIGURE 10.1   Protocol for model application (modified from Ref. [3]).

10.3.2 GRID DESIGN

In a numerical model, the continuous problem domain is replaced by a discretized domain consisting of an array of nodes and associated finite difference blocks or finite cells. The nodal grid forms the framework of the numerical model. The conceptual model and the selection of model type determine the overall dimensions of the grid. The Grid Module of Visual MODFLOW allows the user to define and discretize the model domain. The user can design a suitable grid, add or delete grid lines, change cell or layer elevations or remove cells from the computations. The grid cells outside the model boundaries can be designated as *inactive* or *no flow cells*.

These inactive grid cells are ignored by the model and are not used in any of the calculations of flow or contaminant transport. A scale of 500 m length on the field with respect to the dimension of a single grid is quite reasonable in groundwater modeling studies. The model layers are decided based on the conceptual model of the study area. The elevation of the top and bottom of different layers can be assigned by importing them to the model from a MS-Excel database. Similarly, the location of the pumping wells, observation wells and weekly groundwater levels can be imported to the model from MS-Excel databases.

10.3.3 ASSIGNMENT OF BOUNDARY AND INITIAL CONDITIONS

Correct selection of boundary conditions is a critical step in model design. Under steady state conditions, the boundaries largely determine the flow pattern. Boundary conditions influence transient solutions, when the effects of the transient stress reach the boundary. Setting boundary conditions is a step in model design, which is sometimes subjected to serious error. The boundaries can be physical boundaries or hydraulic boundaries. For example, the groundwater division and the streamline boundaries are hydraulic boundaries, but the river is a physical boundary. The hydrogeologic boundaries can be of three basic types:

a. Specified head boundaries (Dirichlet boundary conditions),
b. Specified flux boundaries (Neumann boundary conditions) and
c. Head dependent flux boundaries (Cauchy boundary conditions).
   MODFLOW has separate subroutines or packages to handle boundary

conditions such as constant head, river, stream, drain, evapotranspiration and recharge, etc.

Initial conditions refer to the head distribution everywhere in the system
at the beginning of the simulation and thus are boundary conditions in time.
It is a standard practice to select the initial condition a steady state head solution generated by a calibrated model.

10.3.4   MODEL PARAMETERS

The numerical model requires assignment of model parameters: aquifer
properties, sources and sinks, groundwater level distribution, and spatial
and temporal distribution of recharge, evapotranspiration, etc. The aquifer
parameters mostly include hydraulic conductivity and specific storage in
case of confined aquifer and specific yield in case of unconfined aquifer.
The pumping test analysis is the most ideal method of estimation of model
parameters on a regional scale. In case of absence of pumping test data, grain
size analysis and textural information can be used for estimation of model
parameters like hydraulic conductivity and specific storage. However, as the
estimation of aquifer parameters can not be guaranteed to be 100% accurate, further refinement of these values can be done during the course of
calibration. It is a standard practice in groundwater modeling that whenever
only horizontal hydraulic conductivity data is available, the K_h to K_v ratio is
assumed as 10 for alluvial aquifer systems [23].

The Well Package of MODFLOW software is designed to simulate
inflows and outflows through recharge wells and pumping wells, respectively. The location of the pumping and observation wells can be imported,
added or deleted using this package. Well-screen intervals, pumping schedules and observed groundwater head data can be provided to each pumping
well either by direct assigning or importing them from MS-Excel files. The
day-wise pumping extraction data is generally not recorded in developing
countries like India. Therefore, questionnaire survey of farmers can be done
to obtain the historical record of pumping data. The agricultural crop coverage and electric consumption can be used as indirect indicators for obtaining
pumping extraction data. Recharge is another input parameter to the model,
which needs to be estimated. Recharge package of MODFLOW is designed
for zone wise and layer wise assignment of the parameter. No single recharge
estimation method can be guaranteed to be 100% accurate. Therefore, two or

three different methods should be used before arriving at a recharge value. Simplest methods of recharge estimation are empirical methods with respect to the monsoon rainfall. Further refinement of the recharge value can be done during the calibration of the model. Evapotranspiration is another input parameter to the model, which can be estimated from the land use map and crop coverage of the study area.

10.3.5 MODEL CALIBRATION

Calibration of the numerical model refers to the demonstration that the model is capable of producing field measured heads and flows. Calibration is accomplished by estimating a set of parameters, boundary conditions, and stresses that produce simulated heads and fluxes that match field-measured values within a pre-established range of error (known as 'calibration target'). Analysis of the difference between observed and simulated heads gives an indication as to where adjustment of calibration parameters may be necessary in order to minimize the difference. Finding this set of values amounts to solving what is known as the inverse problem. Generally, hydraulic conductivity, specific storage and recharge are considered as calibrated parameters for groundwater flow simulation models.

Model calibration can be performed using steady state or transient data sets. Most calibrations are performed under steady state conditions, which may also involve a second calibration to a transient data set. After the calibration, the model is validated using another set of data. There are basically two ways of finding model parameters to achieve calibration, i.e., of solving the inverse problem: trial and error calibration (also known as 'manual calibration') and automated calibration.

In *the trial-and-error calibration*, parameter values are initially assigned to each node or element in the grid. During calibration, parameter values are adjusted in sequential model runs to match simulated heads and flows to the calibration targets. This method is generally very time consuming, cumbersome and influenced by modeler's expertise.

Automated inverse modeling is performed using specially developed codes that use either a direct or indirect approach to solve the inverse problem. An inverse code automatically checks the head solution and adjusts parameters in a systematic way in order to minimize an objective function, an example of which will be to minimize the sum of the squared residuals, i.e., differences between simulated and observed heads. The automated inverse modeling may not be subjective and is not influenced by the modeler. However, it suffers from being complicated and computer intensive.

10.3.6 PERFORMANCE EVALUATION OF THE MODEL

In groundwater modeling studies, different criteria of evaluation like mean error (ME), mean absolute error (MAE), root mean squared error (RMSE), correlation coefficient (r) and Nash-Sutcliffe efficiency (NSE) are generally used during calibration process.

10.3.6.1 Mean Error (ME)

The mean error is a measure of the average residual value defined by the equation:

$$Bias = \frac{1}{N}\sum_{i=1}^{N}(h_{si} - h_{oi}) \tag{1}$$

where, h_{oi} = observed groundwater level of the ith data [L], h_{si} = simulated/predicted groundwater level of the ith data, and N = number of observations.

The positive values of mean error indicate overall over-prediction by the model, while the negative values indicate overall under-prediction by the model.

10.3.6.2 Mean Absolute Error (MAE)

The mean absolute error is similar to the mean error except that it is a measure of the average absolute residual value defined by the equation:

$$MAE = \frac{1}{N}\sum_{i=1}^{N}|h_{si} - h_{oi}| \tag{2}$$

Mean absolute error measures the average magnitude of the residuals, and therefore provides a better indication of calibration than the mean error.

10.3.6.3   Root Mean Squared Error (RMSE)

The *RMSE* is a widely accepted performance evaluation index, and is defined by the following equation:

$$RMSE = \sqrt{\frac{\sum_{i=1}^{N}(h_{si} - h_{oi})^2}{N}} \tag{3}$$

10.3.6.4   Correlation Coefficient (r)

$$r = \frac{\sum_{i=1}^{N}\left(h_{oi} - \overline{h_o}\right)\left(h_{si} - \overline{h_s}\right)}{\sqrt{\sum_{i=1}^{N}\left(h_{oi} - \overline{h_o}\right)^2 \sum_{i=1}^{N}\left(h_{si} - \overline{h_s}\right)^2}} \tag{4}$$

where, $\overline{h_o}$= mean of observed groundwater levels [L], and $\overline{h_s}$= mean of simulated groundwater levels [L].

Correlation coefficient determines whether two ranges of data move together, i.e., whether large values of one data set are associated with large values of the other data set, whether small values of one data set are associated with large values of the other data set, or whether values in both data sets are unrelated.

10.3.6.5   Nash-Sutcliffe Efficiency (NSE)

The Nash-Sutcliffe efficiency is another widely used performance evaluation index for hydrological models and is defined by the following equation:

$$NSE = 1 - \frac{\sum_{i=1}^{N}(h_{oi} - h_{si})^2}{\sum_{i=1}^{N}\left(h_{oi} - \overline{h_o}\right)^2} \tag{5}$$

The best-fit between observed and simulated groundwater levels under ideal conditions would yield ME = 0, MAE = 0, RMSE = 0, r = 1 and NSE = 1.

10.3.7 SENSITIVITY ANALYSIS

Sensitivity analysis is another component of model evaluation, which addresses uncertainty in modeling results. Due to the uncertainties in estimating aquifer parameters, stresses and boundary conditions, a sensitivity analysis is an essential step in modeling studies. This is particularly important when many parameters are to be optimized during calibration. The main objective of a sensitivity analysis is to understand the influence of various model parameters and hydrological stresses on the aquifer system and to identify the most sensible parameter(s), which will need a special attention in future studies. Sensitivity analysis is also simultaneously done during the calibration of the model. Hydraulic conductivity and recharge are generally found to be more sensitive and specific storage or storage coefficient is normally a less sensitive parameter.

10.3.8 SIMULATION OF GROUNDWATER MANAGEMENT SCENARIOS

The calibrated and validated model can be used for a variety of management and planning studies. In a predictive simulation, the parameters optimized during calibration are used to predict the system response to future events. Different type of predictive simulation can be done to study the response of the aquifer to different management scenarios. This may include response of the aquifer to different pumping levels and to simulate groundwater levels in the long run under existing or different management options. Based on the results of this study, management strategies could be formulated for the efficient utilization of water and land resources in the study area.

10.4 CASE STUDY

A groundwater flow simulation model was developed for the Kathajodi-Surua Inter-basin within the Mahanadi deltaic system of eastern India (Figure 10.2). The study area is a river island surrounded by the Kathajodi River and its branch Surua on all sides. It is a part of the Mahanadi Delta, which is located around the confluence of the Mahanadi River with the Bay of Bengal along the eastern coast of India. The area is characterized as a tropical humid climate with an average annual rainfall of 1650 mm,

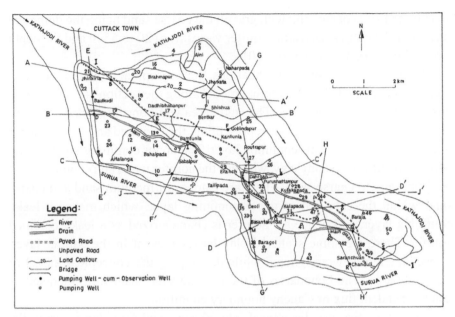

FIGURE 10.2 Location of observation and pumping wells and geologic cross sections in the study area [11].

of which approximately 80% occurs during June to October. Agriculture is the major occupation of the inhabitants and groundwater is the major source of water. At present there are about 70 functioning production wells, which supply major chunk of water required for irrigation and domestic purposes. Monitoring of groundwater levels in the study area was done by selecting nineteen tube-wells in such a way that they represent approximately four west-east and four north-south cross-sections of the study area (Figure 10.2). Weekly groundwater-level data at the nineteen sites was monitored from February 2004 to October 2007, which was used for studying the groundwater characteristics in the study area and calibration of groundwater-flow simulation model.

The drillers log charts available at 70 different sites in the study area were collected from the office of *Orissa Lift Irrigation Corporation* (OLIC), Cuttack, Orissa. The log charts were analyzed by drawing geologic profiles and performing stratigraphic analysis across different cross-sections to characterize the subsurface formation. The analysis showed that the river basin is underlain by a confined/semi-confined aquifer, which contributes a major source of groundwater. The aquifer consisted of coarse sand, medium to coarse sand and coarse sand with gravel; the coarse sand being the dominant formation. The thickness of the aquifer varies from 20 to 55 m and the depth from 15 to 50 m over the basin [10].

Based on the lithologic investigations, the conceptual model of the study area was developed. The conceptual model along the section J-J' of the river basin is shown in Figure 10.3. The study area was conceptualized as a two-layer system with the lower one representing, the confined aquifer and upper one representing the semi-confining layer. The upper confining layer mostly consists of clay and sandy clay, whereas the aquifer material is comprised of medium sand to coarse sand. There are patches of medium sand and coarse sand within the clay bed of upper confining layer, which makes it leaky confining layer, and hence the aquifer is characterized as a leaky confined aquifer. There are some scattered clay lenses present in the aquifer layer. These clay lenses were ignored while developing the conceptual model of the study area. The boundaries of the groundwater basin were modeled as head-dependent flux or Cauchy boundary condition.

The study area was discretized into 40 rows and 60 columns using the Grid module of Visual MODFLOW software (Figure 10.4). This resulted in 2400 cells, each having a dimension of approximately 222 m × 215 m.

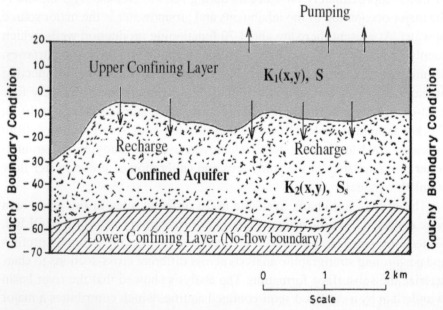

FIGURE 10.3   Conceptual model of the study area [11].

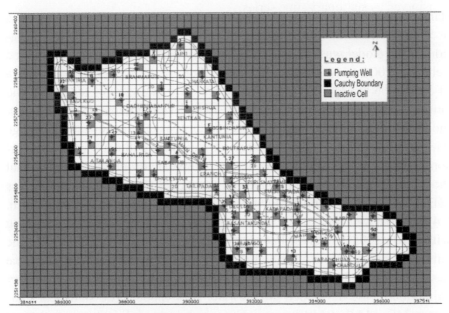

FIGURE 10.4   Design of finite difference grid of Kathajodi-Surua Inter-basin with boundary conditions and location of pumping nodes.

The cells lying outside the study area were assigned as inactive cells. The hydrogeologic setting of the study area as conceptualized earlier was divided into two model layers with the lower one representing the confined aquifer. The data on surface elevation, bottom elevation of the top layer and bottom elevation of the aquifer layer at available 19 sites were imported to the MODFLOW software from the database prepared using MS-Excel files. Similarly, the location of pumping wells, observation wells and weekly groundwater levels of the model period were also imported. The position and length of the screen in the pumping and observation wells was assigned using the model.

The river heads were assigned as varying head boundary conditions using the 'River Package' of Visual MODFLOW software (Figure 10.4). The hydraulic conductivity and specific storage values estimated by pumping test analysis were assigned to the model. The pumping test analysis at 9 sites distributed over the study area showed that the hydraulic conductivity values varied form 11.3 to 96.8 m/day, whereas the specific storage values varied form 1.43×10^{-4} to 9.9×10^{-4}.

The developed groundwater-flow simulation model was firstly calibrated for the steady-state condition and then for the transient condition using standard procedure [3, 26]. The solution of the steady-state calibration was used as an initial condition for the transient calibration. Transient calibration of the model was done using the groundwater level data of the period February 2004 to May 2006; and validation of the model using the data of June 2006 to May 2007. A combination of trial and error technique and automated calibration code PEST was used to calibrate the developed flow model by adjusting the hydraulic conductivity, specific storage and recharge within reasonable ranges. The statistical indicators and the visual comparison of observed and simulated groundwater level hydrographs indicated that there is reasonably good calibration of the model. The statistical indicators ME, MAE, RMSE, r and NSE during the calibration period were -0.063 m, 0.478 m, 0.62 m, 0.957 and 0.915, respectively. The scatter diagram along with 1:1 line, 95% interval lines and 95% confidence interval lines for the validation period is shown in Figure 10.5. The figure shows that the 1:1 line lies within the 95% confidence interval lines which indicates satisfactory validation of the developed groundwater flow model. The statistical indicators ME, MAE, RMSE, r and NSE during the validation period were 0.044 m, 0.489 m, 0.632 m, 0.958 and 0.914, respectively.

The sensitivity analysis of the developed model showed that the model is more sensitive to river stage followed by recharge and hydraulic conductivity; it was very less sensitive to changes in specific storage. This indicated the importance of river in maintaining high groundwater levels in alluvial aquifer systems. The findings suggested that the uncertainty associated with hydraulic conductivity will significantly affect the model's ability to make reliable predictions. On the other hand, the uncertainty in specific storage parameter will have little impact on the model's predictive abilities. The scenario analysis was done to simulate groundwater level after 15 years keeping all the existing conditions constant.

It was observed that with the continuation of existing management practices, there is no significant change in groundwater level even after 15 years of pumping at all the sites. The Kathajodi-Surua Inter-basin is a complete river island surrounded by two rivers and due to this, the effect of the boundary conditions (rivers) on groundwater levels has been found very significant. The water that is pumped from the aquifer is being replenished by the river, and hence there is no significant change in groundwater levels even in the long run. Thus, if the existing conditions continue, there is no threat to the groundwater lowering in the study area for coming 15 years.

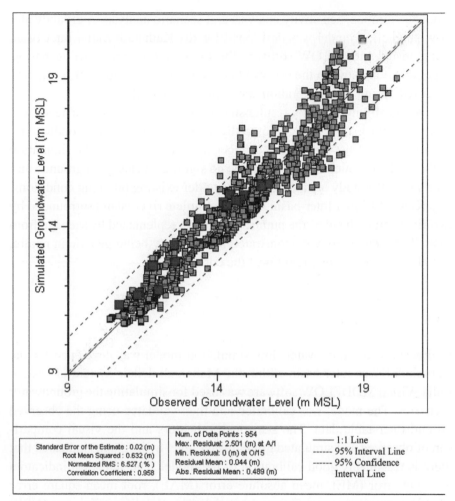

Standard Error of the Estimate : 0.02 (m)	Num. of Data Points : 954	—— 1:1 Line
Root Mean Squared : 0.632 (m)	Max. Residual: 2.501 (m) at A/1	------ 95% Interval Line
Normalized RMS : 6.527 (%)	Min. Residual: 0 (m) at O/15	------ 95% Confidence
Correlation Coefficient : 0.958	Residual Mean : 0.044 (m)	Interval Line
	Abs. Residual Mean : 0.489 (m)	

FIGURE 10.5 Scatter diagram of observed versus simulated groundwater levels for the pooled data during validation period [12].

10.5 CONCLUSIONS

The concept of groundwater flow modeling and steps involved in the modeling procedure has been described in the paper. The major steps involved in the design and development of a groundwater flow simulation model are development of conceptual model, discretization of the basin and grid design, assignment of initial and boundary conditions, input of model parameters, model calibration, sensitivity analysis and scenario analysis. A groundwater flow simulation model was developed for the Kathajodi-Surua Inter-basin using Visual MODFLOW software for simulating groundwater scenarios. A good match between the observed and simulated groundwater levels was observed during both calibration and validation period.

The sensitivity analysis results showed that the groundwater level was more sensitive to river stage followed by recharge and hydraulic conductivity whereas it was least sensitive to changes in specific storage. The simulation of the model revealed that there will be no significant change in groundwater levels over the study area in near future under existing pumping conditions. The Kathjodi-Surua Inter-basin being a complete river island surrounded by two rivers on both sides, the pumped water are replenished by seepage from river. The methodology demonstrated in this study being generic in nature, can also useful for other regions of the country.

10.6 SUMMARY

In this study, a groundwater flow simulation model was developed for the Kathajodi-Surua inter-basin within the Mahanadi deltaic system of eastern India. Visual MODFLOW software was used for simulating the groundwater scenarios. The model was calibrated and then validated using the observed groundwater table data. The statistical indicators and the visual comparison of observed and simulated groundwater level hydrographs indicated that there is reasonably good calibration of the model. The statistical indicators of mean error (ME), mean absolute error (MAE), root mean square error (RMSE), correlation coefficient (r) and Nash –Sutcliffe efficiency (NSE) during the calibration period were -0.063 m, 0.478 m, 0.62 m, 0.957 and 0.915, respectively showing good calibration. Similarly, the statistical indicators ME, MAE, RMSE, r and NSE during the validation period were 0.044 m, 0.489 m, 0.632 m, 0.958 and 0.914, respectively which indicates validation to be strong. Further, sensitivity analysis results showed that the groundwater level was more sensitive to river stage followed by recharge and hydraulic conductivity whereas it was least sensitive to changes in specific storage. The simulation of the model revealed that there will be no significant change in groundwater levels over the study area in near future under existing pumping conditions.

KEYWORDS

- **Aquifer**
- **Calibration**
- **Correlation coefficient**
- **Groundwater**
- **Hydraulic conductivity**
- **Hydrograph**
- **Mean absolute error**
- **Mean error**
- **Model**
- **Model efficiency**
- **Modflow**
- **Nash-sutcliffe efficiency**
- **Recharge**
- **River stage**
- **Root mean square error**
- **Sensitivity analysis**
- **Simulation**
- **Specific storage**
- **Validation**

REFERENCES

1. Ahmed, I., and Umar, R. (2009). Groundwater flow modeling of Yamuna-Krishni inter-stream, a part of central Ganga Plain, Uttar Pradesh. *Journal of Earth System Science*, 118(5), 507–523.
2. Al-Salamah, I. S., Ghazaw, Y. M., and Ghumman, A. R. (2011). Groundwater modeling of Saq Aquifer Buraydah Al Qassim for better water management strategies. *Environmental Monitoring and Assessment*, 173(1–4), 851–860.
3. Anderson, M. P., and Woessner, W. W. (1992). *Applied Groundwater Modeling: Simulation of Flow and Advective Transport*. Academic Press Inc., San Diego, California, 381 pp.
4. Bocanegra, E., Hérnandez, M., and Usunoff, E., (editors) (2005). *Groundwater and Human Development. IAH SP Publication 6, Taylor & Francis*, Abingdon, U.K., 278 pp.
5. CGWB, (2011). *Dynamic Groundwater Resources of India*. Central Ground Water Board (CGWB), Ministry of Water Resources, New Delhi, India.

6. Elango, L., Brindha, K., Kalpana, L., Sunny, F., Nair, R. N. and Murugan, R. (2012). Groundwater flow and radionuclide decay-chain transport modeling around a proposed uranium tailings pond in India. *Hydrogeology Journal*, 20, 797–812.

7. Foster, S., Chilton, J., Cardy, F., Schiffler, M., and Moench, M., (2000). *Groundwater in Rural Development: Facing the Challenges of Supply and Resource Sustainability*. World Bank Technical Paper No. 463, the World Bank, Washington D.C.

8. Lin, Y. C., and Medina, M. A. (2003). Incorporating transient storage in conjunctive stream–aquifer modeling. *Advances in Water Resources*, 26(9):1001–1019.

9. McDonald, M. G., and Harbaugh, A. W. (1988). *A Modular Three Dimensional Finite Difference Groundwater Flow Model (MODFLOW). USGS Scientific Software Group*, Washington D.C.

10. Mohanty, S., Jha, M. K., Kumar, A., and Jena, S. K. (2012). Hydrologic and hydrogeologic characterization of a deltaic aquifer system in Orissa, Eastern India. *Water Resources Management,* 26(7), 1899–1928.

11. Mohanty, S., Jha, M. K., Kumar, A., and Brahmanand, P. S. (2013a). Optimal development of groundwater in a well command of eastern India using integrated simulation and optimization modeling. *Irrigation and Drainage*, 62, 363–376.

12. Mohanty, S., Jha, M. K., Kumar, A., and Panda, D. K., (2013b). Comparative evaluation of numerical model and artificial neural network for simulating groundwater flow in Kathajodi-Surua Inter-basin of Odisha, India. *Journal of Hydrology*, 495, 38–51.

13. Panagopoulos, G. (2012). Application of MODFLOW for simulating groundwater flow in the Trifilia karst aquifer, Greece. *Environmental Earth Sciences*, 67(7), 1877–1889.

14. Rahnama, M. B., and Zamzam, A. (2013). Quantitative and qualitative simulation of groundwater by mathematical models of Rafsanjan aquifer using MODFLOW and MT3DMS. *Arabian Journal of Geosciences*, 6(3), 901–912.

15. Raul, S., Panda, S. N., Hollander, H., and Billib, M. (2011). Integrated water resources management in a major canal command in eastern India. *Hydrological Processes*, 25, 2551–2562.

16. Reeve, A. S., Warzocha, J., Glaser, P. H., and Siegel, D. I. (2001). Regional groundwater flow modeling of the Glacial Lake Agassiz Peatlands, Minnesota. *Journal of Hydrology*, 243(2), 91–100.

17. Rejani, R., Jha, M. K., Panda, S. N., and Mull, R. (2008). Simulation modeling for efficient groundwater management in Balasore coastal basin, India. *Water Resources Management*, 22, 23–50.

18. Rodell, M., Velicogna, I., and Famiglietti, J. S. (2009). Satellite-based estimates of groundwater depletion in India. *Nature*, 460, 999–1002.

19. Rushton, K. R., (2003). *Groundwater Hydrology: Conceptual and Computational Models*. John Wiley and Sons, Chichester, U.K., 416 pp.

20. Sarwar, A., and Eggers, H., (2006). Development of a conjunctive use model to evaluate alternative management options for surface and groundwater resources. *Hydrogeology Journal,* 14:1676–1687.

21. Ting, C. S., Zhou, Y., Vries, J. J. De., and Simmers, H. (1998). Development of a preliminary groundwater flow model for water resources management in the Pingtung Plain, Taiwan. *Groundwater,* 35(6), 20–35.

22. Todd, D. K., and Mays, L. W., (2005). *Groundwater Hydrology*. 3rd edition, John Wiley & Sons, NJ.

23. WHI, (2005). *Visual Modflow Professional Edition User's Manual*. Waterloo Hydrogeologic Inc., Ontario, Canada, 611 pp.

24. World Water Assessment Program, (2003). *The United Nations World Water Development Report 1: Water for People Water for Life*. UNESCO, Paris and Berghahn Books, 575 pp.

25. Zektser, I. S. (2000). *Groundwater and the Environment: Applications for the Global Community*. Lewis Publishers, Boca Raton, Florida, 175 pp.

26. Zheng, C., and Bennett, G. D. (2002). *Applied Contaminant Transport Modeling*. 2nd Edition, John Wiley & Sons, Inc., New York.

27. Zume, J., and Tarhule, A. (2008). Simulating the impacts of groundwater pumping on stream-aquifer dynamics in semi-arid north-western Oklahoma, USA. *Hydrogeology Journal*, 16, 797–810.

CHAPTER 11

ROLE OF CARBON SEQUESTRATION IN RESTORING DEGRADED SOILS: REVIEW

ANIRBAN RAY[1] and SAMPAD GHOSH[2]

Agricultural and Food Engineering Department, Indian Institute of Technology, Kharagpur, West Bengal, Kharagpur 721302, India. Tel.: +91 9231695435, E-mail: anirbanrayiitkgp@gmail.com, anirban@agfe.iitkgp.ernet.in

Department of Chemistry, Indian Institute of Technology, Kharagpur, West Bengal, Kharagpur 721302, India. Tel.: +91 9434238986, E-mail: justsampad@gmail.com

CONTENTS

11.1   INTRODUCTION

The degradation of soil is defined as an implicit or explicit change in the soil health status resulting in a retrograded capacity of the ecosystem in providing goods and services for its habitats and beneficiaries. It is a serious global environmental problem and exacerbated by the climate change. Soil degradation are closely implicated with loss of organic matter, decline in structural condition, erosion, adverse changes in salinity, acidity or alkalinity, decline in soil fertility, and the effects of toxic chemicals, pollutants or excessive flooding [6].

It is noteworthy to mention that soil degradation implies the deterioration of the physical, chemical and biological properties of soil, which is of grave concern encompassing ecosystem sustainability, and livelihood that can potentially snowball into greater international issues including food insecurity leading to inflation of food prices, soil erosion causing sedimentation of dams and floods, reduced revenue, reduced return to investments, increased expenditures, and population migration creating resource and ecological imbalances and so on. Actually, soil is a dynamic system that emerges through a unique balance and interaction of its biological, chemical, and physical components. Hence any remediation work has to be built on a holistic approach to soil rehabilitation and restoration of soil processes using fundamental agro-technologies.

Degraded soils suffer from a multitude of shortcomings in terms of restoring as a functioning system. Foremost problem in degraded soils is poor soil structure that leads to higher bulk density, lower aggregate water stability, lower retention of water useful to plants, and air permeability [13, 24, 25, 28]. Weak aggregate stability leads to crust formation and aggregate breakdown under rainfall thereby making soil susceptible for degradation and erosion washing off crucial nutrients from topsoil [16]. This, in turn, poses economic losses as reduced crop yields and harm to the environment [3, 10, 25]. Conventionally restoration of aggregate stability requires a long-term application of natural and organic fertilizers at high rates, combined with NPK fertilization, liming, and crop rotation [1, 2, 4, 7]. Soil amendments have been shown to be effective in restoring soil physical properties that include various soil polymers and biochars.

A critical point is that action for restoring degraded land has to be integrated across all sectors. The development areas that mandate coordination, cooperation and innovation to highlight a systems approach by applying ecological principles should be taken in account. Therefore, it is necessary to generate ecosystem services and restore soil physical, chemical and biological productivity.

This chapter addresses strategic research challenges by developing a sound understanding of the precincts and embodying inherent solutions.

The key areas, that authors have aimed for to rehabilitate degraded lands, are by promoting soil physical, chemical, microbial properties, nutrient and water use efficiency, sequestering C and thereby mitigating the effects of climate change.

11.2   POTENTIAL OF CARBON SEQUESTRATION IN RESTORING DEGRADED SOILS

Bulk of the required increase in global food production in the future will need to come from intensified farming practices or the world risks high inflation and renewed attack on natural resources. Indian subcontinent in particular already faces the challenge of finding ways to effectively preserve its natural resources, its limited forested areas, and also its soil and water supply. On the other hand aggressive agricultural policies through decades have led to degraded land and the menace continues. The long-term effect of climate change endangers the critical balance of soil processes and ecosystem services further.

Considering the soil degradation in the face of climate change and enhanced accumulation of carbon in air, production of biochar from organic waste and its application in soil can be a promising option in mitigating the problem of climate change by decreasing nitrous oxide (N_2O) and methane (CH_4) emissions. Biochar can be an accessible input to rehabilitate degraded land, improve soil fertility and play a major role in sequestering atmospheric carbon dioxide to mitigate climate change effects [15, 18, 19, 29]. The use of biochar will sequester C in soil and will potentially improve soil structure, soil moisture and soil physico-chemical properties. In particular, amendment of soils with biochar (in the range of 0.5–135 t ha^{-1}) has been shown to increase plant yield, improve chemical and physical properties [11].

Addition of biochar has been shown to increase microbial biomass and activity [20], microbial community composition, and abundance [12, 14, 21, 23, 26, 30, 31]. Incorporation of biochar into soil has been reported to alter nitrification rates and increase N mineralization in forest soils [5, 9]. However, influences within agricultural and grassland soils appear to be uncertain [8, 9]. Biochar is the product of thermal degradation of organic materials in the absence of air [17] that has been promoted by the UN Convention to Combat Desertification for land degradation. Due to its large specific surface area, porous physical structure and aromatic composition [22, 27], biochars have been suggested to increase cation exchange capacity and retain dissolved organic matter as well as generate greater biological activity [22]. The augmentation of microbial activity by biochar amendments has also been attributed to its porous structure providing a habitat for microbial colonization [26], protection from soil micro-arthropods and enhanced availability of C substrates [32].

Application of biochar can provide a long term solution for the large acres of degraded land through innovation in land use, crop system, and use of soil amendments while enhancing production efficiency, improving environmental quality while preserving natural resources and enhancing human well-being. The essence of this chapter echoes well with India's National Agricultural Policy (NAP) that has lays emphasis on management and conservation of resources by promoting sustainable agriculture. Human well-being is intimately related to soil ecosystem services, this chapter strives to develop this critical balance with outcomes oriented towards conservation of resources, rehabilitation of lands and development of an overall stronger economy.

Sustainable agriculture into the future will require research coordination, cooperation and innovation in the face of finite resources, climate change and disease and pest pressure. Key areas for innovation in the face of finite resources and climate change are greater nutrient use efficiency, multiple land uses and reduction of pest pressure by generating ecosystem services. Climate change is likely to compound the challenges in agriculture including more erratic climatic conditions; new plant pests and animal diseases; increased biotic and abiotic stresses on plants. Hence a key challenge for research is to develop an agro-technology that can rejuvenate past degraded lands and provide judicious land use and ecosystem services. Therefore, there should be an concerted efforts towards innovations to configure restoration of degraded land aiming at increased resource use efficiency coupled with higher productivity, preservation of ecosystem services that enhances productivity by enriching soil and reducing environmental costs and mitigating climate change effects by conserving soil moisture and promoting C sequestration.

11.3 SUMMARY

A healthy and resilient soil and water system is the hallmark of successful environmental management, extremely vital to the society and the economy.

Degraded lands are major constraints for global economy and the natural resource system. This problem is likely to be compounded further due to climate change effects and is a priority issue that needs to be addressed. Restoring soil quality in degraded lands is of fundamental importance and is a major focus for environmental management, food security and responsible ecological stewardship. Climate change represents a serious threat to the natural environment because of the adverse effects it can cause to ecosystems, biodiversity, and ecosystem services to society. Actions are needed to ensure ecosystem resilience comprising of sequestering carbon (C) in soils, conservation of water, restoration of nutrient cycles and microbial activities that form the functional basis for natural resource management. These adaptation and mitigation measures are inherent for restoration of degraded lands and countering climate change effects.

KEYWORDS

- **Agricultural policy**
- **Agriculture**
- **Biochar**
- **Carbon sequestration**
- **Climate change**
- **Degraded soil**
- **Ecosystem**
- **Environmental management**
- **Microbial activity**
- **Nutrient cycles**
- **Productivity**

REFERENCES

1. Annabi, M., S. Houot, C. Francou, M. Poitrenaud, and Y. Le Bissonnais (2007). Soil aggregate stability improvement with urban composts of different maturities. *Soil Sci. Soc. Am. J.,* 71, 413–423.
2. Arriaga, F. J. and Lowery, B. (2003). Soil physical properties and crop productivity of an eroded soil amended with cattle manure. *Soil Sci.,* 168, 888–899.

3. Bakker, M. M., G. Covers, R. A. Jones, and M. D. A. Rounsevell (2007). The effect of soil erosion on Europe's crop yields. *Ecozys.,* 10, 1209–1219.

4. Becher, H. H. (2005). Impact of the long-term straw supply on loess-derived soil structure. *Int. Agrophysics.,* 19, 199–202.

5. Berglund, L., T. DeLuca, and O. Zackrisson (2004). Activated carbon amendments to soil alters nitrification rates in Scots pine forests. *Soil Biol. Biochem.,* 36, 2067–2073.

6. Charman, P. and Murphy, B. (1991). *Soils: Their Properties and Management.* Third edition. Oxford Press.

7. Cox, D., D. Bezdicek, and M. Fauci (2001). Effects of compost, coal ash, and straw amendments on restoring the quality of eroded Palouse soil. *Biol. Pert. Soils.,* 33, 365–372.

8. Deenik, J. L., T. McClellan, G. Uehara, N. J. Antal, and S. Campbell (2010). Charcoal volatile matter content influences plant growth and soil nitrogen transformations. *Soil Sci. Soc. Am. J.,* 74, 1259–1270.

9. DeLuca, T. H., M. D. MacKenzie, M. J. Gundale, and W. E. Holben (2006). Wildfire-produced charcoal directly influences nitrogen cycling in ponderosa pine forests. *Soil Sci. Soc. Am. J.,* 70, 448–453.

10. Den Biggelaar, C., R. Lal, K. Wiebe, and V. Breneman (2001). Impact of soil erosion on crop yields in North America. *Adv. Agron.,* 72, 1–53.

11. Glaser, B., J. Lehmann, and W. Zech (2002). Ameliorating physical and chemical properties of highly weathered soils in the tropics with charcoal: a review. *Biol. Pert. Soils.,* 35, 219–230.

12. Grossman, J. M., B. E. O'Neill, S. M. Tsai, B. Liang, E. Neves, J. Lehmann, and J. E. Thies (2010). Amazonian anthrosols support similar microbial communities that differ distinctly from those extant in adjacent, unmodified soils of the same mineralogy. *Microbial Ecol.,* 60, 192–205.

13. Jankauskas, B., G. Jankauskiene, and M. A. Fullen (2008). Soil erosion and changes in the physical properties of Lithuanian Eutric Albeluvisols under different land use systems. *Soil Plant Sci.,* 58, 66–76.

14. Kim, J. S., S. Sparovek, R. M. Longo, W. J. De Melo, and D. Crowley (2007). Bacterial diversity of terra preta and pristine forest soil from the Western Amazon. *Soil Biol. Biochem.,* 39, 648–690.

15. Laird, D. A., (2008). The charcoal vision: a win-win-win scenario for simultaneously producing bioenergy, permanently sequestering carbon, while improving soil and water quality. *Agron. J.,* 100, 178–181.

16. LeBissonnais, Y. (2006). *Aggregate breakdown mechanisms and erodibility.* In: Lai R. (Ed.). Encyclopedia of Soil Science. Taylor Francis Press, Boca Raton, FL, USA.

17. Lehmann, J. and Joseph, S. (2009). *Biochar for environmental management: an introduction. In: Lehmann, J., and S. Joseph. (Eds.), Biochar for Environmental Management: Science and Technology.* Earthscan, London, pp. 1–12.

18. Lehmann, J. (2007a). Bio-energy in the black. *Front. Econ. Env.,* 5, 381–387.

19. Lehmann, J., J. Gaunt, and M. Rondon (2006). Biochar sequestration in terrestrial ecosystems – a review. *Mitig. Adapt. Strategies Glob. Chang.,* 11, 403–427.

20. Lehmann, J., M. C. Rillig, J. Thies, C. A. Masiello, W. C. Hockaday, D. Crowley (2011). Biochar effects on soil biota -A review. *Soil Biol. Biochem.,* 43, 1812–1836.

21. Liang, B., J. Lehmann, D. Solomon, J. Kinyangi, J. Grossman, B. O'Neill, J. O. Skjemstad, J. Thies, F. J. Luizao, J. Petersen, and E. G. Neves. (2006). Black carbon increases cation exchange capacity in soils. *Soil Sci. Soc. Am. J.,* 70, 1719–1730.

22. Liang, B., J. Lehmann, S. P. Sohi, J. E. Theis, B. O'Neill, L. Trujillo, J. Gaunt, D. Solomon, J. Grossman, E. G. Neves, and F. J. Luizao (2010). Black carbon affects the cycling of non-black carbon in soil. *Org. Chem.*, 41, 206–213.

23. O'Neill, B., J. Grossman, M. T. Tsai, J. E. Gomes, J. Lehmann, J. Peterson, E. Neves, and J. E. Thies (2009). Bacterial community composition in Brazilian anthrosols and adjacent soils characterized using culturing and molecular identification. *Microbial Ecol.*, 58, 23–35.

24. Paluszek, J., and W. Zembrowski (2008). Improvement of the soils exposed to erosion in a loessial landscape. *Acta Agrophysica*, 164, 1–160.

25. Papiernik, S. K., T. E. Schumacher, D. A. Lobb, M. J. Lindstrom, M. L. Lieser, A. Eynard, and J. A. Schumacher (2009). Soil properties and productivity as affected by topsoil movement within an eroded landform. *Soil Till. Res.*, 102, 67–77.

26. Pietikainen, J., P. Kiikkila, and H. Fritze (2000). Charcoal as a habitat for microbes and its effect on the microbial community of the underlying humus. *Oikos.*, 89, 231–242.

27. Schmidt, M. W. I. and Noack, A. G. (2000). Black carbon in soils and sediments: analysis, distribution, implications, and current challenges. *Global. Biogeochem. Cy.*, 14, 777–793.

28. Shukla, M. K., and R. Lal (2005). Erosional effects on soil physical properties in an on-farm study on Alfisols in West Central Ohio. *Soil Sci.*, 170, 445–456.

29. Sohi, S., E. Krull, E. Lopez-Capel, and R. Bol (2010). A review of biochar and its use and function in soil. *Adv. Agron.*, 105, 47–82.

30. Steinbeiss, S., G. Gleixner, and M. Antonietti (2009). Effect of biochar amendment on soil carbon balance and soil microbial activity. *Soil Biol. Biochem.*, 41, 1301–1310.

31. Yin, B., D. Crowley, G. Sparovek, W. J. De Melo, and J. Borneman (2000). Bacterial functional redundancy along a soil reclamation gradient. *Appl. Environ. Microb.*, 66, 4361–4365.

32. Zackrisson, O., M. C. Nilsson, and D. A. Wardle (1996). Key ecological function of charcoal from wildfire in the boreal forest. *Oikos.*, 77, 10–19.

CHAPTER 12

ANALYSIS AND PREDICTION OF WATER QUALITY USING PRINCIPAL COMPONENT ANALYSIS AND NEURAL NETWORK

MRUNMAYEE M. SAHOO,[1] K. C. PATRA,[2] J. B. SWAIN,[3] and K. K. KHATUA[2]

[1]*Retired Principal and Scientist of ICAR-IIWM, Bhubaneswar-751023, India. E-mail: narayansahoo65@yahoo.in*

[2]*Department of Civil Engineering, National Institute of Technology, Rourkela, India, E-mail: kcpatra@nitrkl.ac.in, kkkhatua@nitrkl.ac.in*

[3]*Department of Civil Engineering, National Institute of Technology, Rourkela, India, E-mail: Jnkballav.2009@gmail.com*

CONTENTS

12.1   INTRODUCTION

The fresh water on our planet is very unevenly distributed. Of the liquid surface fresh water, 87% is contained in lakes, 11% in swamps, and only 2% in rivers. In India, surface water is a major source of water for the industries (41%), followed by groundwater (35%) and municipal water (24%). The use of municipal water is limited to industries located in urban or peri-urban areas. A vast majority of industries use surface and groundwater in conjunction with groundwater being relied as a source when surface water availability is on a decline or is impacted by water pollution bound to have an impact on the industrial process. In recent years, the rate of discharge of pollutants and effluents into the environment is continuously increasing due to rapid growth of population, urbanization and industrialization. It causes variety of soluble inorganic, soluble organic and organic compounds. In addition to all these, water can carry large amounts insoluble materials that held in suspension. The polluted water endangers not only the valuable human life but also causes considerable biological disorder in the organisms.

Surface water contamination and its management have become important issues because of far reaching impact on human health. Assessment of water quality measures the analysis of physical, chemical and biological characteristics of water. Water quality indices (WQI) aim at giving a single value to the water quality of a source reducing great amount of parameters into a simpler expression and enabling easy interpretation of monitoring data [4]. The index helps in interpreting the water quality in to a single numerical value. WQI is strongly dependent on various correlated parameters taken for the study. Also, identification of the suitability of the parameters is critical for accurate evaluation of WQI.

Water quality is generally ascertained based on guidelines by agencies such as the World Health Organization (WHO) and the Bureau of Indian Standards (BIS). However, the interdependency of these water quality parameters makes the evaluation inaccurate. Therefore, it is important to study the correlation among the parameters. The quality of surface water and ground water is studied earlier by various researchers. Among them, Akkaraboyina et al. [1] made a comparative study of water quality indices of Godavari River in India.

The principal component analysis technique (PCA) was employed to evaluate the seasonal correlations of water quality parameters, while principal factor analysis technique was used to extract the parameters that are most important in assessing seasonal variations of river quality [6]. Khandelwal and Singh [5] predicted the chemical parameters like sulfate, chlorine, chemical oxygen demands, total dissolved solids and total suspended solids in mine water using artificial neural network (ANN) by incorporating the pH, temperature and hardness. The prediction by ANN is also compared with multivariate variance analysis (MVRA). Stewart [8] used feed forward neural network modeling techniques. Two neural network models are proposed in series:

- The simulated daily mean dissolved oxygen concentration,
- Superimposed any daily periodic signals.

The final calibrated neural network models predicted the dissolved oxygen concentration with acceptable accuracy, producing high correlations between measured and predicted values [8].

This chapter discusses application of empirical approach for classification of water samples using Q-mode of principal component analysis (PCA) and prediction of water quality by artificial neural network (ANN). A case study is also included.

12.2 MATERIALS AND METHODS

Along the River Brahmani of Odisha – **India,** the gaging stations are located. The stations are at the down streams of Rourkela steel plant, NALCO Smelter Plant and Captive Power Plant, Mahanadi Coal field Limited and chromites mines. The river water at these locations is contaminated heavily resulting in acidity, toxicity, presence of heavy metal and microbes. During monsoon, generally these industries and mines are filled with water, which contaminates the river water and gradually disperses, making unsuitable for use. In addition, fertilizers used for agricultural purpose affect pH and nitrate content of river water. Hence, evaluation of water quality indices (WQI) of

Brahmani River is most important at these gaging stations close to industries and mines to prepare remedial measures.

Five gaging stations were selected on the basis of mining and industrial activities prevalent nearby as shown in Figure 12.1 namely: Panposh downstream, Talcher up-stream, Kamalanga downstream, Aul and Pottamundai. From these five gaging stations, the data were sampled during 2003–2012 for monsoon season. From these five gaging stations, the data are sampled during 2003–2012 in the monsoon season. Eleven water quality parameters were used for the further analysis, such as:

- pH,
- Dissolved oxygen in mg/L (DO),
- Biological Oxygen Demand in mg/L (BOD),
- Conductivity in mmho/cm,
- Nitrogen as Nitrate in mg/L (Nitrate-N),
- Total Coli-form in MPN/100 ml (TC),
- Fecal Coli-form in MPN/100 ml (FC),
- Chemical Oxygen Demand in mg/L (COD),
- Ammonia as Nitrogen in mg/L (NH_4-N),
- Total Alkali as $CaCO_3$ in mg/L (TA as $CaCO_3$),
- Total Hardness as $CaCO_3$ in mg/L (TH as $CaCO_3$).

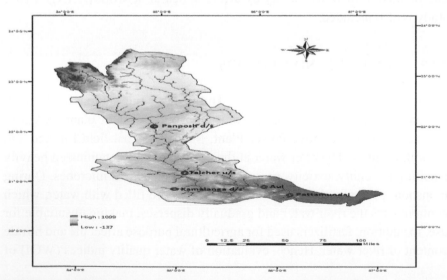

FIGURE 12.1   Map shows the Locations Gaging Stations of Brahmani River.

TABLE 12.1 Correlation Matrix for Water Quality Parameters in Brahmani River Basin

	pH	DO	BOD	Cond	Nitrate-N	TC	FC	COD	NH$_4$-N	TA as CaCO$_3$	TH as CaCO$_3$
pH	1.00										
DO	0.13	1.00									
BOD	0.07	-0.22	1.00								
Conductivity	0.10	-0.35	0.30	1.00							
Nitrate-N	0.16	-0.04	0.30	-0.01	1.00						
TC	0.11	-0.25	0.30	0.33	-0.05	1.00					
FC	0.11	-0.17	0.43	0.36	-0.01	0.68	1.00				
COD	0.10	-0.19	0.75	0.32	0.21	0.32	0.40	1.00			
NH4-N	0.12	0.05	0.02	0.18	-0.16	0.12	0.12	0.11	1.00		
TA as CaCO$_3$	0.07	-0.14	0.13	0.21	0.05	0.05	0.04	0.16	-0.03	1.00	
TH as CaCO$_3$	0.13	0.03	0.34	0.20	0.37	0.18	0.23	0.46	0.08	0.28	1.00

Pearson correlation matrix was prepared within the studied parameters for the data in monsoon season of nine years and shown in Table 12.1. It is observed that parameters such as DO, Conductivity, Nitrate-N, TC, FC, COD, NH_4-N and TH as $CaCO_3$ exhibit slight correlation with pH varies from 0.10 to 0.16. BOD shows slight and moderate correlation with Nitrate-N, TC and TH as $CaCO_3$ varies from 0.30 to 0.43 where as it shows significant correlation with COD of 0.75. Conductivity exhibits slight correlation with TC, FC and COD varies from 0.32 to 0.36. Nitrate-N shows slight correlation with TH as $CaCO_3$ of 0.37. As TC and FC are more related to each other, shows a moderate correlation of 0.68 and TC exhibit a slight correlation with COD of 0.32. FC shows a moderate correlation of 0.40 with COD. COD exhibit a moderate correlation if 0.46 with TH as $CaCO_3$. DO show slight negative correlation with Conductivity of −0.35. The strong correlation between some parameters could be due to changes in land use, mining, industrial activity and improper effluent discharge in the River.

12.2.1 DETERMINATION OF WATER QUALITY INDEX (WQI)

In the calculation of water quality for river water, the importance of various water quality parameters depends on the intended use of water and for the point of view of suitability for domestic purposes. The standards (permissible values of various water quality parameters) for drinking water are recommended by *Indian Council of Medical Research* (ICMR). When ICMR standards for water quality are not available, quality parameters are used from the relation: the standards of *United States Public Health Services* (USPHS), *World Health Organization* (WHO), *Indian Standard Institution* (ISI) and European Economic Community (EEC) are considered. The water quality rating q_i for the i-th water is given below:

$$q_i = 100 \ (v_i/s_i) \tag{1}$$

where, v_i = value of the *i*-th water quality parameter at a given sampling station, and s_i = standard permissible value of *i*-th water quality parameter.

This equation ensures that $q_i = 0$ when a pollutant (the *i*th water quality parameter) is absent in the water while $q_i = 100$ if the value of this parameter is just equal to its permissible value for drinking water. Thus, the larger the value of q_i, the more polluted is the river water with the *i*-th pollutant. However, water quality ratings for pH and DO require special handling and care. The permissible range of pH for the drinking water is 7.0–8.5. Water quality rating for pH can be written as:

$$q_{pH} = 100[(v_{pH} - 7)/(8.5 - 7.0)] \tag{2}$$

where, v_{pH} = value of pH ~ 7 that implies the numerical difference between v_{pH} and 7.0 ignoring algebraic sign.

Equation (2) ensures that the $q_{pH} = 0$ for pH = 7.0. In contrast to other water quality parameters, the case of DO is slightly complicated because the quality of water is enhanced if it contains more DO. Therefore, the water quality rating q_{DO} has been calculated as below:

$$q_{DO} = 100[(14.6 - v_{DO})/(14.6 - 5)] \tag{3}$$

where, v_{DO} = value of observed DO; the value 14.6 is the solubility of oxygen (mg/L) in distilled water at 0°C and 5.0 mg/L is the standard for drinking water.

Equation (3) gives $q_{DO} = 0$ when DO = 14.6 mg/L and $q_{DO} = 100$ when $v_{DO} = 5.0$ mg/L. The more harmful a given water quality parameter is, the smaller is its permissible value for drinking water. So the 'weights' for various water quality parameters are assumed to be inversely proportional to the standards recommended by ICMR for the corresponding water quality parameters, that is:

$$W_i = \frac{K}{S_i} \tag{4}$$

where, W_i = unit weight for the ith water quality parameter (i = 1, 2, 3, . . ., 11), K = constant of proportionality which is determined from the condition and $K = 1$ for sake of simplicity. So the sum of unit weight of 11 water quality parameters can be given as:

$$\sum_{i=1}^{11} W_i = 1 \tag{5}$$

To calculate the WQI, first the sub index (SI)$_i$ corresponding the ith water quality parameter is calculated as the product of the quality rating q_i and the unit weight W_i of the ith parameter given as:

$$(SI)_i = q_i W_i \tag{6}$$

The overall WQI of river Brahmani is then calculated by aggregating these sub indices (SI) linearly. Thus, WQI can be written as:

$$WQI = [\sum_{i-1}^{11} q_i W_i / \sum_{i=1}^{11} W_i] = \sum_{i=1}^{11} q_i W_i \qquad (7)$$

For domestic purposes, water quality can be categorized into five classifications depending on WQI values of the parameters:

Quality	WQI value
Excellent	0–25
Good	26–50
Poor	51–75
Very poor	76–100
Unsuitable	100

12.2.1.1   Generation of Euclidean Distance Matrix

The parameters collected from the River Basin are normalized to reduce the scaling effect. A simple normalization technique of separating selected variables by their maximum values was used here, after which all data vary from 0 to 1. To apply the factor analysis, one requires correlation matrix obtained from Euclidean distance matrix. Euclidean distance is defined as the sum of squares of the difference between the values of attributes of two samples. Mathematically, it may be given as:

$$d(x, y) = \sqrt{\sum_i (x_i - u_i)^2} \qquad (8)$$

where, $d(x, y)$ = Euclidean distance, $x = x_1, x_2, ..., x_m$; and $y = y_1, y_2, ..., y_m$ represent m attribute values of two samples.

If the distance is zero, both the parameter samples are similar. If it is above zero, the Euclidean distance indicates the intensity of dissimilarity between two samples. An entry in the matrix denotes Euclidean distance between the p-th row and the $(p+1)$-th row of the water samples. The Euclidean distance matrix is thus generated for the use in Q-mode PCA.

12.2.1.2   Q-Mode Principal Component Analysis

Principal component analysis (PCA) is the most widely used method for transforming a given set of interrelated variables into a new set of variables called principal components. The components generated are orthogonal to each other; hence correlation between them is 0. They are linear combinations of original variables for the total variance of the data. The principal components are generated sequentially in an ordered manner with decreasing values of variance. This property of variance shows that the data points are rigorously separated into distinct clusters projected into a space by first few principal components, which are called factors. These factors are achieved by dimension reduction objective of the factor analysis.

Based on application, PCA can be broadly classified into two categories; R-mode and Q-mode. If PCA is used to develop a structure among large variables, the analysis is named as R-mode PCA. When PCA is used in-group cases, the analysis is called as Q-mode PCA. Rotation method is used to transform factors; after rotation each variable will be related to one of the components; and each component will have high correlation with only a small set of variables. Q-mode of PCA has been widely used in classification of water quality and gene regulatory [2, 7, 9].

Euclidean distance matrix explains the correlation between each pair of sample data. PCA framework is used for grouping the sample data into separate independent clusters. In PCA, the initial clusters are extracted out by the eigenvalue-eigenvector analysis of the similarity coefficient matrix as given

$$(S - I\lambda_i)Y_i = 0 \text{ for } i = 1, 2, \dots, P \tag{9}$$

where, S is a $P \times P$ Euclidean distance matrix, I is the identity matrix, λ_i are the eigen-values and Y_i are the corresponding eigen-vectors.

Equation (9) is an eigen-value eigen-vector equation; and $\lambda_1 \geq \lambda_2 \geq \dots \lambda_p$ are the real, nonnegative roots of the determinant polynomial of degree P given as:

$$I(S - I\lambda_i)I = 0 \tag{10}$$

The above equation is solved for λ_i and the Y_i can be calculated using the values of λ_i in Eq. (9).

The eigenvectors computed represent the unique set of P independent components of the data set, which maximize the variance [2]. According to PCA, each *P* independent principal components can be written as linear combination of original variables, with the elements of the *P* eigenvectors as the coefficients of linear combinations.

The elements of these eigenvectors reflect the degree of association between each component and the sample data, and are called the factor

TABLE 12.2   Water Quality Index (WQI) Values for All Samples

Stations	WQI	Comments	Stations	WQI	Comments
pa.03	43.34	Cluster 2, G	pa.08	47.25	Cluster 2, G
ta.03	23.75	Cluster 1, Ex	ta.08	20.27	Cluster 1, Ex
ka.03	14.57	Cluster 1, Ex	ka.08	23.35	Cluster 1, Ex
aul.03	16.02	Cluster 1, Ex	aul.08	15.71	Cluster 1, Ex
po.03	14.66	Cluster 1, Ex	po.08	17.42	Cluster 1, Ex
pa.04	6.56	Cluster 1, Ex	pa.09	36.59	Cluster 2, G
ta.04	9.44	Cluster 1, Ex	ta.09	15.93	Cluster 1, Ex
ka.04	14.58	Cluster 1, Ex	ka.09	18.97	Cluster 1, Ex
aul.04	26.41	Cluster 2, G	aul.09	17.14	Cluster 1, Ex
po.04	26.51	Cluster 2, G	po.09	14.15	Cluster 1, Ex
pa.05	46.69	Cluster 2, G	pa.10	42.79	Cluster 2, G
ta.05	26.72	Cluster 2, G	ta.10	19.51	Cluster 1, Ex
ka.05	26.73	Cluster 2, G	ka.10	26.18	Cluster 2, G
aul.05	43.17	Cluster 2, G	aul.10	33.73	Cluster 2, G
po.05	38.82	Cluster 2, G	po.10	3.97	Cluster 1, Ex
pa.06	35.11	Cluster 2, G	pa.11	46.08	Cluster 2, G
ta.06	21.21	Cluster 1, Ex	ta.11	29.72	Cluster 2, G
ka.06	5.46	Cluster 1, Ex	ka.11	29.02	Cluster 2, G
aul.06	22.98	Cluster 1, Ex	aul.11	29.17	Cluster 2, G
po.06	11.55	Cluster 1, Ex	po.11	24.99	Cluster 1, Ex
pa.07	41.68	Cluster 2, G	pa.12	50.02	Cluster 3, P
ta.07	15.62	Cluster 1, Ex	ta.12	31.54	Cluster 2, G
ka.07	9.92	Cluster 1, Ex	ka.12	29.30	Cluster 2, G
aul.07	12.91	Cluster 1, Ex	aul.12	21.86	Cluster 1, Ex
po.07	17.79	Cluster 1, Ex	po.12	37.18	Cluster 2, G

loadings of the samples on the ith principal component. Each of the P independent principal components represents a cluster. The WQI values calculated by SPSS are given in Table 12.2.

12.2.2 ARTIFICIAL NEURAL NETWORK (ANN) BY MATLAB

The artificial neural network (ANN) has the ability to learn from the pattern acquainted before. Once the sufficient numbers of sample data sets are trained by the neural network, it can predict on the basis of its previous learning about the output related to input data set of similar pattern. The data were collected from the gaging stations of River Brahmani during monsoon season of 2003–2011. The data were normalized and used as input in Principal Component Analysis as described in section of Principal Component Analysis (PCA). Then the principal components extracted by SPSS Version 20 were normalized and used as an input to ANN. The output for each data is the WQI calculated as per procedure given in section of calculation and formulation of WQI. Every network can be defined using three fundamental components: transfer function, network architecture and learning law. A network should be trained before interpreting some information. Back propagation algorithm is the most efficient learning procedure for multilayer neural network. The feed forward back propagation neural network (BPNN) consists of three layer: input layer, hidden layer and output layer. Each layer consists of neurons, which send information from the input layer to output layer through hidden layer. Output layers compute output vectors corresponding to the solution.

12.2.2.1 Architecture and Basic Learning Rules of ANN

Data is processed from the input layer through hidden layer to output layer (forward pass). Output values are compared with the measured output that is the true output and the errors between them are processed back through network (backward pass) for updating the individual weights of the connections and the biases of the individual neuron. The process is repeated for all training input and output data set, until the network error is converged to a threshold minimum defined by a cost function; usually Root Mean Squared

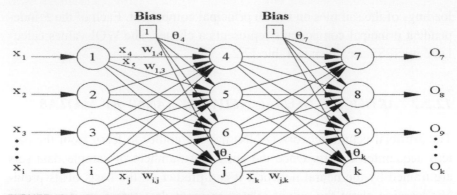

FIGURE 12.2 Back Propagation Neural Network.

Error (RMS) or Summed Squared Error (SSE). In Figure 12.2, the jth neuron is connected to number of inputs:

$$x_i = (x_1, x_2, x_3 \ldots \ldots x_n) \tag{11}$$

The input values in the hidden layers will be:

$$Net_j = \sum_{i=1}^{n} x_i W_{ij} + \theta_j \tag{12}$$

where, x_i = Input values, W_{ij} = Weight on the connection point of ith and jth neuron, θ_j = Bias neuron, and n = Number of input units.

Net output from the hidden layer is calculated using a logarithmic sigmoid function given as:

$$O_j = f(Net_j) = 1/1 + e^{-(Netj + \theta j)} \tag{13}$$

The total input to the *k*th unit is given as:

$$Net_k = \sum_{j=1}^{n} W_{jk} O_j + \theta_k \tag{14}$$

where, θ_k = bias neuron and W_{jk} = weight between jth neuron and kth output.

Total output from one unit will be given as:

$$\theta_k = f(Net_k) \tag{15}$$

The network computes its own output pattern using its weights and thresholds from the given input pattern and corresponding output pattern. Now, actual output is compared with the desired output. Hence, the error in any output layer k can be given as:

$$e_l = t_k - O_k \tag{16}$$

where, t_k= desired output and O_k= actual output.

The total error function is given as:

$$E = 0.5 \sum_{k=1}^{n} (t_k - O_k)^2 \tag{17}$$

Training of the network is the process to get an optimum weight space of the network. The descent down error surface is calculated by the following rule:

$$\nabla W_{jk} = -\eta(\delta E / \delta W_{jk}) \tag{18}$$

where, η = learning rate parameter and E = error function.

The update of the weights for the $(n+1)^{th}$ pattern is given as:

$$W_{jk}(n+1) = W_{jk}(n) + \nabla W_{jk}(n) \tag{19}$$

This network patterns are repeated for each pair of training network. Each pass through all training patterns is called cycle or epoch. The epochs are repeated to minimize the error.

All input and output parameters are scaled between 0 to 1 to utilize the most sensitive part of neuron. As the output neuron is sigmoid, it can range its value from 0 to 1. To test and validate the neural network model, a new data set is chosen. The results of the testing and validation give the network performance, which is the correlation coefficient between predicted and observed values. Training was done using hidden layer. There was no danger of over fitting problems; hence network was trained with epochs.

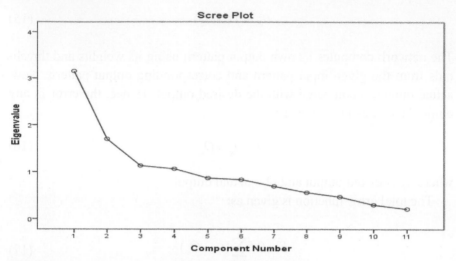

FIGURE 12.3   Scree plot for selection of number of Clusters.

12.3   RESULTS AND DISCUSSION

The corresponding eigenvalues and eigenvectors of the water samples for the Euclidean distance matrix were calculated using SPSS 20 software. Varimax rotation is applied to obtain the optimal distribution of variances in various components. The number of factors or Clusters are selected on the basis of principal components showing eigenvalues greater than 1 or number of principal components forming the cliff in scree plot or Akaike's information criterion (AIC) [3]. It can be observed from the plot in Figure 12.3.

Four factors are extracted from the scree plot as well as from the eigenvalues calculated by SPSS 20. These four factors contribute 64.112% of variances. In order to compare the WQI values, the water quality can be categorized into four classifications depending on WQI values. Water quality can be excellent, good, poor and very good if WQI lies in the ranges 0–25, 26–50, 51–75 and 76–100, respectively.

To assign the water samples into various Clusters, the absolute values of the elements of the eigenvectors (the factor loadings) are used to identify the Clusters for the water samples. Rotated factor matrix for the water quality parameters is calculated to obtain the four principal components values of respective stations are given in Table 12.3. For example, for NH_4-N, the factor loading in Cluster 4 is 0.702, which is higher as compared to loadings in other Clusters, hence a stronger relationship with Cluster 4 rather than with

TABLE 12.3 Loadings for Varimax Factor Matrix of Four Factor Model

Variable	Factors			
	1	2	3	4
pH	0.119	−0.315	0.029	0.014
DO	−0.101	0.189	−0.336	0.515
BOD	0.230	0.233	−0.073	−0.093
Conductivity	0.102	−0.137	0.392	0.017
Nitrate-N	−0.012	0.387	−0.095	−0.201
TC	0.352	−0.133	−0.118	−0.015
FC	0.371	−0.071	−0.166	0.032
COD	0.194	0.235	0.000	0.062
NH_4-N	0.030	−0.115	0.091	0.702
TA as $CaCO_3$	−0.233	0.055	0.684	0.075
TH as $CaCO_3$	−0.014	0.332	0.146	0.290

Cluster 1, 2 and 3. From Table 12.2, it can be seen that WQI of samples of the gaging stations from 2003 to 2012 varies from excellent to good in most of cases in Brahmani River.

Therefore, water samples belong to cluster 1 (principal component 1 or PC1), cluster 2 (principal component 2 or PC2), cluster 3 (PC3) and cluster 4 (PC4) are treated as of excellent, good, poor and very poor quality, respectively. The same procedure is repeated for all samples to find out their respective clusters. In the same argument, sample of the Panposh downstream station on 2012 belongs to cluster 3, poor quality.

After WQI analysis by Q-mode PCA, Artificial Neural Network (ANN) is applied for its prediction and analysis. Feed forward network is adopted here as architecture and is reported as a suitable tool for problem identification. The target time steps used for neural network analysis are; 30% for training 35% each for validation and testing. The estimated mean squared errors (MSE) for training, validation and testing are 1.07539e-1, 84.96309e-1, and 47.45446e-1 respectively. The response output and error (MSE) after target time steps for the given input and output is shown in Figure 12.4.

Furthermore, ANN provides a computationally elegant method with lesser dependence on choice of parameters. After the complete iteration by ANN model the predicted data is generated by the model and is recorded

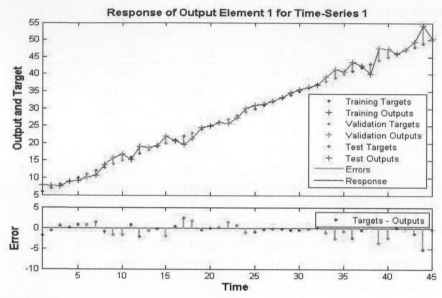

FIGURE 12.4 Response Output by ANN model.

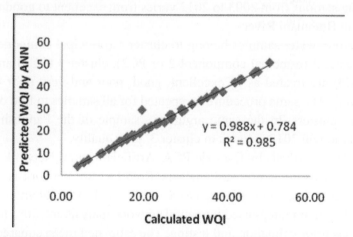

FIGURE 12.5 Correlation between Actual and Predicted WQI.

for correlation analysis for the validation of input data and the output of the model. The correlation graph is shown in Figure 12.5.

However, quality is a vague term that cannot be easily described as having pH value of 7.0 or above. Above all, water quality can be best described based on its degree of portability and potential uses rather than expressing its constituents in numerical terms. Therefore, the non-parametric empirical method like Q-mode PCA and clustering was proposed in this chapter. The method can take any number of parameters and also classifies water samples into proper groups. The performance of Q-mode of clustering is improved if the size of the data is increased.

12.4   CONCLUSIONS

In this study, PCA base classification has been proposed to know the quality of sample collected. It has been seen that the methodology; Q-mode PCA efficiently classifies into various clusters. It can be used in the field and laboratory due to easy accessibility and availability of statistical packages like SPSS. Classification of WQI by Q-mode PCA has resulted; out of 50 water samples, 27 belong to cluster 1, 22 belong to cluster 2 and only 1 belong to cluster 3. It means mostly the quality of water samples varies from excellent to good during monsoon season. There are several other advantages like; the physicochemical analysis of any collected water sample can be determined without any sophisticated laboratory set up.

After classification of water samples by Q-mode PCA, the data are normalized by PCA to use as input in artificial neural network for further error analysis and prediction of WQI. The feed forward back propagation neural network (BPNN) is used to build neurons as a connection link between hidden and output layer. The correlation analysis by ANN of the calculated WQI and predicted WQI shows strong correlation of correlation coefficient of 0.985. The results of this ANN and Q-mode PCA study have shown that, with combination of computational efficiency, Statistical measures and ability of input parameters which describe physical behavior of water quality parameters, improve the model predictability and is possible in artificial neural network study.

12.5   SUMMARY

The Brahmani river basin is contaminated heavily due to acidity, alkalinity and effluent discharge of nearby industries at Panposh downstream and Talcher upstream. Evaluation of quality of water is extremely important to prepare its remedial measures. In this research work water samples from Panposh downstream and Talcher upstream stations in monsoon seasons of nine different years are collected to assess the quality of water.

This paper presents application of empirical approach for classification of water samples, i.e., Q-mode of principal component analysis and prediction of water quality by artificial neural network. The water quality parameters of the river are uncorrelated using Q-mode of principal component analysis with Euclidean distance matrix and varimax rotation. The water samples are categorized considering the parameters like pH, DO, BOD, Conductivity, Nitrogen as ammonia (NH_4-N), Nitrogen as nitrate (NO_3-N), TC, FC, TH of $CaCO_3$, TA of $CaCO_3$. The non-parametric method by Q-mode of PCA proposed here efficiently assesses water quality index for classification of water quality.

ACKNOWLEDGEMENTS

Authors sincerely acknowledge the Central Water Commission, Bhubaneswar, and the Odisha Pollution Control Board, Bhubaneswar in providing the required information for preparing this chapter. They also thank National Institute of Technology Rourkela to provide the well-instrumented environmental laboratory for testing and validating the water quality indicator collected.

KEYWORDS

- Artificial neural network
- Akaike's information criterion
- Back propagation algorithm
- Calibration
- Central water commission
- Chemical composition
- Cluster
- Eigen value
- Empirical method
- Epoch
- Euclidean distance matrix
- Factors
- Factor loadings
- Feed forward back propagation network
- Gauging station
- Hidden layers
- Mean squared errors
- Neuron
- Non-parametric method
- Normalized
- Pearson correlation coefficients
- Physiochemical analysis
- Principal component analysis

- Q-mode of principal component analysis
- Quality rating
- River Brahmani
- Rotated factor matrix
- Root mean squared error
- Scree plot
- Sigmoid
- Sub Index
- Threshold
- Validation
- Variance
- Varimax
- Water quality index
- Weights

REFERENCES

1. Akkaraboyina, M. K., and Raju, B. S. N. (2012). A comparative study of water quality indices of river Godavari. *International Journal of Engineering Research and Development*, 2(3), 29–34.
2. Albadawia, Z., Bashirb, H. A., and Chenc, M. (2005). A mathematical approach for the formation of manufacturing cells. *Computers & Industrial Engineering,* 48(1), 3–21.
3. Basilevsky, A. 1994. *Statistical Factor Analysis and Related Methods*. John Wiley & Sons, New York.
4. Bharti, N. and Katyal, D. (2011). Water quality indices used for surface water vulnerability assessment. *International Journal of Environmental Sciences*, 2(1), 154–173.
5. Khandelwal, M. and Singh, T. N. (2005). Prediction of mine water quality by physical parameters. *Journal of Scientific & Industrial Research*, 64, 564–570.
6. Ouyang, Y., Nkedi-Kizza, P., Wu, Q. T., Shinde, T. and Huang, C. H. (2006). Assessment of seasonal variations in surface water quality. *Journal of Water Resources*, 40(20), 3800–10.
7. Park, S., Choi, D., and Jun, C. H. (2001). A clustering method for discovering patterns using gene regulatory processes. *Genome Inform*, 12, 249–251.
8. Stewart, A. R. (2002). Development of neural network model for dissolved oxygen in the Tualatin River, Oregon. *Proceedings of the Second Federal Interagency Hydrologic Modeling Conference*, Las Vegas.
9. Singh, B., Sudhir, D., Sandeep, J., Garg, V. K., and Kushwaha, H. S. (2008). Use of fuzzy synthetic evaluation for assessment of groundwater quality for drinking usage: a case study of southern Haryana. *Indian Environ. Geol.*, 54, 249–255.

PART IV

RAINFALL ANALYSIS FOR CROP PLANNING

COMPARISON OF IMD AND CMIP5 RAINFALL PRODUCTS FOR ANNUAL DAILY EXTREME PRECIPITATION: INDIA

SOUMEN MAJI,[1] TRUSHNAMAYEE NANDA,[2] and
BANAMALI PANIGRAHI[3]

[1]*Department of Civil Engineering, Central Institute of Technology, Kokrajhar, Assam-783370, India. E-mail: s.maji@cit.ac.in*

[2]*Department of Agricultural and Food Engineering, Indian Institute of Technology, Kharagpur, West Bengal, India, 721302, Tel.: +91-8895705688, E-mail: nanda.trushnamayee@yahoo.com*

[3]*Department of Regulatory Sciences, GVK Informatics Pvt. Ltd., Hyderabad, Telangana, India, Tel.: +91-9440942065 E-mail: banamali.panigrahi25@gmail.com*

CONTENTS

13.1   INTRODUCTION

Rainfall is an important climatic element for global climate change indicator. It is part of the hydrological cycle, which gives information on water availability, long-term trends, variability, and extreme events. Extreme events produce very serious impacts on different parts such as: urban areas, hilly areas, agricultural areas, and also river basins [1, 7] and hence, are of interest for water resource planning and management. Moreover, the information regarding changes in spatial pattern of rainfall over a region is important from water distribution point of view. Precipitation data may be obtained either from networks of rain gage stations or from remotely sensed satellite data, and even from climate model simulations. The *India Meteorological Department* (IMD) is the authority, which provides rainfall data for rain gage stations in India. Daily gridded rainfall dataset, developed by IMD [16], is available to the research community for non-commercial purposes. Though the satellite and climate model data sources include certain deficiencies [11], these are alternatives to gage based rainfall data. In the current climate change scenarios, the global climate model (GCM) simulations are useful for long-term climate change studies. Moreover, Taylor et al. [12–14] mentioned that the Fifth Phase *Coupled Model Inter-comparison Project* (CMIP5) established by the World Climate Research Program (WCRP) can provide global datasets of area-averaged and time-integrated precipitation based on all suitable observation techniques.

Hibbard et al. [5] added that CMIP5 strategy includes two types of climate change modeling experiments such as long-term integrations and near-term integrations. Rain gauge measurements are point measurements, which are locally influenced, and are contaminated by systematic measuring errors but they give the best estimate of the true precipitation amount, if the network density is sufficient. Satellite data precipitation estimates are based on indirect approaches and need to be calibrated and validated.

There is an increase in extreme rainfall events in a contiguous region extending from the north-western Himalayas in Kashmir through most of the Deccan Plateau in the southern peninsula region of India and they also found decrease in extreme rainfall events in eastern part of the Gangetic plain and parts of Uttaranchal [10]. Trends in extreme rainfall indices for the period 1901–2000 for 100 stations over India were examined by [6] and it was found that most of the extreme rainfall indices show positive trends in extreme rainfall over the west-coast and north-western parts of peninsula. 1° x 1° daily gridded rainfall dataset for the period 1951–2000 were used [3] and it was observed that there was an increasing trend in the frequency of extreme rainfall events and increase in the intensity of precipitation over the central India. The spatial variation in the trends in frequency and intensity of extreme rainfall indicated that the north and central India show decreasing trend in the frequency and intensity of extremes, while the coastal regions of peninsular India and eastern India have shown increasing trends [8]. Decrease in frequency of heavy rainfall events in major parts of central and north India and increasing trend in peninsular, east and north-east India have been found [4]. Performance of CMIP5 models in simulating the climatology of the Greater Horn of Africa was also evaluated by [9]. Model (CMIP5) can capture the dominant features of the geographical distribution of temperature and precipitation [15].

In this study, authors have identified, if any there is any change in spatial distribution in daily extreme rainfall magnitude before and after the significant warming trend in temperature observed in seventies using both observed and modeled data. They also present frequency analysis with annual daily extreme rainfall values for 5 years, 25 years and 50 years return periods. The IMD and CMIP5 rainfall data of 53 years were divided into two successive periods of 24 years (1951–1974) and 29 years (1975–2003), respectively. Frequency analysis of annual daily extreme rainfall was performed for the two sets of rainfall data separately and percentage changes in annual daily extreme rainfall depth were compared over India.

13.2 METHODOLOGY

13.2.1 INDIA METEOROLOGICAL DEPARTMENT (IMD) GRIDDED DATA

Daily gridded rainfall data at spatial resolution of 1^0 x 1^0 latitude/longitude was obtained from IMD [16]. IMD has rainfall record of 6329 rain gauge stations in India for the period 1951–2003. However, only 1803 stations having minimum 90% data availability during the analysis period were used for preparation of the gridded data (Figure 13.1). Methods proposed by [17] with directional effects were considered for the interpolation of station data

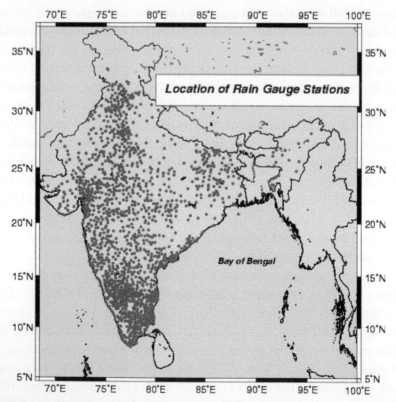

FIGURE 13.1   Location of rain gauges over India [16].

into 1° x 1° regular grids. This method is based on the weights calculated from the distance between the station and the grid point. The distribution of rain gauges in India is shown in Figure 13.1 [16].

13.2.2 FIFTH PHASE COUPLED MODEL INTER-COMPARISON PROJECT (CMIP5) DATA

Under CMIP5, the recorded data is discoverable under CMIP5 activity. Model records consist of collections of atomic datasets. Atomic dataset is a subset of the output saved from a single model run which is uniquely characterized by a single activity, product, institute, model, experiment, data sampling frequency, modeling realm, variable name, MIP table, ensemble member, and version number. The activity that is selected for the study is CMIP5. The four options available in product are: "output," "output1," "output2," and "unsolicited." For CMIP5 data from the requested variable list was assigned a version and placed in either "output1" or "output2. Variables not specifically requested by CMIP5 remained designated "unsolicited." The version number indicates the date of publication. Institute identifies the institute responsible for the model results.

In CMIP5, the rcp45 refers to a particular experiment in which a representative concentration pathway (RCP) has been specified which leads to an approximate radioactive forcing of 4.5 W.m^{-2}, whereas historical refers to an experiment which contains data from 1850 to 2012. Frequency indicates the interval between individual time-samples in the atomic dataset. Thirty-nine models are available for historical observation (1850–2012), out of which EC-EARTH is used. The data is 1.125° x 1.125° daily gridded rainfall data. EC-EARTH model is selected due to its nearness in resolution with the observed rainfall data. The data contains the entire spatio-temporal domain, with values reported at each included time and location. Out of this global precipitation record besides choosing Indian region by fixing latitude and longitude of India, the temporal domain 1951–2003 is also fixed for the analysis of modeled data. Details of the currently available CMIP5 historical experiments and their forcing are available at: http://cmip-pcmdi.llnl.gov/cmip5/.

For both IMD and CMIP5, daily rainfall data for 53 years from 1951–2003 for each grid points were obtained. For each year for a particular grid point, the highest of the 365 (366) rainfall values was selected. For handling the data, frequency analysis was done through programming using MATLAB software (MATLAB R, 2010a). The study undertaken herein was for the spatial coverage over India and on daily temporal scale. The annual daily extreme precipitation depth was found using Extreme Gumbel distribution or Value Type I (EVI) distribution method in both the IMD observed and CMIP5 model simulated rainfall data.

13.2.3 COMPUTATION OF ANNUAL DAILY MAXIMUM PRECIPITATION DEPTH

The $1^0 \times 1^0$ daily gridded rainfall data taken from India Meteorological Department (IMD) and $1.125^0 \times 1.125^0$ daily gridded rainfall data taken from EC-EARTH model were used as reference. The observed rainfall data is for India region and is available from the year 1951–2003 and the model rainfall data included the entire spatio-temporal domain, from which India region with year 1951–2003 is extracted. The rainfall data from both the products were divided into two sections: the first 24 years (1951–1974) covers the first section and the last 29 years (1975–2003) covers the second section. Annual daily maximum precipitation depth was computed by using frequency analysis for 5-year, 25-year and 50-year return period. The rainfall distribution characteristics of India were studied by using EVI probability distribution as follows [2]:

$$F(x) = \exp\left[-\exp\left(-\frac{x-u}{\alpha}\right)\right] \quad -\alpha \leq x \leq \alpha \tag{1}$$

where, x is the rainfall magnitude; u is mode of distribution, and α is a parameter that is to be determined and u are determined from the formula as given below:

$$\alpha = \frac{\sqrt{6}}{\Pi} s \tag{2}$$

$$u = \bar{x} - 0.5772\alpha \tag{3}$$

where, s is the standard deviation and $\bar{x}$ is the mean of annual daily maximum rainfall.

Let n be the total number of rainfall event and x_i be the rainfall magnitude in i^{th} event, then:

$$\overline{x} = \frac{1}{n}\sum_{i=1}^{n} x_i \qquad (4)$$

$$s = \sqrt{\frac{1}{n}\sum_{i=1}^{n}\left(x_i - \overline{x}\right)^2} \qquad (5)$$

The reduced variate y is defined as

$$y = \frac{x - u}{\alpha} \qquad (6)$$

Substituting in Eq. (1), we get:

$$F(x) = \exp\left[-\exp(-y)\right] \qquad (7)$$

Solving for y

$$y = -\ln\left[\ln\left(\frac{1}{F(x)}\right)\right] \qquad (8)$$

Let an extreme event occur if a random variable x_e greater than or equal to some level x_T, probability (P) of occurrence of an event is the inverse of its return period (T):

$$\frac{1}{T} = P\left(x_e \geq x_T\right) \qquad (9)$$

$$F(x_T) = \frac{T-1}{T} \qquad (10)$$

Substituting in Eq. (8), we get:

$$y_T = -\ln\left[\ln\left(\frac{T}{T-1}\right)\right] \qquad (11)$$

From Eq. (6), annual daily maximum precipitation depth for T years return period is:

$$x_T = u + \alpha\, y_T \tag{12}$$

13.2.4   COMPUTATION OF PERCENTAGE CHANGE IN ANNUAL DAILY MAXIMUM PRECIPITATION

Let annual daily maximum precipitation depth for 5 year return period for the time period of 24 years (1951–1974) and the next 29 years (1975–2003) of rainfall for a single grid are x_1 and x_2, respectively. Therefore, change in annual daily maximum precipitation depth is = $(x_2 - x_1)$ and percentage change in annual daily maximum precipitation depth = $(x_2 - x_1)/x_1 \times 100$. In this way percentage change in annual daily maximum precipitation depth for all the grids and for 25 year and 50 year return periods were estimated. Similar procedure was repeated for CMIP5 rainfall data to find percentage change in annual daily maximum rainfall depth over the two time periods. Isohyetal maps were prepared for percentage change in annual daily maximum precipitation depth for both the sets of rainfall data for 5 years, 25 years and 50 years return periods.

13.3   RESULTS AND DISCUSSION

13.3.1   BASED ON OBSERVED (IMD) DATA AND CMIP5 DATA

In this context, spatial variation of extreme rainfall depth for 5 years, 25 years and 50 years return period of the observed and CMIP5 rainfall data was investigated. The data were divided in the time range 1951–1974 and 1975–2003 for comparison. Then, percentage change in rainfall depth for the same return period was described. Frequency analysis were performed using Extreme value type I distribution over both observed and modeled annual daily maximum rainfall data for 5 year, 25 year and 50 year return periods for the period 1951–1974 and 1975–2003. The contours of rainfall were plotted to study the spatial variation of rainfall depth within Indian region.

13.3.1.1 Analyses of Spatial Variation IMD and CMIP5 Rainfall for 1951–1974

Figure 13.2 shows the spatial variation for rainfall from IMD and CMIP5 in case of 5 years return period. From Figure 13.2, it can be envisaged that the rainfall values are under-predicted by the climate models (CMIP5) than the observed rainfall by IMD, for the CMIP5 simulations. Despite underestimation, the CMIP5 rainfall shows almost similar pattern as that of IMD in different parts of India. In Figure 13.2a, highest rainfall is in west-coast and in some parts of north-east, east-coast and central India, whereas southern peninsula, north-west and some portion of northern and north-east India have shown low rainfall values. Also it can be envisaged from Figure 13.2b that north-east, east-coast and west-coast has high rainfall depth, and south peninsula, north-west and some part of central India have low rainfall depth. Rainfall in north-east region and east-coast is high compared to other zones and is clearly visible from the these contours which is not the case in 5 years return periods of observed data.

Similarly, Figures 13.3 and 13.4 show the variation of annual daily maximum rainfall over 25 and 50 years return period, respectively for IMD as well as CMIP5 over India. From Figure 13.3a, it is envisaged that for 25 years return period, IMD rainfall data shows highest rainfall in west-coast and in

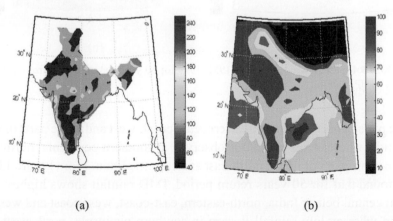

(a)   (b)

FIGURE 13.2 Precipitation depth for 1951–1974 for 5 years return period (rainfall in mm) from: (a) IMD and (b) CMIP5 rainfall data.

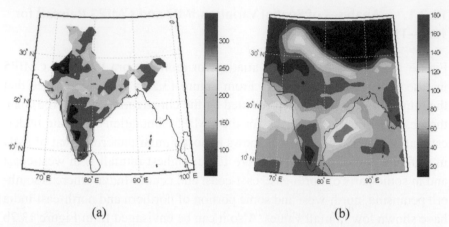

FIGURE 13.3 Precipitation depth of 1951–1974 for 25 years return period from: (a) IMD and (b) CMIP5 rainfall.

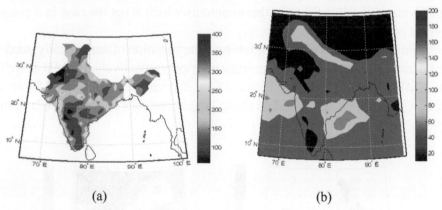

FIGURE 13.4 Precipitation depth of 1951–1974 for 50 years return period from: (a) IMD and (b) CMIP5 rainfall.

some parts of northern, north-eastern, east-coast, west and some parts of central India whereas low rainfall is observed in southern peninsula, northwest and in some parts of north, north-east and eastern region. From Figure 13.4a, it is found that for 50 years return period, IMD rainfall shows highest values in central belt of India, north-eastern, east-coast, west-coast and western region whereas low rainfall is seen in southern peninsula, north-west and in some parts of north and northeast India. Figure 13.4b shows that northeast, east-coast, west-coast has high rainfall depth and also south peninsula, northwest and some part of central India have low rainfall depth according to CMIP5 model simulated rainfall data. The spatial change in CMIP5 rainfall pattern for 25 years return period is same to that for the 5 years return period except that the rainfall magnitude is increased in the later. Highest precipitation depth is seen in 50 years return period at northeast, east-coast and west-coast and lowest precipitation depth is seen in northwest and south. In case of IMD data for 50 years return period (Figure 13.4a) for the same time range, high precipitation depth is seen in west-coast, north-east and some parts of central India and east while in modeled data high rainfall depth is seen in north-east and in some parts of east-coast and west-coast.

13.3.1.2 Analysis of Spatial Variation of IMD and CMIP5 Rainfall for 1975–2003

Frequency analysis was conducted with rainfall dataset from both the IMD and CMIP5 estimates for 1975–2003. Figure 13.5 shows the pattern of rainfall from both the sources over India for 5 years return period. For 5 years return period, IMD data shows high rainfall in north-east, east-coast, west-coast, west and in some parts of central India and low rainfall is seen in southern peninsula, North-west and in some parts of north and north-east India. However, for modeled data, north-east, east-coast, west-coast has high rainfall while northwest and southern India has low rainfall value. High rainfall depth in north-east part is very clear from Figure 13.5 and IMD data show clear contours of high rainfall depth in west-coast.

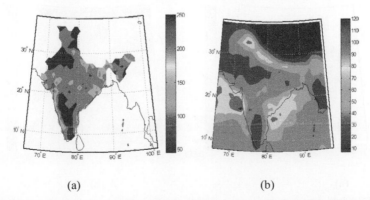

(a)   (b)

FIGURE 13.5 Precipitation depth of 1975–2003 for 5 years return period from: (a) IMD and (b) CMIP5 rainfall.

Figures 13.6 and 13.7 shows the spatial pattern of annual daily maximum rainfall depth for 25 and 50 years return period, respectively. From Figure 13.6a it is understood that for 25 years return period IMD data gives high precipitation depth in west-coast, west, north-east, east-coast and in some parts of central India and rainfall depth is low in southern peninsula, north-west and in some parts of northern and north-eastern region. Very few parts are under high rainfall depth for 25 years return period for modeled data (Figure 13.6b). North-east, east-coast, west-coast has high rainfall while North-west, some parts of central India and southern India has low rainfall value. For 50 years return period high precipitation depth is

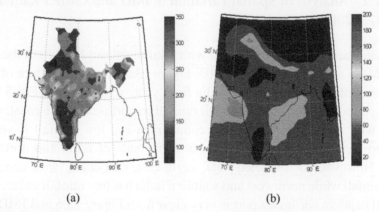

(a)   (b)

FIGURE 13.6   Precipitation depth of 1975–2003 for 25 years return period from: (a) IMD and (b) CMIP5 rainfall.

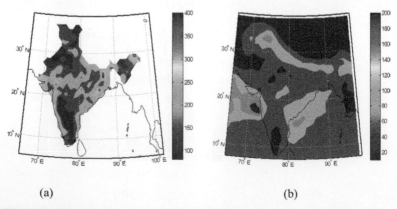

(a)   (b)

FIGURE 13.7   Precipitation depth of 1975–2003 for 50 years return period from: (a) IMD and (b) CMIP5 rainfall.

visible (Figure 13.7a) in west-coast, west, central belt, north-east, east-coast and low precipitation depth is seen in southern peninsula, north-west, north and in some parts of north-east and east-coast. The contour of second set of rainfall data for 25 years and 50 years of return periods is same except the magnitude of precipitation. Figure 13.7b shows that using CMIP5 data, the rainfall contour for 50 years return period has high values in north-east, east-coast, west-coast and a small part of central India while northwest and southern India has low rainfall values.

13.3.2 PERCENTAGE CHANGE IN ANNUAL DAILY MAXIMUM RAINFALL DEPTH

Annual daily maximum rainfall depth estimated for 5, 25 and 50 years return periods from both IMD and CMIP5 were analyzed for change in rainfall over two historical periods: 1951–1974 and 1975–2003. From the percentage change in annual daily maximum precipitation depth, one can distinguish the areas under high, constant and low rainfall depth over the periods. The changes undergone in precipitation depth over the periods are plotted in Figure 13.8. Most of the parts in west-coast have no change in rainfall depth for the 5 years return period in spite of a region receiving high rainfall whereas the western part which receives low rainfall has shown (Figure 13.8a) increase in percentage change of rainfall. In Figure 13.8b, increase in percentage change in rainfall value is seen mostly in central belt of India. North and north-eastern region have decrease in rainfall amount for 25 years return periods. Most parts have no change in precipitation depth. North-east region that receives high rainfall has negative change in precipitation depth in some parts.

Western region that receives low rainfall has shown positive percentage change in rainfall value. For 50 years return period it is seen (Figure 13.8c) that most of the regions have no change in precipitation depth. North, north-east and some parts of central India, west-coast has decreased in precipitation depth. East-coast, west, east and some part of north-east has increased rainfall depth. From the contour map of Figure 13.8, it is clear that annual daily maximum precipitation depth is increased in eastern, east-coast, western, northwest, central and north-east India. Decrease in annual daily maximum precipitation depth is observed in most parts of northern India, southern-peninsula and in some parts of east-coast, central and north-east India. Light blue colored contour shows no change in annual daily maximum precipitation depth.

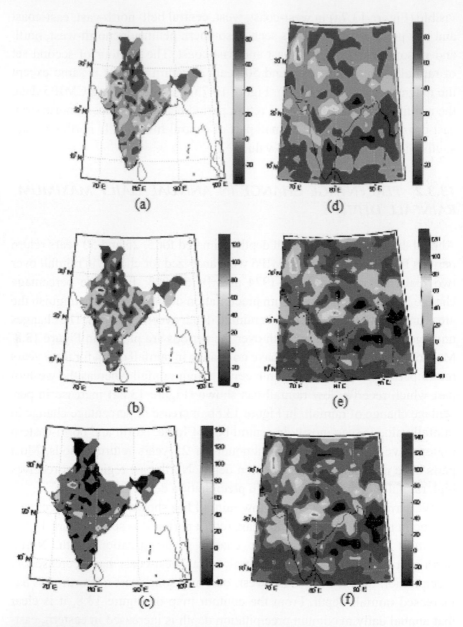

FIGURE 13.8 Percentage change in annual daily maximum precipitation depth using IMD data for (a) 5, (b) 25 and (c) 50 years return period; using CMIP5 data for (d) 5 years return period, (e) 25 years return period, (f) 50 years return period.

It is clearly shown in Figure 13.8d that annual daily maximum precipitation depth is increased in east-coast, north-west, central and north-east India. Decreased in annual daily maximum precipitation depth is occurred in most parts of northern India, southern-peninsula and in some parts of central and north-east India. Light blue colored contour in Figure 13.8 shows no change in annual daily maximum precipitation depth. Major parts of India have shown no change in rainfall depth. From the contour above (Figure 13.8e), it is clear that annual daily maximum precipitation depth is increased in north-east, central India, north-west and a small part of west-coast. There is also decrease in annual daily maximum precipitation depth in south, east, north and some part of central India. Most parts of India have no change in rainfall depth.

From the contours of percentage change (Figures 13.8a–13.8c) in rainfall depth, it is found that precipitation depth is increased in east-coast, west, North-west and some parts of north-east and central India whereas it is decreased in north and some parts of west-coast and central India. It is understood from Figures 13.8d–13.8f that percentage change in rainfall is increased in north-east, northwest, east-coast and some parts of central India. Percentage change in rainfall is decreased in south, west-coast, north and in some parts of central India. Hence, there are some similarity about percentage change in rainfall between IMD and CMIP5 data.

13.4   CONCLUSIONS

In India, spatio-temporal variation of rainfall across the country is very high. Seasonality affects various decisions making in hydrology and agriculture. This study uses two sets of rainfall data from: IMD data from India Meteorological Department (IMD); and another is modeled rainfall data from Coupled Model Inter-comparison Project Phase Five (CMIP5). Frequency analysis was performed on both datasets for 5 years, 25 years, and 50 years return periods using extreme value type I (EVI) distribution.

The results of IMD rainfall show percentage increase in rainfall depth in east-coast, west, north-west, north-east and some parts of central India and percentage decrease in north, south and some parts of northeast and central India. Similarly, the CMIP5 rainfall shows percentage increase in rainfall depth in east-coast, west, north-west, north-east and some parts of central India and percentage decrease in rainfall depth in north, south and some parts of northeast and central India. Overall, comparison of the two rainfall datasets concluded that CMIP5 data shows almost similar pattern of annual daily maximum rainfall depth as IMD data for most of the regions over India.

In this study, only simple comparisons between model and IMD rainfall are presented, there is no in depth analysis of resolution of the model itself, its physical process or other processes neither comparison among models in simulating better rainfall output has been done. Future study will focus on how to adopt a more rational and scientific approach to extracting useful information from the model simulation results and how to use better simulation results for detecting and attributing climate change and comparisons of those results should be done with observed one in order to know the reliability of the model output.

13.5 SUMMARY

In this study, observed rainfall data from India Meteorological Department (IMD) and modeled rainfall data from Coupled Model Inter-comparison Project Phase Five (CMIP5) were taken as reference data. Frequency analysis is performed on both datasets for 5 years, 25 years and 50 years return period and their results are plotted as contours. The results of IMD rainfall shows percentage increase in rainfall depth in east-coast, west, north-west, north-east and some parts of central India and percentage decrease in rainfall depth in north, south and some parts of northeast and central. While the results of CMIP5 rainfall shows almost same as IMD result in percentage change in annual daily maximum precipitation depth. Annual daily maximum precipitation depth may increase by 92% (IMD) and 73% (CMIP5) for 5 year return period, 132% (IMD) and 108% (CMIP5) for 25 year return period, 144% (IMD) and 122% (CMIP5) for 50 year return period. By comparing the two results it is concluded that CMIP5 and IMD rainfall data show almost similar annual daily maximum rainfall depth for most of the regions.

However in the current study, only simple comparisons between model and IMD rainfall data are presented. There is no in-depth analysis of resolution of the model itself, its physical process or other processes neither comparison among models in simulating better rainfall output has been done.

KEYWORDS

- **Agriculture**
- **Climate change**
- **CMIP5**
- **Contour**
- **Extreme precipitation**
- **Extreme value type I**
- **Frequency analysis**
- **Global climate change**
- **Gumbel distribution**
- **Hydrology**
- **Hydrologic cycle**
- **IMD**
- **IMD rainfall**
- **Mean**
- **Probability distribution**
- **Rain gage**
- **Rainfall**
- **Reduced variate**
- **Return period**
- **Seasonality affect**
- **Simulation**
- **Spatio-temporal variation**
- **Standard deviation**

REFERENCES

1. Booij, M. J. (2002). Extreme daily precipitation in Western Europe with climate change at appropriate spatial scales. *International Journal of Climatology*, 22, 69–85.
2. Chow, V. T., Maidment, D. R., and Mays, L. W. (1988). Indicate full title. *Applied Hydrology*, 380–415.
3. Goswami, B. N., Venugopal, V., Sengupta, D., Madhusoodanan, M. S., and Xavier, P. K. (2006). Increasing trend of extreme rain events over India in a warming environment. *Science*, 314, 1442–1445.

4. Guhathakurta, P., Srojith, O. P., and Menon, P. A. (2011). Impact of climate change on extremes rainfall events and flood risk in India. *Indian Academy of Sciences, Vol.* 3, 359–373.

5. Hibbard, K. A., Meehl, G. A., Cox, P., and Friedlingstein, P. (2007). A strategy for climate change stabilization experiments. *Eos, Transactions American Geophysical Union, Vol.* 88(20), 217–221.

6. Joshi, U. R., and Rajeevan, M. (2006). Trends in precipitation extremes over India. In: *National Climate Change* by India Meteorological Department.

7. Kharin, V. V., Zwiers, F. W., Zhang, X., and Hegerl, G. C. (2007). Changes in temperature and precipitation extremes in the IPCC ensemble of global coupled model simulations. *Journal of Climate,* 20(8), 1419–1444.

8. Krishnamurthy, C. K. B., Lall, U., and Kwon, H. H. (2009). Changing frequency and intensity of rainfall extremes over India from 1951 to 2003. *Journal of Climate*, Vol. 22(18), 4737–4746.

9. Otieno, V. O. and Anyah, R. O. (2012). CMIP5 simulated climate conditions of the Greater Horn of Africa (GHA). Part 1: contemporary climate. *Climate Dynamics*, Vol. 41(7–8), 2081–2097.

10. Roy, S. and Balling, R. C. (2004). Trends in extreme daily precipitation indices in India. *International Journal of Climatology*, Vol. 24(4), 457–466. doi*: 10.1002/joc.995.*

11. Rudolf, B., Hauschild, H., Rüth, W., and Schneider, U. (1996). Comparison of rain gage analyzes, satellite-based precipitation estimates and forecast model results. *Advances in Space Research*, Vol. *18*(7), 53–62.

12. Taylor, K. E., Balaji, V., Hankin, S., Juckes, M., Lawrence, B., and Pascoe, S. (2012). *CMIP5 Data Reference Syntax (DRS) and Controlled Vocabularies*, 16 pp.

13. Taylor, K. E., Stouffer, R. J., and Meehl, G. A. (2009). A summary of the CMIP5 experiment design. *PCDMI Rep.*, Vol. 33, 1–45.

14. Taylor, K., Stouffer, R. J., and Meehl, D. A. (2012). An Overview of CMIP5 and the Experiment Design. *American Meteorological Society,* 485–498.

15. Ying, X. U. and Chong-Hai, X. U. (2012). Preliminary assessment of simulations of climate changes over China by CMIP5 multi-models. *Atmospheric and Oceanic Science Letters*, Vol. 5(6), 489–494.

16. Rajeevan, M., Bhate, J., Kale, J. D., and Lal, B. (2005). Development of a high resolution daily gridded rainfall data for the Indian region. *Met. Monograph Climatology No. 22,* India Meteorological Department, pp. 26.

17. Shepard, D. (1968). A two-dimensional interpolation function for irregularly spaced data. Proc. 1968 *ACM Nat. Conf.,* 517–524.

CHAPTER 14

SUSTAINABILITY CONCEPT TO EVALUATE PERFORMANCE OF RAINFALL TIME SERIES

DEEPESH MACHIWAL,[1] DEVI DAYAL,[2] and MADAN KUMAR JHA[3]

Central Arid Zone Research Institute, Regional Research Station, Kukma – 370105, Bhuj, Gujarat, India. Tel.: +91-2832-271238; Fax: +91-2832-271238, E-mail: dmachiwal@rediffmail.com, devidayal.cazri@yahoo.co.in

Agricultural and Food Engineering Department, Indian Institute of Technology, Kharagpur – 721302, West Bengal, India, E-mail: madan@agfe.iitkgp.ernet.in

CONTENTS

14.1 INTRODUCTION

Climate change has emerged as the single most-pressing problem globally, and accordingly, the same has received increasing attention across the world [4]. The rainfall and temperature are two most important climatic variables clearly evidencing the occurrence of the climate change. Rainfall is particularly more susceptible to the impacts of climate change as the changing rainfall patterns may ultimately lead to extreme hydrological events, i.e., floods/droughts in different regions [60]. In rainfed agriculture in semi-arid and arid regions, variability of the rainfall due to climate vulnerability significantly affects the spatial and temporal water availability [13]. It is also revealed from the literature that number of studies exploring spatial and temporal variability of the rainfall time series at different time scales is currently increasing [5, 7, 14, 20, 39, 50]. Understanding spatial and temporal variability of the rainfall is further important for the arid lands, which are spreading over 61 million km^2 worldwide (46% of the global area) [16].

In the arid lands, the water deficit scenario is quite common due to relatively less occurrence of rainfall and its low magnitudes. The arid lands in India, extending over 50.8 million ha (15.8% of the country's geographical area) [46, 48], can be further sub-divided into hot arid and cold arid zones. A large portion of the hot arid zone in the country, i.e., 32 million ha, exists in western Rajasthan consisting of 62% of country's hot arid zone, and another 19.6% of country's total arid land is situated in Gujarat [28].

Time series modeling is considered as a comprehensive technique for exploring temporal patterns of hydrologic time series through identification/ determination of all important time series characteristics, i.e., normality, stationarity, homogeneity, presence/absence of trend and persistence [1, 37–39, 56]. From the statistical perspective, the term homogeneity determines whether or not the entire data in the time series belong to one population, and if true, the time series should have a time-invariant mean. The non-homogeneity may be introduced in the time series due to anthropogenic factor when a change occurs in the method of data collection and/ or due to natural factors when the environment, in which data collection is done, is changed [17]. The term stationarity determines whether the statistical parameters of the time series, i.e., mean, standard deviation, etc. computed from different samples change only due to sampling variations or due to any other reason. Presence of trend in a time series is observed when a significant relation is found between the observations and time as revealed from the values of correlation coefficient. Trends in hydrologic time series are incurred due to natural as well as anthropogenic activities [56]. The term persistence reflects the tendency for the magnitude of an event (or data) to be dependent on the magnitude of the previous event(s) or data.

Sustainability index (SI) is another useful tool that can be used to evaluate performance of the hydrologic time series based on reliability-resilience-vulnerability approach. The SI makes it possible to evaluate and compare hydrologic time series of the different stations with respect to their sustainability. The concept of SI was first defined by Loucks [35] using reliability (R_y), resilience (R_e) and vulnerability (V_y) as the performance criteria with an aim to evaluate and compare water management policies. Thereafter, the index has been utilized by researchers for the scientific use [45, 51, 55]. It is revealed from the extensive literature search that the SI has never been applied to evaluate the performance of the hydrologic stations based on their time series records.

This chapter aims at highlighting role of time series modeling in identifying vital characteristics of the hydrologic time series. Firstly, fundamental characteristics of both time series analysis and sustainability criteria are defined and/or described. Then, the chapter summarizes theoretical procedures for applying sustainability concept by explaining its historical development and recent applications in hydrology and water resources engineering. Thereafter, a case study is presented demonstrating a comprehensive methodology for analyzing the rainfall time series in an arid region of western India. This study introduces the sustainability approach, for the first time, to analyze a hydrologic time series, and demonstrates its novel applicability to rainfall time series of arid region.

14.2 TIME SERIES MODELING

Time series modeling is performed by investigating a temporally sequence of dataset and is explained by the synthesis of a model for making predictions wherein time is an independent variable. Sometimes, time is not actually used as the independent variable to predict the magnitude of a random variable such as peak runoff rate, but the data are ordered by time in a series. The goal of time series modeling is to detect and describe quantitatively each of the generating processes underlying a given sequence of observations [56].

Hydrologic time series model accomplishes multiple tasks. In literature, time series modeling is employed mainly to detect a trend in several hydrologic variables especially in precipitation and temperature datasets. The time series analysis also help developing and calibrating a model that describes the time-dependent characteristics of a hydrologic variable. Additionally, the time series models may be used to predict future values of a time-dependent hydrologic variable. Besides the modeling of time-dependent hydrologic data series, the concept of time series modeling may also be used for space-dependent data series of hydrologic systems, which are known as 'spatial data series'. In spatial data series, the independent variable is site and the dependent variable is a hydrologic parameter that may have different values over the space at any time. Many of the time series modeling methods can be adequately used for spatial data series [56].

According to Rogers et al. [52], time series modeling is a four-step process involving detection, analysis, synthesis, and verification. The first step detects the systematic components, e.g. trend, periodicity, etc. of the time series. In detection step, physical and statistical significance of the systematic components is also decided. In the third step, the systematic components are analyzed to identify their characteristics including magnitudes, form and their duration over which the effects exist. In the synthesis step, information from the analysis step is accumulated to develop a time series model and to evaluate goodness-of-fit of the developed model. In the final step, the developed time series model is evaluated to verify using independent sets of data. Elaborated text on time series modeling can be found in the specialized books on time series analysis such as [6, 10, 11, 38, 54].

14.2.1   TIME SERIES CHARACTERISTICS

There are a set of key assumptions, which needs to be satisfied prior to use of any hydrologic time series for many statistical analyzes involved in water resources studies. These assumptions include the time series is homogenous, stationary, free from trends and shifts, non-periodic with no persistence [1]. However, either none of them or only few criteria are checked to confirm that the time series follow the conditions.

14.2.1.1   Normality

Normality of a hydrologic time series indicates whether the distribution of the hydrologic data in the series follows or not the normal distribution. Many statistical tests used for time series modeling are based on the assumptions that the data in the series were sampled from a normal distribution. This assumption is very critical to test reliability of the test especially of parametric tests which depend upon the parameters of the data distribution, i.e., mean and standard deviation, among others. For a normally-distributed hydrologic time series, value of skewness coefficient should be zero and value of kurtosis should be three. Otherwise, the curve of the probability density function on the normal probability plot will be either left-skewed/right-skewed or platykurtic/leptokurtic.

14.2.1.2   Homogeneity

Homogeneity is the term, which checks whether any subset(s) of the entire time series belong to the one population. If a time series has time invariant mean then homogeneity is supposed to be present in the time series.

The factors/causes responsible for arising the non-homogeneity in a time series can be anthropogenic, e.g. due to changes in the method of data collection, and/or natural such as change occurring in environment in which data collection is done [17].

14.2.1.3 Stationarity

Stationarity in a time series is considered to be present if values of the statistical parameters such as mean and standard deviation do not significantly change for different samples of the series. Some changes in the statistical parameters of the time series may be due to sampling variations, which are not accounted while testing stationarity of the time series. The stationarity in a time series may be of two types; strict stationarity if statistical properties of the time series do not vary with changes of time origin, and weak stationarity or second-order stationarity when the first- and second-order moments depend only on time differences [9]. In nature, it is rare to find strict stationarity in a time series, and a time series with weakly stationarity is practically considered as stationary time series.

14.2.1.4 Presence of Trends

Trends in a time series indicate some kind of change occurring over time. A change in the hydrologic time series can place in many ways. For example, a change can occur gradually over the time where it is difficult to locate a clear-cut point of change. Such kind of change is known as trend. On the contrary, a change in the hydrologic time series may be abrupt over an instant time, which is known as step change or jump. The trend may also take more complex form completely different from gradual and abrupt changes [56]. A trend is determined by a unidirectional and gradual change in the mean value of a hydrologic variable that may be either falling or rising [56].

Trend present in the time series may or may not be statistically significant, which may be confirmed by testing strength of relationship (positive or negative) between the observed values and time. Usually, trends and shifts in a hydrologic time series are incurred due to gradual natural or human-induced changes in the hydrologic environment [19, 53]. Factors causing gradual or natural changes in hydrologic variables can be global or regional climate change, gradual urbanization in an area surrounding the monitoring site, changes in the method of measurement at the monitoring site, or by moving the monitoring site even a short distance away. On the other hand, step changes or jumps in a time series usually result from catastrophic natural events such as earthquakes, tsunami, cyclones, or large forest fires, which quickly and considerably alter the hydrologic regime of an area. In addition, the anthropogenic changes such as the closure of a new dam, the start or end of groundwater pumping, or other such developmental activities may also cause jumps in some hydrologic time series [19]. Similar to trends, the jumps can also be either positive or negative.

14.2.1.5 Periodicity

Periodicity of hydrologic time series explains a steady or oscillatory form of movement that recurrently occurs over a fixed time interval [56]. The periodicity in the hydrologic time series is generally introduced by astronomic cycles, e.g. earth's cyclic rotations made around the sun [19, 30]. Impact of the astronomic cycles is clearly observed in most of the hydrologic time series such as rainfall, temperature, evapotranspiration, stream flow, groundwater levels, seawater levels, soil moisture, etc. [19]. In addition to annual-scale periodicity, there may be periodicity at lesser scales of time such seasonal, monthly and weekly periodicity. The seasonality in the hydrologic time series may be caused due to seasons. For example, rainfall in the northwest part of the India falls during four-month period (June to September) when southwest monsoon sets in the region. Hence, the rainfall will concentrate in rainy season and there will be negligible rainfall during the dry period. Accordingly, stream flow in seasonal rivers will exist during monsoon and post-monsoon seasons but the stream may be completely dry during pre-monsoon or summer season.

The seasons may also affect the groundwater level time series. In monsoon season, there will be adequate availability of the surface water, and negligible quantities of the groundwater will be extracted for agricultural purposes. However, in response to negligible rainfall received during the post-monsoon or rabi season, large groundwater withdrawals will be made and that may lower down the groundwater levels in the aquifer system. Even weekly cycles may also be observed in the water-use data of domestic, industrial, or agricultural sectors; many times the water-use time series contain both annual and weekly periodicities [19]. In order to identify the periodicity in the hydrologic time series, it is suggested in the literature that the time scale should be less than a year, e.g. monthly or seasonal [38].

14.2.1.6   Persistence

Persistence in the hydrologic time series remains present when successive members of a time series are linked in some dependent manner [56]. Persistence can also be defined as a memory effect or the tendency by which magnitude of a hydrologic event remains dependent on the magnitude of its previous event(s); for example, the tendency for low rainfall to follow low rainfall and that for high rainfall to follow high rainfall. Consequently, persistence is considered identical to autocorrelation [49]. Persistence, for the first time, was described in a comprehensive manner in the studies on a reservoir design across the Nile River [25, 26]. At that time, the persistence was defined by a parameter known as 'Hurst's coefficient', having average value of approximately 0.73 for a time series with very large samples/dataset. Capodaglio and Moisello [8] suggested that its theoretical value is 0.5 for an independent Gaussian process to which hydrologic series are assimilated. When the theoretical and the observed values of Hurst's coefficient do not match each other, then it is known as 'Hurst's phenomenon'. Almost all the stochastic models proposed to represent hydrologic processes attempt to include the persistence. However, it is virtually impossible to identify any long-term persistence in the hydrologic time series with the time series records commonly available in hydrology [8].

14.2.1.7   Stochastic Component

A time series model consists of a systematic pattern explained in terms of two components, i.e., trend and seasonality, and a stochastic component. The stochastic component is usually makes the pattern difficult to be identified. The systematic pattern is deterministic in nature, whereas the stochastic component accounts for the random error. In general, the stochastic component contains a dependent part that can be described by a p-order autoregressive (AR) and q-order moving average (MA) model abbreviated as ARMA(p,q), and an independent part that can only be described by some sort of probability distribution function. When p = 0, the ARMA(p,q) represents an MA(q) model, and when q = 0, it represents an AR(p) model.

14.3   CONCEPT OF SUSTAINABILITY

It is revealed from the literature that the concept of sustainable develop-ment was introduced, for the first time, by the *World Conservation Strategy*. Several researchers have attempted to define and develop methodologies for assessing the sustainability of water resources systems [3, 15, 27, 34, 35, 40, 47, 57, 58, 59, 61, 62].

Later on, the sustainability index (SI) was proposed to initially evaluate the performance of alternative policies from the perspective of water users and the environment [35]. The SI can also be defined as a measure of a system's adaptive capacity to reduce its vulnerability. For example, if imple-menting a policy makes a system more sustainable then the SI will suggest that the system has larger adaptive capacity. Thus, the concept of SI described by Ref. [35] considered three performance criteria, i.e., reliability (R_y), resilience (R_e) and vulnerability (V_y) in order to evaluate and compare dif-ferent water management policies. Afterwards, the SI has been utilized in many scientific studies by researchers [45, 51, 55]. In literature, the R-R-V based sustainability concept is mainly used to evaluate performance of the water resources systems [2, 29, 35, 55]. It is also revealed from the literature that the sustainability concept has not applied for any other purpose except to evaluate performance of water resources systems, e.g. to assess sustain-ability of the storage reservoirs.

14.3.1   SUSTAINABILITY OF WATER RESOURCES SYSTEMS

Before defining sustainability of the water resources systems, few impor-tant issues need to be considered. Loucks [36] identified the most important issues as change, scale, technology, risk and training. The first four issues are said to be of direct importance to methodologies for assessing sustain-ability and, the same are discussed in this section.

14.3.1.1   Change

In nature, stationarity seldom occurs in most of the hydrologic processes and environmental systems due to changing conditions over the course of time. It is well-known fact that the natural, economic, environmental and social subsystems are interrelated, and therefore, change in any of the system(s) will have an effect on the other system(s) and, this will have an effect on the entire system. In addition, the management objectives might change over time and simply guessing about the future may result in wrong predictions. In order to handle with such situations and to cope up with dynamic natural systems, adaptive management was introduced as a tool in natural resources management [36]. It is suggested that a method used for assessing sustainability should also be able to work within an adaptive management framework.

14.3.1.2 Scale

The sustainability should be assessed by considering the appropriate scale, over both time and space. When finalizing the temporal scale, both the planning horizon and the duration of the time steps used in the analysis need to be considered. One of the key elements of the sustainability is need of future or predicted events (water supply), however there is no guideline as to how many future events should be considered for the analysis. The appropriate duration of time steps may be decided by taking into account the variability of the water supply systems. The extreme events, such as occurrence of droughts and floods, are naturally occurring in the water cycle, but they can temporarily put at risk the efforts to achieve sustainable development [32]. Therefore, the assessment of sustainability for the water resources system should adopt a time scale making the inclusion of possible extreme events. Loucks [35] suggested that the duration of the time steps used for the sustainability analysis should be such that natural variation in a resource, like water, is averaged out over the period. Thereafter, many researchers recommended various time lengths for the planning horizon and time steps [3, 24, 33, 35, 36, 40, 41, 43, 44, 52, 59].

14.3.1.3 Risk

There are external causes such as extreme events (floods and droughts) and degraded water quality, etc., which may result in failure of water resources systems completely. These influencing factors should be given careful due considerations while planning and designing of the water resource systems. In most developing countries, water resources systems are not dependable due to political and economic constraints, however a sustainable water resources system should definitely experience diminishing frequency and less severity of failures over time [36]. Conventionally, the water resources systems, e.g. reservoirs, were designed to have a high degree of reliability or low probability of failure. However, following the suggestions and recommendations of Refs. [18, 22], generous efforts have been devoted to the two additional risk criteria, i.e., resilience defined as likelihood of return to normal operation after a failure, and vulnerability explaining likely magnitude of failure. A sustainable system should have a high degree of resilience and low vulnerability [15].

14.3.1.4 Technology

Advent of sophisticated technologies along with advancement of computer-based modeling work resulted in rapid development of tools for sustainability assessment [23, 62]. One of the advanced developed modeling system tools for sustainability assessment is Decision Support Systems (DSS), which are supposed to assist the decision-makers in making informed decisions. It is well-understood fact that participation of the stakeholders is a key component in achieving success in the water resources management [63], and hence, the developed DSS should have relatively simple models with an easy to understand graphical user interface enhancing the user-friendly possibilities for achieving a useful shared-vision model set-up [36].

14.3.2 *SUSTAINABILITY INDEX FOR HYDROLOGIC TIME SERIES*

The sustainability concept to evaluate performance of the hydrologic time series based on R-R-V approach is applied for the first time for rainfall time series [39]. The performance criteria, initially defined and applied to water resources systems, are slightly modified in order to apply them to assess the sustainability of hydrologic time series. The estimators of reliability, resilience, vulnerability and sustainability index along with their explanations are described ahead.

14.3.2.1 Reliability

Reliability of water demand in the water resources systems is defined as the probability at which the available water supply meets the water demand during the period of simulation [22, 31]. For a hydrologic time series, the 'reliability' is expressed as ratio of the number of data in a satisfactory (successful) state to the total number of data in the time series. It is considered that the satisfactory state for a particular hydrologic variable will be such that its value in the entire time series x_n with n sample size remains equal to or greater than the mean threshold x^T, and then the reliability of the hydrologic time series can be expressed as [39]:

$$R_y = f_{SE}/n \tag{1}$$

where, R_y = reliability; f_{SE} = number of successful events or satisfactory values in hydrologic time series (x_n), when $x_t \geq x^T$ (t=1, 2, ... n); and n = sample size of time series.

14.3.2.2 Resilience

Resilience of a water resources system is defined as its capacity to adapt to changing conditions [64]. The 'resilience' of a hydrologic time series is described as the probability or the changes of occurrence that if value of a hydrologic variable in a time series is in an unsatisfactory (failure) state at any time, the next state will be satisfactory (successful). It may also be explained as the probability of having a satisfactory value or successful event in time period t-1, given an unsatisfactory value or failure event in any time period t. The resilience of the hydrologic time series can be expressed as shown below [39]:

$$R_e = f_{FE-SE}/f_{FE} \tag{2}$$

where, R_e = resilience of time series; f_{FE-SE} = number of times a satisfactory value (successful event) follows an unsatisfactory value (failure event); and f_{FE} = number of times an unsatisfactory value occurs in the time series.

14.3.2.3 Vulnerability

The vulnerability is defined as the probable value of deficits, if they occur [22]. In other words, the vulnerability can be explained as a measure of the extent of the differences between the threshold value and the failure events among hydrologic data series. Thus, vulnerability is a probabilistic measure, which is also known as expected values, maximum observed values, and probability of exceedance to vulnerability measures. Considering an expected value measure of vulnerability is to be used, vulnerability of the hydrologic time series can be expressed as follows [39]:

$$V_y = \sum_{i=1}^{n} \text{difference}\left(x^T - x_t\right) \Big/ f_{FE} \quad , \text{for } t = 1, 2, \dots n \qquad (3)$$

where, V_y = vulnerability of hydrologic time series; and $\sum_{i=1}^{n} \text{difference}$ $(x^T - x_t)$ = sum of positive values of $(x^T - x_t)$.

14.3.2.4   Sustainability Index

It is revealed from the literature review that sustainability index was originally developed by [35] in order to evaluate different water management policies by making quantitative measures of sustainability of water resources systems. The quantitative sustainability index facilitates comparison of the different water management policies. The sustainability index (SI) depends upon quantitative values of reliability, resilience and vulnerability, and is expressed by the following equation [35]:

$$SI = R_y \times R_e \times \left(1 - V_y\right) \qquad (4)$$

Value of the SI range from 0 to 1, and it becomes zero if value of any of three performance parameters, i.e., reliability, resilience and vulnerability is zero.

14.4   APPLICATION OF SUSTAINABILITY CRITERIA IN HYDROLOGY

The sustainability criterion has rarely been applied to hydrologic studies as revealed from the extensive literature search. This clearly reflects that there is vast scope for applying the SI concept to several types of hydrologic time series. Definitely, the SI concept may emerge as effective tool to measure, evaluate and compare the sustainability of the hydrologic stations/sites with respect to different hydrologic variables.

In this section, a case study is presented demonstrating a comprehensive methodology for the evaluation of performance of the hydrologic stations based on rainfall time series in an arid region of India. The study was conducted in Kachchh district of Gujarat, India.

14.5　OVERVIEW OF STUDY AREA

Kachchh (study area), the second largest district of the country, is situated in Gujarat State and experiences hot and arid climate over the entire 100% occupied land [12, 21]. It encompasses an area of 45,612 km^2 and is situated from 22°44'08" to 24°41'30" north latitudes and 68°07'23" to 71°46'45" east longitude. The study area comes under sensitive seismic zones of the country with very high vulnerability of occurring earthquakes; one of the major earthquakes occurred in January 2001. Rainfall in the study area is highly erratic and unpredictable in nature. Scarcity of surface water resources is a common phenomenon in the study area and groundwater resources are mostly unusable due to deeper availability and considerably high salinity levels mainly due to coastal location.

14.6　METHODOLOGY

Annual rainfall data for a period of 34 years (1980–2013) of ten rain-gage sites (i.e. Naliya, Anjar, Bhachau, Bhuj, Gandhidham, Dayapar, Mandvi, Mundra, Nakhatrana and Rapar) were collected from Revenue Department, Gandhinagar, Gujarat, India. At Gandhidham, the rain gage was installed in the year 1998, and therefore, the data were available afterwards.

In this study, the sustainability concept was applied to annual rainfall time series of ten sites. The SI based on the reliability-resilience-vulnerability concept was computed for the rainfall time series of ten sites and then it was compared to each other in order to find the most sustainable rainfall series with respect to mean threshold rainfall value over the space. The vulnerability value of the rainfall time series was further divided by the mean threshold rainfall in order to make them range between 0 and 1. It is worth-mentioning that rainfall in a year was considered as success if the annual rainfall in that year exceeded the mean threshold rainfall value. On the other hand, a failure event indicated that the rainfall in a particular year did not exceed the mean threshold value.

14.7 RESULTS AND DISCUSSION

Values of reliability (R_y), resilience (R_e) and vulnerability (V_y) for rainfall time series of ten sites are shown in Figure 14.1. It is revealed from this figure that value of R_y for annual rainfall time series of two rain-gage sites, i.e., Mundra and Mandvi is 0.50, which is relatively high compared with to that of other sites. On the other hand, the value of R_y for annual rainfall series of Bhachau and Dayapar sites is the lowest ($R_y \leq 0.35$) among all sites. Whereas, the R_y may be considered as moderately low for the annual rainfall of five sites, i.e., Naliya, Anjar, Bhuj, Nakhatrana and Rapar.

It is seen from Figure 14.1 that the value of R_e is the highest (0.47) for Naliya and Mundra sites, which indicates that the annual rainfall time series of these two sites is the most resilient for a period of 34 years. However, value of $R_e=0.32$ for Dayapar site renders it as the least resilient for the annual rainfall time series. The R_e values for the annual rainfall of the other sites is moderately low.

Furthermore, it is apparent from Figure 14.1 that the value of V_y is the lowest for two rain-gage sites, i.e., Bhachau ($V_y=0.36$) and Rapar ($V_y=0.37$), which suggests that the annual rainfall of these two sites is less vulnerable among others. On the contrary, the annual rainfall of two stations, i.e., Mandvi ($V_y=0.59$) and Bhuj ($V_y= 0.541$) is the most vulnerable among other sites. A peculiar observation of the sustainability approach is the least vulnerability for annual rainfall series of Bhachau site ($V_y=0.36$), which has the least reliability ($R_y=0.32$) and moderate resilience ($R_e=0.43$). Likewise,

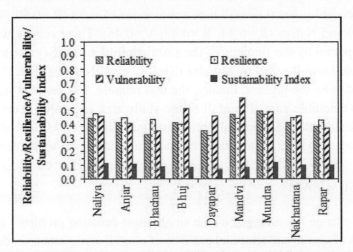

FIGURE 14.1   Reliability, resilience and vulnerability of annual rainfall time series for 10 sites.

the annual rainfall of Rapar rain-gage site is relatively less vulnerable (V_y=0.37) even with less reliability (R_y=0.38) and less resilience (R_e=0.43) of the annual rainfall time series. The site having the most dependable and sustainable rainfall time series is Mundra (R_y=0.50; R_e=0.47; V_y=0.49) followed by Naliya (R_y=0.44; R_e=0.47; V_y=0.46).

Moreover, two most sustainable rainfall time series as revealed from value of the SI are Mundra (SI=0.12) and Naliya (SI=0.112) (Figure 14.1). The sustainability of annual rainfall time series decreases in the order of:

Mundra>Naliya>Anjar>Rapar>Nakhatrana>Bhachau>Mandvi>Bhuj> Dayapar.

The least three sustainable rain-gage sites are Mandvi, Bhuj and Dayapar, where conservation and management of rainwater should be among the top priorities for sustainable water resources management. Overall, it is understood the annual rainfall of the area is little less reliable, less resilient and moderately vulnerable, and there is an urgent need to manage the rainwater adequately to meet the water demands on sustainable basis in case of droughts or failure of monsoon.

14.8 CONCLUSIONS

In this study, sustainability concept based on reliability, resilience and vulnerability approach is applied to evaluate the performance of rainfall time series in an arid region of India. Based on integrated values of three-performance indicator (i.e., R_y, R_e and V_y), the results indicated that the most sustainable and dependable annual rainfall time series is for Mundra (R_y=0.50; R_e=0.47; V_y=0.49) and Naliya (R_y=0.44; R_e=0.47; V_y=0.46). These results were further supported by the findings of the sustainability index (SI) whose values for annual rainfall series of Mundra (SI=0.12) and Naliya (SI=0.112) were observed to be the highest. Finally, the less reliable, less resilient and moderately vulnerable annual rainfall of the study area suggested necessity for adopting suitable management options on sustainable basis for conserving the rainwater to meet escalating water demands during drought years.

14.9 SUMMARY

Climate change has emerged as the single most-pressing problems globally. Analysis of spatial and temporal patterns of rainfall is important for detecting climate change or variability especially in arid and semi-arid regions where the abrupt changes in the rainfall are more likely to occur. In general, the spatio-temporal variability of a rainfall time series is determined by employing time series modeling techniques for detecting presence/absence of trends, testing normality, examining persistence, identifying change points, computing drought indices, etc. by employing statistical analysis/techniques. Thus, time series modeling offers a comprehensive tool to investigate rainfall variability by detecting all important characteristics of a time series. It is revealed from the extensive literature search that time series modeling has been widely used to explore variability of rainfall characteristics in humid and/or semi-arid regions. However, very few attempts have been made to analyze rainfall of arid regions of the world, which may be likely due to relatively less occurrences and low magnitudes of rainfall events in arid regions.

The concept of sustainability was originally developed for evaluating performance of the water resources systems to policy changes. Thereafter, many performance criteria and indices to measure sustainability of the systems have been reported in the literature. Among those criteria and indices, sustainability index comprised of three performance indicators of reliability, resilience and vulnerability has been widely used and discussed in the past studies. It is learnt from the literature that the sustainability index has rarely been used for evaluating performance of the hydrologic time series.

This chapter aims at highlighting role of time series modeling in identifying vital characteristics of the hydrologic time series. Firstly, fundamental characteristics of both time series analysis and sustainability criteria are defined and/or described. Then, the chapter summarizes theoretical procedures for applying sustainability concept by explaining its historical development and recent applications in hydrology and water resources engineering. Thereafter, a case study is presented demonstrating a comprehensive methodology for analyzing the rainfall time series in an arid region of western India. This study introduces the sustainability approach, for the first time, to analyze a hydrologic time series, and demonstrates its novel applicability to rainfall time series of arid region.

KEYWORDS

- **Abrupt change**
- **Anthropogenic activities**
- **Arid region**

- **Climate change**
- **Drought Indices**
- **Homogeneity**
- **Humid region**
- **Hurst's coefficient**
- **Hydrologic time series**
- **Normality**
- **Performance indicator**
- **Periodicity**
- **Persistence**
- **Probabilistic measure**
- **Rain-gage**
- **Reliability**
- **Resilience**
- **Seasonality**
- **Spatio-temporal variability**
- **Stationarity**
- **Step change**
- **Stochastic component**
- **Sustainability**
- **Time series modeling**
- **Trend**
- **Vulnerability**
- **Water resources systems**

REFERENCES

1. Adeloye, A. J., and Montaseri, M. (2002). Preliminary stream flow data analyzes prior to water resources planning study. *Hydrological Sciences Journal*, 47(5), 679–692.
2. Ajami, N. K., Hornberger, G. M., and Sunding, D. L. (2008). Sustainable water resource management under hydrological uncertainty. *Water Resources Research*, 44(W11406), 1–10.
3. Baan, J. A. (1994). Evaluation of water resources projects on sustainable development. In: *Water Resources Planning in a Changing World, Proc.* Int. UNESCO Symp.. Karlsruhe, Germany, 28–30 June, IV, 63–72.

4. Berrang-Ford, L., Ford, J. D., and Paterson, J. (2011). Are we adapting to climate change? *Global Environmental Change*, 21, 25–33.
5. Beven, K. J. (2001). *Rainfall-Runoff Modeling. The Primer*. Wiley, Chichester.
6. Bras, R. L., and Rodriguez-Iturbe, I. (1985). *Random Functions and Hydrology*. Addison-Wesley, Reading, M. A.
7. Cannarozzo, M., Noto, L. V., and Viola, F. (2006). Spatial distribution of rainfall trends in Sicily (1921–2000). *Physics and Chemistry of Earth*, 31, 1201–1211.
8. Capodaglio, A. G., and Moisello, U. (1990). Simple stochastic model for annual flows. *Journal of Water Resources Planning and Management*, ASCE, 116(2), 220–232.
9. Chen, H. L., and Rao, A. R. (2002). Testing hydrologic time series for stationarity. *Journal of Hydrologic Engineering,* ASCE, 7(2), 129–136.
10. Clarke, R. T. (1998). *Stochastic Processes for Water Scientists: Development and Applications.* John Wiley and Sons, New York.
11. Cryer, J. D. (1986). *Time Series Analysis*. PWS Publishers, Duxbury Press, Boston, MA.
12. Dayal, D., Ram, B., Shamsudheen, M., Swami, M. L., and Patil, N. V. (2009). *Twenty Years of CAZRI, Regional Research Station, Kukma-Bhuj*. Regional Research Station, Central Arid Zone Research Institute, Kukma-Bhuj, Gujarat, pp. 35.
13. De Luis, M., Raventós, J., González-Hidalgo, J. C., Sánchez, J. R., and Cortina, J. (2000). Spatial analysis of rainfall trends in the region of Valencia (east Spain). *International Journal of Climatology,* 20(12), 1451–1469.
14. Delitala, A. M. S., Cesari, D., Chessa, P. A., and Ward, M. N. (2000). Precipitation over Sardinia (Italy) during the 1946–1993 rainy seasons and associated large scale climate variations. *International Journal of Climatology*, 20, 519–541.
15. Duckstein, L., and Parent, E. (1994). System engineering of natural resources under changing physical conditions: a framework for reliability and risk. In: L. Duckstein and E. Parent (editors), *Engineering Risk in Natural Resources Management*, Kluwer Academic Publishers, The Netherlands.
16. FAO-AGL. (2003). FAO Terrastat Database. *http://www.fao.org/ag/agl/agll/terrastat/wsrout.asp?wsreport=2a&region=8&search=Display+statistics+%21*.
17. Fernando, D. A. K., and Jayawardena, A. W. (1994). Generation and forecasting of monsoon rainfall data. In: *20th WEDC Conference on Affordable Water Supply and Sanitation* (Colombo, Sri Lanka), 310–313. Accessed 27 Feb 2008). *http://wedc.lboro.ac.uk/conferences/pdfs/20/Fernandd.pdf* (only available online)
18. Fiering, M. B. (1982). Alternative indices of resilience. *Water Resources Research*, 18(2), 33–39.
19. Haan, C. T. (1977). *Statistical Methods in Hydrology*. Iowa State University Press, Iowa, 378p.
20. Haigh, M. J. (2004). Sustainable management of head water resources: the Nairobi head water declaration (2002) and beyond. *Asian Journal of Water, Environment and Pollution,* 1(1–2), 17–28.
21. Harsh, L. N., and Tewari, J. C. (2007). Agroforestry Systems in Arid Regions of India. In: Puri, S. and Panwar, P. (editors), *Agroforestry: Systems and Practices*, New India Publishing Agency, New Delhi, India, 647p.
22. Hashimoto, T., Loucks, D. P., and Stedinger, J. (1982). Reliability, resilience and vulnerability for water resources system performance evaluation. *Water Resources Research,* 18(1), 14–20.

23. Hersh, M. A. (1999). Sustainable decision making: the role of decision support systems. *IEEE Transaction on Systems, Man and Cybernetics – Part C: Applications and Reviews,* 29(3), 395–408.

24. Hoekstra, A. Y. (1998). *Perspectives on Water – An Integrated Model Based Exploration of the Future.* International Books, Utrecht, The Netherlands.

25. Hurst, H. E. (1951). Long term storage capacity of reservoirs. *Transactions, ASCE,* 116, 770–800.

26. Hurst, H. E. (1957). A suggested statistical model of some time series which occur in nature. *Nature,* 180(4584), 494.

27. Jordaan, J., Plate, E. J., Prins, E., and Veltrop, J. (1993). *Water in Our Common Future.* Committee on Water Research (COWAR), UNESCO International Hydrology Program, Paris.

28. Kar, A., Garg, B. K., Singh, M. P., and Kathju, S., (editors) (2009). *Trends in Arid Zone Research in India.* Central Arid Zone Research Institute, Jodhpur. 481 p.

29. Kay, P. A. (2000). Measuring sustainability in Israel's water system. *Water International,* 25(4), 617–623.

30. Kite, G. (1989). Use of time series analyzes to detect climatic change. *Journal of Hydrology,* 111, 259–279.

31. Klemes, V., Srikanthan, R., and McMahon, T. A. (1981). Long-memory flow models in reservoir analysis: what is their practical value? *Water Resources Research,* 17(3), 737–751.

32. Kundzewicz, Z. W. (1999). Flood protection – Sustainability issues. *Hydrological Sciences Journal,* 44(4), 559–571.

33. Lane, M. E., Kirshen, P. H., and Vogel, R. M. (1999). Indicators of impacts of global climate change on U. S. water resources. *Journal of Water Resources Planning and Management,* 125(4), 194–204.

34. Loucks, D. P. (1994). Sustainability implications for water resources planning and management. *Natural Resources Forum,* 18(4), 263–274.

35. Loucks, D. P. (1997). Quantifying trends in system sustainability. *Hydrological Sciences Journal,* 42(4), 513–530.

36. Loucks, D. P. (2000). Sustainable water resources management. *Water International,* 25(1), 3–10.

37. Machiwal, D., and Jha, M. K. (2008). Comparative evaluation of statistical tests for time series analysis: Application to hydrological time series. *Hydrological Sciences Journal,* 53(2), 353–366.

38. Machiwal, D., and Jha, M. K. (2012). *Hydrologic Time Series Analysis: Theory and Practice.* Springer, Germany and Capital Publishing Company, New Delhi, India, 303 p.

39. Machiwal, D., Kumar, S., and Dayal, D. (2015). Characterizing rainfall of hot arid region by using time-series modeling and sustainability approaches: a case study from Gujarat, India. *Theoretical and Applied Climatology, DOI 10.1007/s00704-015-1435-9.*

40. Makoni, S. T., Kjeldsen, T. R., and Rosbjerg, D. (2001). Sustainable reservoir development – a case study from Zimbabwe. In: A. H. Schumann, M. C. Acreman, R. Davis, M. A. Marino, D. Rosbjerg and Xia Jun (editors), *Regional Management of Water Resources,* IAHS Publication No. 268, 17–23.

41. Matheson, S., Lence, B., and Furst, J. (1997). Distributive fairness considerations in sustainable project selection. *Hydrological Sciences Journal,* 42(4), 531–548.

42. McCuen, R. H. (2003). *Modeling Hydrologic Change: Statistical Methods.* Lewis Publishers, CRC Press LLC, Florida, 433 pp.

43. McLaren, R. A., and Simonovic, S. P. (1999a). Data needs for sustainable decision making. *International Journal of Sustainable Development and World Ecology*, 6, 103–113.

44. McLaren, R. A., and Simonovic, S. P. (1999b). Evaluating sustainability criteria for water resource decision making: Assiniboine Delta Aquifer case study. *Canadian Water Resources Journal*, 24(2), 147–163.

45. McMahon, T. A., Adeloye, A. J., and Sen-Lin, Z. (2006). Understanding performance measures of reservoirs. *Journal of Hydrology*, 324, 359–382.

46. MoEF. (2001). *India: National Action Program to Combat Desertification in the Context of United Nations Convention to Combat Desertification (UNCCD)*. Volume I: Status of Desertification. Ministry of Environment & Forests, Government of India, New Delhi. 294 p.

47. Nachtnebel, H. P. (2001). Irreversibility and sustainability in water resources systems. In: by J. J. Bogardi and Z. W. Kundzewicz (editors), *Risk, Reliability, Uncertainty and Robustness of Water Resources Systems*, Cambridge University Press, Cambridge, UK.

48. NBSS&LUP. (2001). *Agro-ecological Sub-regions of India for Planning and Development*. NBSS&LUP Publication, Nagpur.

49. O'Connel, P. E. (1977). ARIMA models in synthetic hydrology. In: T. A. Ciriani, U. Maione and J. R. Wallis (editors), *Mathematical Models for Surface Water Hydrology*, John Wiley and Sons, Inc., New York.

50. Oguntunde, P. G., Friesen, J., van de Giesen, N., and Savenije, H. H. G. (2006). Hydroclimatology of Volta River Basin in West Africa: Trends and variability from 1901 to 2002. *Physics and Chemistry of Earth*, 31, 1180–1188.

51. Ray, P. A., Vogel, R. M., and Watkins, D. W. (2010). Robust optimization using a variety of performance indices. *Proceedings of the World Environmental and Water Resources Congress*, ASCE, Reston, VA.

52. Rogers, P., Jalal, K. F., Lohani, B. N., Owens, G. M., Yu, Chang-Ching, Dufournaud, C. M. and Bi, J. (1997). *Measuring Environmental Quality in Asia*. Harvard University Press, USA.

53. Salas, J. D. (1993). Analysis and Modeling of Hydrologic Time Series. In: D. R. Maidment (editor-in-chief), *Handbook of Hydrology*, McGraw-Hill, Inc., USA, pp. 19.1–19.72.

54. Salas, J. D., Delleur, J. W., Yevjevich, V., and Lane, W. L. (1980). *Applied Modeling of Hydrologic Time Series*. Water Resources Publications, Littleton, CO.

55. Sandoval-Solis, S., McKinney, D. C., and Loucks, D. P. (2011). Sustainability index for water resources planning and management. *Journal of Water Resources Planning and Management*, ASCE, 137, 381–390.

56. Shahin, M., Van Oorschot, H. J. L., and De Lange, S. J. (1993). *Statistical Analysis in Water Resources Engineering*. A. A. Balkema, Rotterdam, the Netherlands, 394 p.

57. Shamir, U. (1996). Sustainable management of water resources. *Proc. of the International Conference on Water Resources & Environment Research*, Kyoto, Japan, Oct. 29–31, Vol. II, 15–29.

58. Simonovic, S. P. (editor). (1997). *Hydrological Sciences Journal*, 42(4). Special issue on Sustainable Development of Water Resources.

59. Simonovic, S. P., Burn, D. H., and Lence, B. J. (1997). Practical sustainability criteria for decision making. *International Journal of Sustainable Development and World Ecology*, 4, 231–244.

60. Takeuchi, K., Hamlin, M., Kundzewicz, Z. W., Rosbjerg, D., and Simonovic, S. P., (editors). (1998). *Sustainable Reservoir Development and Management*, IAHS Publication No. 251.

61. UNESCO. (1999). *Sustainability Criteria for Water Resources Systems.* Cambridge University Press, Cambridge, UK.
62. World Bank. (1993). *Water Resources Management.* A World Bank Policy Paper, Washington DC, USA.
63. World Health Organization (WHO). (2009). *Summary and policy implications Vision 2030: The resilience of water supply and sanitation in the face of climate change,* Geneva.
64. Yevjevich, V. M. (1972). *Stochastic Processes in Hydrology.* Water Resources Publications, Fort Collins, CO.

CHAPTER 15

RESERVOIR INFLOWS IN A TROPICAL HUMID CLIMATE: THAILAND

PROLOY DEB[1] and NIRAKAR PRADHAN[2]

[1]Center for Water, Climate and Land Use (CWCL), Faculty of Science and Information Technology, School of Environmental and Life Sciences, University of Newcastle, Callaghan, Newcastle, Australia, 2308 E-mail: debproloy@gmail.com; Proloy.Deb@uon.edu.au; Phone: +61-478744219

[2]Environmental Engineering and Management Program, Asian Institute of Technology, Pathumthani, 12120, Thailand

CONTENTS

15.1 INTRODUCTION

Thailand's major economy relies on agriculture and it is expected to be seriously affected by the adverse impacts of climate change due to the high sensitivity of agriculture on climatic variables including water resources [3]. Studies indicate that among Southeast Asia, Thailand has the highest per capita water use and 94% to total water use is accounted for agricultural sector [2]. The vulnerability of freshwater resources attributed to climate change is undoubtedly negligible in the region, since the development of the region depends on water resources [13]. Moreover, the observed changes in water regime driven by climatic factors have not only affected agriculture [3, 4] but also the energy production in past decades [7].

Although not many researches focusing on climate change have been conducted in Chi River basin of Thailand, yet the existing findings illustrate the presence of ambiguous and increasing trends in precipitation and temperature respectively [1]. An increase of 1.2 to 1.9°C in temperature is projected by 2050s relative to historical period under climate change in Thailand [8]. Further studies on climate change has illustrated that the shifts in precipitation pattern are not coherent and therefore it has its implications on regional scale [12, 18, 19]. The observed and projected changes in the climatic variables will have significant influence on the stream flow and watershed hydrology [20]. The alteration of the rainfall pattern will certainly influence in the seasonal reservoir inflows and therefore shift in the reservoir operations are necessary. Although on a global scale majority of studies have mainly focused on the downstream beneficial interests in large river systems, yet merely a handful of studies have focused on the climate change implications on the reservoir inflows [10, 14].

General Circulation Models (GCMs) are tools, which provide the future climatic data for a given greenhouse emission scenario. However, due to the coarse spatial resolution, it is not suitable to apply the outputs at basin scale or sub-grid level hydrological assessment studies. Statistical or dynamic downscaling [regional climate models (RCMs)] methods are generally applied for refining the climatic data for catchment modeling. Even though some studies have applied RCM data directly in impact assessment studies, yet globally in many basins output of 20–50 km resolution are not sufficient to represent the true climate of the regions at station level and hence further downscaling or bias correction are suggested [16, 22].

In order to complement these problems, the present study was conducted to forecast the future reservoir inflows using bias corrected future climate data and hydrological model for IPCC special report on emission scenarios (SRES) A2 and B2 for Ubolratana dam located in Chi river basin, Northeast Thailand. The main objectives are to investigate: (1) The future climate change in the upstream of the Ubolratana reservoir and (2) Response of the climate change on the reservoir inflows for future periods.

15.2 STUDY AREA

The Chi River Basin is located in the north-east of Thailand extending from 15°30'–17°30' N latitude and 101°30'-104°30' E longitude and covers an area of 49,477 km^2 in twelve provinces extending about 360 km from the east to west and 210 km from the north to south. Figure 15.1 shows the location of the study area in the basin. The climate is moderately tropical with average annual temperature ranging from 26.6–27.8°C. Further, the region is dominated by two monsoon seasons namely the southwest which influences from mid-May to mid-October with heavy showers and the northeast monsoon extending from mid-November to mid-February which accumulates 1700 mm of average annual rainfall. South China Sea contributes to the tropical cyclone in the region from August to September. Long-term observations suggest the average annual runoff of the Chi river basin is 11,244 MCM which is composed of 9638 MCM for wet and 1606 MCM for dry season [15].

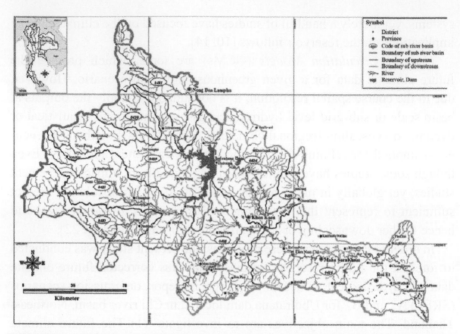

FIGURE 15.1 The location of the study area showing the upstream and downstream of the Ubolratana dam.

15.2.1 CHARACTERISTICS OF UBOLRATANA DAM

The Ubolratana is a multipurpose dam with a catchment area of 12,000 km² for development of electricity, irrigation, flood control, transportation, fisheries and tourism. The dam is located on the Nam Pong River at Kok Soong Sub-district, Ubolratana District of Khonkaen province. The study area is consisted of an earth core rock-fill dam and was constructed in 1984 with a height of 32 m, crest length of 885 m, crest width of 6 m. Normal Flood Level is 182.00 m (MSL) with a maximum storage capacity of 2,559 MCM. The total catchment area is 12,000 km².

15.2.2 FLOOD PROBLEMS IN UBOLRATANA RESERVOIR

Historical data suggests that the average inflow in the reservoir is 2,481 MCM which is equivalent to the capacity of the reservoir and therefore during the extreme rainfall years the water resource management is a big issue in the reservoir. Figure 15.2 illustrates the annual inflow from 1970 to 2008 in the dam. It can be observed that the inflow was twice the

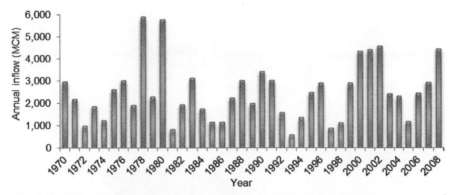

FIGURE 15.2   Record of annual inflow from 1970 – 2008 to Ubolratana dam.

capacity in 1978 and 1980 and therefore due to the safety concerns spill-ways had to operate far beyond the designed discharge which led to flash flood in downstream.

15.2.3   DATA COLLECTION

Two sets of meteorological data (rainfall and temperature) in the Pong river sub-basin were collected for 81 stations lying within and adjacent to the basin according to data availability and frequency. In addition, stream flow data from 26 stations were extracted for the upstream and downstream of the reservoir. The meteorological data was obtained from Thai Meteorological Department (TMD), whereas the stream flow data was retrieved from Royal Irrigation Department (RID).

The future climate data were retrieved from the RCM *Providing Regional Climates for Impact Studies* (PRECIS), which is developed by Hadley Center at UK Met Office (http://cc.start.or.th/). The model has a spatial resolution of 20 km and derives its boundary conditions from the GCM – ECHAM4. A comparative study of the suitability of several GCMs in Ping river basin (an adjacent basin) suggests that ECHAM4 is the best suitable model in the basin in order to represent the precipitation and temperature for the historical climate [17].

15.3   METHODOLOGY

The Figure 15.3 illustrates the methodological flowchart used in this study. First of all the bias correction of the PRECIS dataset for A2 and B2 scenarios

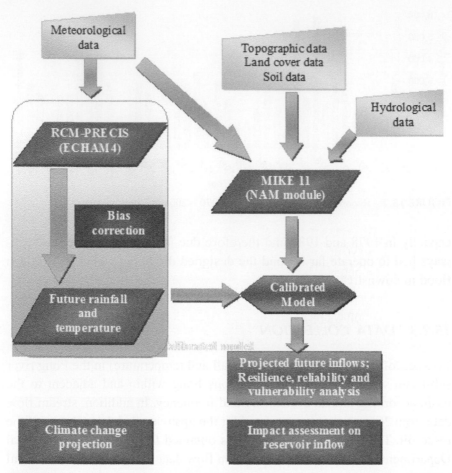

FIGURE 15.3 Schematic representation of the methodology used in assessing the reservoir inflow.

were done at all respective stations contributing to the basin which was followed by set-up of the MIKE 11 rainfall-runoff model (NAM) for the upstream of reservoir. Further, the calibrated and validated hydrological model was used for projection of future reservoir inflows under the considered climate scenarios. Additionally resilience, reliability and vulnerability (RRV) analysis was also conducted for the future inflow.

15.3.1 EXTRACTION OF CLIMATE PROJECTION DATA

Although there were 81 stations considered in the study, however, based on continuity of available dataset 39 rain gauge stations were selected to create the Thiessen interpolation. The selection scheme of rain gauge was done based on the distance of each station, the completion and reliability of rainfall data. Existing gaps in the rain gauge station data were interpolated by using square inverse distance interpolation method from nearby stations, which had equivalent rainfall. Moreover, RCM provides the rainfall data in the form of grids (20 × 20 km) at its center. Data from 42 grids were retrieved which were covering all the upstream and downstream rain gauge stations. In order to get better estimation of the rainfall from the RCM data, weights were determined for each station based on the fraction of the Thiessen polygon area lying on each RCM grid.

15.3.2 BIAS CORRECTION

Due to the spatial dependence of the biases in temperature and precipitation, performing bias correction is necessary for each station. The biases from the temperature were removed by power law transformation theorem where the data is normally distributed. It uses the scaling and shifting of the mean and variance of the dataset [11, 21]. Further the bias correction for precipitation was also done by non-linear method of multi-day window for correction of coefficient of variation (CV). The baseline period considered for correcting the future period dataset was 1976–2005. Future climate projections were done for three time windows 2010–2039 (2020s), 2040–2069 (2050s) and 2070–2099 (2080s).

For correcting rainfall, although the block length for bias correction as recommended by is 5 days [12], yet if the block lengths are chosen too small, there are high chances of correcting the natural variability rather than correcting the systematic model error. Based on this recommendation, in the present study a sensitivity analysis for different block lengths 15, 25, 35, 45 and 65 days was done to represent the best performance of statistic for bias correction. Moreover, the performance of the multi-day analysis was assessed by calculating Root Mean Square Error (RMSE) and Efficiency Index (EI) at monthly scale.

15.3.3 HYDROLOGICAL MODEL

One-dimensional hydrodynamic model MIKE-11 was applied to simulate the flow and the water level of the river. The computational core of MIKE 11 is hydrodynamic simulation engine and complements a wide range of additional modules and extensions. The general rainfall-runoff modules integrated in this model are the Nedbør –Afstrømnings -Model (NAM), the unit hydrograph method (UHM), conceptual continuous hydrological model, a monthly soil moisture accounting model, runoff methods tailored to urban environments (URBAN) and semi-distributed rainfall-runoff-geomorphological approach (DRiFt). For this study, NAM approach was used due to its suitability for large river basins with numerous catchments with complex networks of rivers [5].

The MIKE 11-NAM model represents the various components of the rainfall–runoff process by continuously accounting for the water content in four different and mutually interrelated storages namely snow, surface, lower zone and groundwater storage where the each storage represents different physical elements of the catchment (Figure 15.4). The basic input requirements for the NAM model are model parameters, initial conditions, meteorological data and stream flow data. In the present application, the nine most important parameters of the NAM model are determined by calibration.

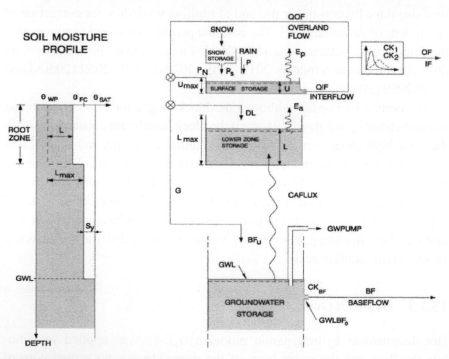

FIGURE 15.4　NAM model structure.

Split-sample testing scheme was used for validation of the model, which suggests calibration of a model based on 3–5 years of data and validation for another period of similar length [9]. Calibration period from 2003 to 2007 was chosen to represent the recent climate, whereas validation period was selected from 1998 to 2002. The selection of the calibration and validation periods considered all the low and extreme flows and therefore better model set up was expected. The future inflow projections were done for the bias corrected three time windows: 2020s, 2050s and 2080s. The changes in the future inflows were analyzed based on the average monthly flows and daily flow duration curves.

In addition to the projection of the change in inflows under climate change, reliability (C_R), resilience (C_{RS}) and vulnerability (C_V) (RRV) analysis was also done in order to evaluate the performance of the inflows. RRV criteria were evaluated under future climate conditions for both A2 and B2 emission scenarios. First a criterion, C is defined for the normal range of inflow, where an unsatisfactory condition occurs when inflow is out of normal range. The normal range of inflow was between 20th percentile and 80th percentile of historical inflow data (1970–2008) [6]. If the inflow is in normal range, we can conclude it to be in a satisfactory (S) state, otherwise in an unsatisfactory (U) state (Eq. 1):

$$Zt = \begin{cases} 1, \text{ if } X_t \in S \\ 0, \text{ if } X_t \in U \end{cases} \qquad (1)$$

where: Z_t is a generic indicator of satisfactory or unsatisfactory. Another indicator, W_t, which represents a transition from S to U states, is defined in Eq. 2:

$$wt = \begin{cases} 1, \text{ if } X_t \in U \text{ and } X_{t+1} \in S \\ 0, \text{ otherwise} \end{cases} \qquad (2)$$

Furthermore, if the periods of X_t is in unsatisfactory state then based on $J_1, \ldots, J_N$ where N is the number of U periods. Then reliability, resilience and vulnerability indices during the total time period (T) were calculated using Eqs. (3)–(5), respectively.

Reliability

$$C_R = \frac{\sum_{t=1}^{T} Z_t}{T}$$
(3)

Resilience

$$C_{RS} = \frac{\sum_{t=1}^{T} W_t}{T - \sum_{t=1}^{T} Z_t}$$
(4)

Vulnerability time

$$C_v = \max\{J_1, J_2, \ldots, J_N\}$$
(5)

15.4 RESULTS AND DISCUSSION

15.4.1 *PERFORMANCE OF BIAS CORRECTION FOR RCM DATA*

The RCM outputs forced by ECHAM5 were bias corrected by applying the power law transformation for rainfall data and the linear approach for temperature data. The observed data for the 30-year period of 1976–2005 were used as a baseline in this study due to available climate data.

The comparison of the observed and raw RCM data along with multi-day analysis for the baseline period and two grids of RCM from the study area are shown in Figures 15.5(a) and 15.(b). From visual observation, all multi-day blocks considered followed similar patterns of monthly rainfall for the grid ID 1681024 except the raw data which is observed to deviate widely relative to the observed values. Further, for grid ID 1661016, a significant deviation in rainfall was observed for May to August where the raw RCM data overestimates the rainfall significantly. However, the multi-day data analysis estimates suggested that larger blocks lead to greater variation especially in the months with higher rainfall. Moreover, low day blocks (15 and 25 days) visually performed well in line with the observed values.

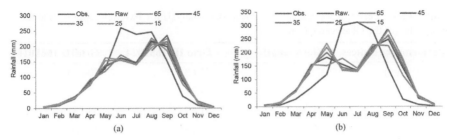

FIGURE 15.5 Comparison of multi-day blocks for two RCM grids (a) Id 1681024 and (b) Id 1661016 at the study site.

Performance indicators (RMSE and EI) were calculated for the grids, and these suggest that 25 days block is the most suitable since it represents the least RMSE and highest EI relative to the other blocks (Table 15.1). As expected, lowest performance is observed for the raw data with highest RMSE of 75.68 and 106.14 mm and poorest EI of -45.3 and -23.17 for two grids. In addition, a relative low performance is also observed for the lowest block (15 days) compared to 25 days block. The correction of the natural climatic variability due to small block size may the attributing factor for the observed low performance. On contrary, 5 day blocks were considered for maximum and minimum temperature. The comparisons of the bias corrected results suggest good capability of the representativeness of the observed temperature at all stations considered.

15.4.2 PROJECTION OF RAINFALL

Projected rainfall under climate change for both scenarios indicates higher intensity of rainfall for all time windows relative to the historical climate (Figure 15.6). Moreover, it is also evident, for the past climate observed average annual rainfall is 1900 mm. However, by 2080s average annual rainfall is expected to rise to 3000 mm for both scenarios. Furthermore, a shift in the probability of being less precipitation is observed which is highest for A2 scenario relative to B2 that increases higher risks of floods. In addition from the annual rainfall analysis, it is clear that under both scenarios, the magnitude of the mean, median and the quintiles of rainfall are expected to rise in the future from 1200 to 1600 mm in the last part of the century for A2 scenario and 1650 mm as per the projection for B2 scenario

TABLE 15.1   Performance Statistics of Multi-day Analysis for Two Selected Grids at the Study Site for the Observed Climate

Performance indices	Block lengths	Grid ID 1681024	GridID 1661016
RMSE (mm)	Raw	75.68	106.14
	65 days	69.31	72.17
	45 days	71.57	64.13
	35 days	55.13	40.32
	25 days	36.14	37.66
	15 days	40.16	37.98
EI	Raw	−45.30	−23.17
	65 days	−1.29	−8.92
	45 days	−0.64	−6.67
	35 days	−0.17	0.16
	25 days	0.62	0.73
	15 days	0.16	0.55

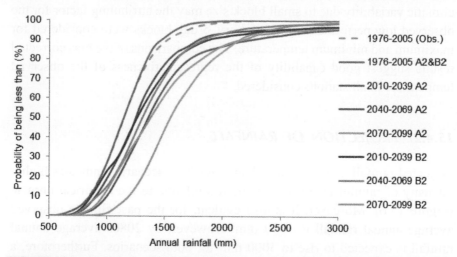

FIGURE 15.6   Cumulative distribution function of projected annual rainfall for each time windows considered under A2 and B2 scenario at upstream of Ubolratana dam.

(Figures 15.7a and 15.7b). However, the median values of annual rainfall are 1550 and 1400 mm, respectively for the corresponding scenarios.

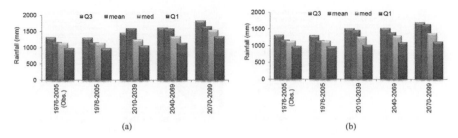

(a)   (b)

FIGURE 15.7   Projected annual rainfall for each period under SRES (a) A2 and (b) B2 at upstream and downstream area of the Ubolratana dam.

15.4.3   PROJECTION OF TEMPERATURE

Bias correction for maximum and minimum temperature suggests an increase in magnitude in future (Figures 15.8a and 15.8b). The highest increase is observed in case of the late century for both scenarios with A2 responds to be severe. Analysis on the change of the maximum and minimum temperature reveals both follows similar trend of shift (Figures 15.9a and 15.9b). The minimum change is observed in case of May whereas, maximum is predominant in July for all the scenarios and time widows considered. Interestingly a significant decline the change is observed for the November and December relative to other months although the magnitude of change is higher relative to May. Nevertheless, it can be summarized that the maximum and minimum temperatures for the basin is expected to increase for all the time periods and scenarios in the future with maximum shift in the A2 scenario and July.

15.4.4   RUNOFF MODELING

15.4.4.1   Model Setup

The hydrological model NAM was calibrated at daily time step with the fine tuning of the parameters as presented in Table 15.2. The model was calibrated by iterating the simulation by changing values of one parameter within the range provided in Table 15.2 and keeping other parameter values constant. Comparison of the simulated and observed discharge in terms of various model evaluation indexes validates the model can simulate the

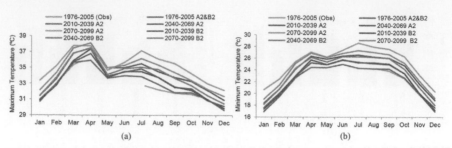

FIGURE 15.8 Projected mean monthly (a) maximum and (b) minimum temperature under A2 and B2 scenarios for future time windows at the study site.

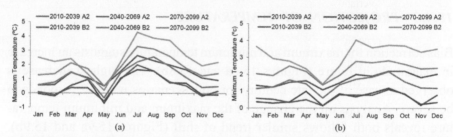

FIGURE 15.9 Projected changes in the mean monthly (a) maximum and (b) minimum temperatures for A2 and B2 scenarios for future time periods in study site.

TABLE 15.2 NAM Model Parameters Calibrated for the Basin

Parameter	Description	Lower limit	Upper limit	Calibrated value
U_{max} (mm)	Maximum water content in the surface storage. This storage can be interpreted as including the water content in the interception storage, in surface depression storages, and in the uppermost few cm's of the soil	0	35	20
L_{max} (mm)	Maximum water content in the lower zone storage. L_{max} can be interpreted as the maximum soil water content in the root zone available for the vegetative transpiration	50	350	300
CQOF (–)	Overland flow runoff coefficient. CQOF determines the distribution of excess rainfall into overland flow and infiltration	0	1	0.297

TABLE 15.2   Continued

Parameter	Description	Lower limit	Upper limit	Calibrated value
TOF (–)	Threshold value for overland flow. Overland flow is only generated if the relative moisture content in the lower zone storage is larger than TOF	0	0.9	0.0000327
TIF (–)	Threshold value for interflow. Interflow is only generated if the relative moisture content in the lower zone storage is larger than TIF	0	0.9	0.86
TG (–)	Threshold value for recharge. Recharge to the groundwater storage is only generated if the relative moisture content in the lower zone storage is larger than TG	0	0.9	0.87
CK_{IF} (h)	Time constant for interflow from the surface storage. It is the dominant routing parameter of the interflow because $CK_{IF} >> CK_{1,2}$	500	1000	560.3
$CK_{1,2}$ (h)	Time constant for overland flow and interflow routing. Overland flow and interflow are routed through two linear reservoirs in series with the same time constant $CK_{1,2}$	3	72	50
CK_{BF} (h)	Baseflow time constant. Baseflow from the groundwater storage is generated using a linear reservoir model with time constant CK_{BF}	500	5000	3999

runoff in good agreement with the observed values in the basin (Table 15.3). Although, higher volumetric error (–11.304 %) is observed in case of validation which is probably due to the inability of the model to capture the extreme high flow observed in case of 2002. Also, higher Efficiency Index (EI) and Coefficient of Determination (R^2) is observed for both calibration and validation reflecting the applicability of the model in the study site.

15.4.4.2   Future Runoff Projection

The comparison of the mean monthly inflow to the reservoir for the historical period and the future suggests an increase in the magnitude of the inflow

TABLE 15.3   Evaluation of Model Performance for Calibration and Validation

Evaluation indexes	Calibration (2003–2007)	Validation (1998–2002)
Volume error (%)	–0.007	–11.304
Mean = |Qsim-Qobs|/Qobs	2.00	1.31
R^2	0.811	0.826
EI	0.809	0.807

for future under both scenarios considered (Figure 15.10). In addition, an insignificant shift in the peak is also noticeable for all the future time windows relative to the historical period. Surprisingly, in case of A2 scenario for 2020s, double peak is observed the first in February and second in August. The maximum peak flow (18,000 m^3/s) can be observed for 2080s under A2 scenario whereas a relative lower magnitude of peak flow (13,700 m^3/s) is observed for the corresponding time period for B2 scenario. Furthermore, a significant increase in the peak flow is also observed for the 2050s time window under both scenarios. The expected increase in the flow under future climate indicates higher intensity of flood under future climate.

The flow duration curve generated based on the simulation results suggests the percentage of time that inflow to the dam is likely to equal or exceed some specified value of interest. The shape of a flow-duration curve in its upper and lower regions is particularly significant in evaluating the flow characteristics. The projected inflows show a very steep curve in the high-flow region, which is expected for rain-caused floods on this basin (Figure 15.11). In the low-flow region, the beginning of 21st century exhibit high percentage of no flow which is relatively higher than the historical time period, until the mid of the century where there are more low flows in each step. In addition, an inclined curve indicates that moderate flows are not sustained throughout the year due to natural inflow regulation, or because a small groundwater capacity cannot sustains the base flow to the stream.

15.4.4.3   RRV Analysis for Future Climate

Annual inflow data of Ubolratana dam from 1970 to 2008 is used as the level of water for baseline period. Low annual inflow or less than 20th percentile of this period is assumed to cause drought whereas high inflow or more than 80th percentile may cause flood. Flow between 20th and 80th percentiles are

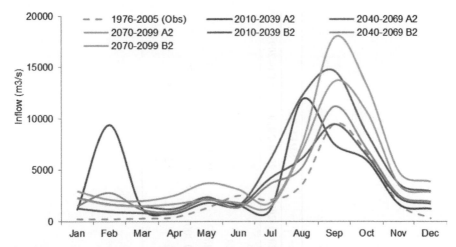

FIGURE 15.10   Inflow to Ubolratana dam for different time windows under future climate.

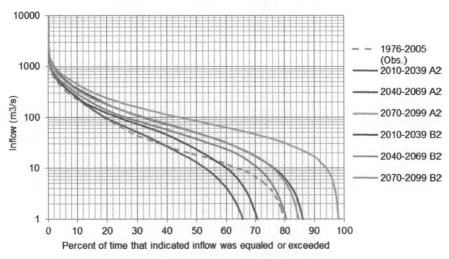

FIGURE 15.11   Flow duration curve of daily inflow to Ubolratana dam.

assumed to be the appropriate inflow for which the dam operators safely. Table 15.4 illustrates the RRV analysis results obtained from the study based on the projected future runoff. Evidently number of low annual inflow (Q20) is expected to decrease in the future. However, the number of high annual inflow (Q80) is projected to increase under both the emission scenarios.

Surprisingly, although for certain years in the future is projected to have high intense rainfall, yet low annual inflows are observed in the

TABLE 15.4 Results Obtained by RRV Analysis for the Inflow to Ubolratana Dam

RRV analysis	Observed	SRES A2			SRES B2		
	1976–2005	2010–2039	2040–2069	2070–2099	2010–2039	2040–2069	2070–2099
Q20 (years)	7	9	1	0	6	5	3
Q80 (years)	7	8	18	26	14	18	17
Reliability, C_R	0.533	0.433	0.367	0.133	0.333	0.233	0.333
Resiliency, C_{RE}	0.643	0.529	0.368	0.115	0.300	0.174	0.350
Vulnerability time, C_V (years)	3	3	7	11	6	12	8

corresponding year. This is probably due to higher rate of evapotranspiration attributing to the high temperature and the contribution of more percolation in the aquifers. In addition our analysis show higher and lower rainfall in the future will decrease resilience and reliability however increases the vulnerability for both A2 and B2 scenarios. The analysis also shows that 2050s time period for B2 scenario is the most vulnerable contributing to vulnerability for 12 years. However, for A2 scenario, 2080s time window is more vulnerable relative to other time periods. Nonetheless, the future is ascertained to be more severe and the reservoir operation rule is necessary to be reviewed for future climate.

15.5 CONCLUSIONS

The present study analyzes the future inflow and the Resilience, Reliability, Vulnerability (RRV) analysis of the flow to Ubolratana dam in Thailand under future climate for 2020s, 2050s and 2080s under A2 and B2 climate scenarios. Climate data were collected from 39 meteorological stations and stream flow data from 26 gauging stations in the upstream of the dam. Bias correction of the climate data was done for 42 grids from upstream and downstream of the dam for the RCM PRECIS. Power law transformation was applied to correct the maximum and minimum temperatures whereas; non-liner method of multi-day window for correction of coefficient of variation was used to correct the precipitation. Further the bias corrected temperature and precipitation was used as input for the hydrological model MIKE-11 NAM to simulate the future inflow. An additional RRV analysis of the simulated inflow is also done to analyze the vulnerability of the dam under future climate.

The results suggest an increase in the precipitation for both scenarios under future climate and all time windows considered. A significant increase of 36, 35 and 42% in average annual rainfall is expected for 2020s, 2050s and 2080s under A2 scenario whereas, 25, 19 and 40% for B2 scenarios for the corresponding time periods. Similarly, climate change is expected to induce higher temperature for the future climate with a magnitude of 0.50, 1.36 and 2.46°C for 2020s, 2050s and 2080s under A2 scenario for mean maximum temperature and 0.51, 1.13 and 1.85°C for the corresponding time windows under B2 scenario. Likewise mean annual minimum temperature is expected to increase by 0.61, 1.71 and 3.13°C for 2020s, 2050s and 2080s respectively under A2 scenario and 0.60, 0.43 and 2.30°C for B2 scenario for corresponding time periods relative to the baseline period of 1976–2005. The simulated runoff changes are driven by combined effects of rainfall changes and their seasonality. Simulated inflows shows increase for all period and both emission scenarios, with the greatest change occurring in period 2080s for A2 emission scenarios. Most of extreme changes are in low and moderate flow quantile ranges. Compared to the historical period, the number of high annual inflow will increase while the number of low annual inflow will decrease. The RRV criteria show that with the increasing rainfall in future will contribute to lower resilience and reliability whereas higher vulnerability. The results of this study show an increase in the volume of inflow for all the projected period which will affect the storage of dam. Therefore, the appropriate planning and management should be ready to counteract this problem for the future.

15.6 SUMMARY

This study analyzes the future climate implications on the reservoir inflows for Ubolratana dam, Thailand. The future climate data of precipitation, maximum and minimum temperature was derived from regional climate model RCM Providing Regional Climates for Impact Studies (PRECIS) for A2 and B2 climate scenarios. Bias correction was performed on the climate data for finer spatial resolution. Future inflow was estimated by the simulation of the future flow by hydrological model MIKE 11 NAM module. The results suggest elevated maximum and minimum temperatures relative to the baseline period. Similarly, higher intense rainfall is also expected in the future for both scenarios considered. Hydrological model simulation results for future climate suggests higher inflows in the future for both scenarios however less intense in case of B2 scenario. Resilience, reliability and vulnerability (RRV) analysis show that with the increasing rainfall in future will contribute to lower resilience and reliability whereas higher vulnerability.

KEYWORDS

- **climate change**
- **climate data**

- **climate scenarios**
- **climate variables**
- **climatic factors**
- **humid climate**
- **hydrological model**
- **MIKE 11**
- **NAM**
- **PRECIS**
- **RCM**
- **reliability**
- **reservoir inflow**
- **resilience**
- **resilience, reliability and vulnerability analysis, RRV**
- **RRV analysis**
- **Southeast Asia**
- **stream flow**
- **Thailand**
- **vulnerability**
- **water resources**
- **water use**
- **watershed**

REFERENCES

1. Artlert, K., Chaleeraktrakoon, C., and Nguyen, V. T. V. (2013). Modeling and analysis of rainfall processes in the context of climate change for Mekong, Chi and Mun river basins (Thailand). *J. Hydro-environ. Res.*, 7(1), 2–17.
2. Chulalongkorn, (2012). *Thailand water account (2005–2007)*. Water Resources System Research Unit. Faculty of Engineering, Chulalongkorn University, Bangkok.
3. Deb, P., Kiem, A. S., Babel, M. S., Chu, S. T., and Chakma, B. (2015). Evaluation of climate change impacts and adaptation strategies for maize cultivation in the Himalayan foothills of India. *J. Water Clim. Change,* doi: 10.2166/wcc.2015.070.
4. Deb, P., Shrestha, S., and Babel, M. S., (2014). Forecasting climate change impacts and evaluation of adaptation options for maize cropping in the hilly terrain of Himalayas: Sikkim, India. *Theor. Appl. Climatol.*, doi: 10.1007/s00704–014–1262–4.
5. DHI Water & Environment (2007). *MIKE 11: A modeling system for rivers and channels.* Reference Manual by DHI Water & Environment.

6. Fowler, H. J., Kilsby, C. G., and O'Connell, P. E., (2003). Modeling the impacts of climatic change and variability in the reliability, resilience and vulnerability of a water resource system. *Water Resource Res.,* doi: 10.1029/2002WR001778.

7. Hunukumbura, P. B., and Tachikawa, Y. (2012). River discharge projection under climate change in the Chao Phraya river basin, Thailand using MRI_GCM3.1S dataset. *J. Meteorol. Soc. Jap.,* 90A:137–150.

8. Intergovernmental Panel on Climate Change 2007. *Climate Change 2007: The physical science basis – Summary of policy makers.* Cambridge, UK and New York: Cambridge University Press.

9. Klemeš, V. (1986). Operational testing of hydrologic simulation models. *Hydrolog. Sci. J.,* 31,13–24.

10. Lauri, H., de Moel, H., Ward, P. J., Räsänen, T. A., Keskinen, M., and Kummu, M. (2012). Future changes in Mekong River hydrology: impact of climate change and reservoir operation on discharge. *Hydrol. Earth Syst. Sci.,* 16, 4603–4619.

11. Leander, R., and Buishand, T. A. (2007). Resampling of regional climate model output for the simulation of extreme river flows. *J. Hydrol.,* 332, 487–496.

12. Manomaiphiboon, K., Octaviani, M., Torsri, K., and Towprayoon, S. (2013). Projected changes in means and extremes of temperature and precipitation over Thailand under three future emissions scenarios. *Clim. Res.,* 58, 97–115.

13. Pradhan, N., Habib, H., Venkatappa, M., Ebbers, T., Duboz, R., and Shipin, O., (2015). Framework tool for a rapid cumulative effects assessment: case of a prominent wetland in Myanmar. *Environ. Monit. Assess.,* 187, 341, doi: 10.1007/s10661-015-4508-4.

14. Raje, D., and Mujumdar, P. P., (2010). Reservoir performance under uncertainty in hydrologic impacts of climate change. *Adv. Water Res.,* 33:312–326.

15. Royal Irrigation Department, RID, (2002). *Study plan to support the development of primary water resources and improvement of irrigation project for 9th economic development planning.* 25th River Basin Status Report, Royal Irrigation Department, Bangkok, Thailand.

16. Sharma, D., and Babel, M. S., (2013). Application of downscaled precipitation for hydrological climate change impact assessment in the Ping River Basin of Thailand. *Clim. Dyn.,* 41, 2589–2602.

17. Sharma, D., Gupta, A. D. and Babel, M. S., (2007). Spatial disaggregation of bias corrected GCM precipitation for improved hydrologic simulation: Ping River Basin, Thailand. *Hydrol. Earth Syst. Sci.,* 11, 373–1390.

18. Shrestha, S., Deb, P., and Bui, T. T. T., (2014). Adaptation strategies for rice cultivation under climate change in Central Vietnam. *Mitig. Adapt. Strateg. Glob. Change,* doi: 10.1007/s11027–014–9567–2.

19. Shrestha, S., Thin, N. M. M., and Deb, P. (2014). Assessment of climate change impacts on irrigation water requirement and rice yield for Ngamoeyeik Irrigation Project in Myanmar. J. Wat. Clim. Change. 5(3), 427–442.

20. Sivakumar, B. (2011). Global climate change and its impacts on water resources planning and management: assessment and challenges. *Stoch. Environ. Res. Risk Assess.,* 25, 583–600.

21. Terink, W., Hurkmans, R. T. W. L., Torfs, P. J. J. F., and Uijlenhoet, R. (2010). Evaluation of a bias correction method applied to downscaled precipitation and temperature reanalysis data for the Rhine basin. *Hydrol. Earth Syst. Sci.,* 14, 687–703.

22. Teutschbein, C., and Seibert, J., (2012). Bias correction of regional climate model simulation for hydrological climate-change impact studies: review and evaluation of different methods. *J. Hydrol.*, 456–457, 12–29.
23. World Bank (2014). *World Bank Database*. Retrieved on January 17 from, http://data.worldbank.org/country/Thailand.

APPENDICES

(Modified and reprinted with permission from: Goyal, Megh R., 2012. Appendices. Pages 317–332. In: *Management of Drip/Trickle or Micro Irrigation* edited by Megh R. Goyal. New Jersey, USA: Apple Academics Press)

APPENDIX A   CONVERSION SI AND NON-SI UNITS

To convert the Column 1 in the Column 2, Multiply by	Column 1 Unit SI	Column 2 Unit Non-SI	To convert the Column 2 in the Column 1 Multiply by

LINEAR

0.621 _____	kilometer, km (10^3m)	miles, mi _____	1.609
1.094 _____	meter, m	yard, yd _____	0.914
3.28 _____	meter, m	feet, ft _____	0.304
3.94×10^{-2} ___	millimeter, mm (10^{-3})	inch, in _____	25.4

SQUARES

2.47 _____	hectare, he	acre _____	0.405
2.47 _____	square kilometer, km^2	acre _____	4.05×10^{-3}
0.386 _____	square kilometer, km^2	square mile, mi^2 _____	2.590
2.47×10^{-4} ___	square meter, m^2	acre _____	4.05×10^{-3}
10.76 _____	square meter, m^2	square feet, ft^2 _____	9.29×10^{-2}
1.55×10^{-3} ___	mm^2	square inch, in^2 _____	645

CUBICS

9.73×10^{-3} ___	cubic meter, m^3	inch-acre _____	102.8
35.3 _____	cubic meter, m^3	cubic-feet, ft^3 _____	2.83×10^{-2}

6.10×10^4 ___ cubic meter, m^3	cubic inch, in^3 _____	1.64×10^{-5}
2.84×10^{-2} ___ liter, L (10^{-3} m^3)	bushel, bu _____	35.24
1.057 _____ liter, L	liquid quarts, qt _____	0.946
3.53×10^{-2} ___ liter, L	cubic feet, ft^3 _____	28.3
0.265 _____ liter, L	gallon _____	3.78
33.78 _____ liter, L	fluid ounce, oz _____	2.96×10^{-2}
2.11 _____ liter, L	fluid dot, dt _____	0.473

WEIGHT

2.20×10^{-3} ___ gram, g (10^{-3} kg)	pound, _____	454
3.52×10^{-2} ___ gram, g (10^{-3} kg)	ounce, oz _____	28.4
2.205 _____ kilogram, kg	pound, lb _____	0.454
10^{-2} _____ kilogram, kg	quintal (metric), q ___	100
1.10×10^{-3} ___ kilogram, kg	ton (2000 lbs), ton .	907
1.10^2 _____ mega gram, mg	ton (US), ton _____	0.907
1.10^2 _____ metric ton, t	ton (US), ton _____	0.907

YIELD AND RATE

0.893 _____ kilogram per hectare hectare	pound per acre _____	1.12
7.77×10^{-2} ___ kilogram per cubic meter	pound per fanega _____	12.87
1.49×10^{-2} ___ kilogram per hectare	pound per acre, _____ 60 lb	67.19
1.59×10^{-2} ___ kilogram per hectare	pound per acre, _____ 56 lb	62.71
1.86×10^{-2} ___ kilogram per hectare	pound per acre, _____ 48 lb	53.75
0.107 _____ liter per hectare	galloon per acre _____	9.35
893 _____ ton per hectare	pound per acre _____	1.12×10^{-3}
893 _____ mega gram per hectare	pound per acre _____	1.12×10^{-3}
0.446 _____ ton per hectare	ton (2000 lb) per _____	2.24

acre
2.24 _____ meter per second   mile per hour _____ 0.447

SPECIFIC SURFACE

10 _____ square meter per   square centimeter _____ 0.1
kilogram   per gram
10^3 _____ square meter per   square millimeter _____ 10^{-3}
kilogram   per gram

PRESSURE

9.90 _____ megapascal, MPa   atmosphere _____ 0.101
10 _____ megapascal   bar _____ 0.1
1.0 _____ megagram per   gram per cubic _____ 1.00
cubic meter   cubic centimeter
2.09×10^{-2} ___ pascal, Pa   pound per square _____ 47.9
                                          feet
1.45×10^{-4} ___ pascal, Pa   pound per square _____ 6.90×10^3
                                          inch

TEMPERATURE

1.00   _____ Kelvin, K   centigrade, °C _____ 1.00
(K-273)   (C+273)
(1.8 C _____ centigrade, °C   Fahrenheit,°F _____ (F-32)/1.8
+ 32)

ENERGY

9.52×10^{-4} ___ Joule J   BTU _____ 1.05×10^3
0.239 _____ Joule, J   calories, cal _____ 4.19
0.735 _____ Joule, J   feet-pound _____ 1.36
2.387×10^5 ___ Joule per square   calories per square ___ 4.19×10^4
meter   centimeter
10^5 _____ Newton, N   dynes _____ 10^{-5}

WATER REQUIREMENTS

9.73×10^{-3} ___ cubic meter	inch acre _____ 102.8
9.81×10^{-3} ___ cubic meter per hour	cubic feet per _____ 101.9 second
4.40 _____ cubic meter per hour	galloon (US) per _____ 0.227 minute
8.11 _____ hectare-meter	acre-feet _____ 0.123
97.28 _____ hectare-meter	acre-inch _____ 1.03×10^{-2}
8.1×10^{-2} ____ hectare centimeter	acre-feet _____ 12.33

CONCENTRATION

1 _____ centimol per kilogram	milliequivalents _____ 1 per 100 grams
0.1 _____ gram per kilogram	percents _____ 10
1 _____ milligram per kilogram	parts per million _____ 1

NUTRIENTS FOR PLANTS

2.29 _____ P	P_2O_5 _____ 0.437
1.20 _____ K	K_2O _____ 0.830
1.39 _____ Ca	CaO _____ 0.715
1.66 _____ Mg	MgO _____ 0.602

NUTRIENT EQUIVALENTS

Column A	Column B	Conversion A to B	Equivalent B to A
N	NH_3	1.216	0.822
	NO_3	4.429	0.226
	KNO_3	7.221	0.1385
	$Ca(NO_3)_2$	5.861	0.171

Column A	Column B	Conversion A to B	Equivalent B to A
	NH_4NO_3	5.718	0.175
	NH_4NO_3	5.718	0.175
N	NH_3	1.216	0.822
	NO_3	4.429	0.226
	KNO_3	7.221	0.1385
	$Ca(NO_3)_2$	5.861	0.171
	$(NH_4)_2SO_4$	4.721	0.212
	NH_4NO_3	5.718	0.175
	$(NH_4)_2HPO_4$	4.718	0.212
P	P_2O_5	2.292	0.436
	PO_4	3.066	0.326
	KH_2PO_4	4.394	0.228
	$(NH_4)_2HPO_4$	4.255	0.235
	H_3PO_4	3.164	0.316
K	K_2O	1.205	0.83
	KNO_3	2.586	0.387
	KH_2PO_4	3.481	0.287
	Kcl	1.907	0.524
	K_2SO_4	2.229	0.449
Ca	CaO	1.399	0.715
	$Ca(NO_3)_2$	4.094	0.244
	$CaCl_2 \times 6H_2O$	5.467	0.183
	$CaSO_4 \times 2H_2O$	4.296	0.233
Mg	MgO	1.658	0.603
	$MgSO_4 \times 7H_2O$	1.014	0.0986
S	H_2SO_4	3.059	0.327
	$(NH_4)_2SO_4$	4.124	0.2425
	K_2SO_4	5.437	0.184
	$MgSO_4 \times 7H_2O$	7.689	0.13
	$CaSO_4 \times 2H_2O$	5.371	0.186

APPENDIX B   PIPE AND CONDUIT FLOW

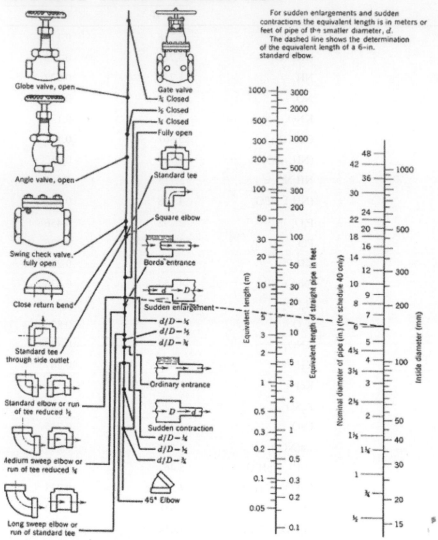

For sudden enlargements and sudden contractions the equivalent length is in meters or feet of pipe of the smaller diameter, *d*. The dashed line shows the determination of the equivalent length of a 6-in. standard elbow.

Friction loss (m per 100 m length of main line) of portable aluminum pipe with couplings: Based on Scobey's formula, for KS = 10 m.

Flow		Pipe diameter, cm						
Liters/s	GPM	7.5	10	12.5	15	17.5	20	25
2.52	40	0.658	0.157	—	—	—	—	—
3.15	50	1.006	0.239	—	—	—	—	—

Flow		Pipe diameter, cm						
Liters/s	GPM	7.5	10	12.5	15	17.5	20	25
3.79	60	1.423	0.339	—	—	—	—	—
4.42	70	1.906	0.449	0.150	—	—	—	—
5.05	80	2.457	0.584	0.193	—	—	—	—
5.68	90	3.073	0.731	0.242	—	—	—	—
6.31	100	3.754	0.893	0.295	0.120	—	—	—
7.57	120	5.307	1.263	0.413	0.170	—	—	—
8.83	140	7.113	1.693	0.560	0.227	—	—	—
10.10	160	9.169	2.182	0.721	0.293	—	—	—
11.36	180	11.47	2.729	0.967	0.366	—	—	—
12.62	200	14.01	3.333	1.102	0.448	0.209	—	—
13.88	220	16.79	3.996	1.321	0.537	0.251	—	—
15.14	240	19.81	4.713	1.558	0.633	0.296	—	—
16.41	260	23.06	5.448	1.814	0.737	0.344	—	—
17.67	280	26.55	6.316	2.089	0.849	0.397	—	—
18.93	300	30.27	7.203	2.381	0.967	0.452	0.235	—
20.19	320	34.22	8.142	2.692	1.094	0.511	0.265	—
21.45	340	38.39	9.137	3.020	1.227	0.573	0.298	—
22.72	360	42.80	10.18	3.366	1.368	0.639	0.332	—
23.98	380	47.43	11.29	3.731	1.516	0.708	0.368	—
25.24	400	52.28	12.44	4.113	1.671	0.781	0.399	0.136
26.50	420	—	13.95	4.513	1.833	0.857	0.445	0.149
27.76	440	—	14.57	4.930	1.988	0.936	0.486	0.163
29.03	460	—	16.23	5.364	2.179	1.019	0.529	0.177
30.29	480	—	17.59	5.815	2.363	1.104	0.573	0.192
31.55	500	—	19.01	6.284	2.554	1.193	0.620	0.208
34.70	550	—	22.79	7.532	3.060	1.430	0.742	0.249
37.86	600	—	26.88	9.886	3.611	1.687	0.876	0.294
41.01	650	—	31.30	10.35	4.204	1.965	1.020	0.342
44.17	700	—	36.03	11.91	4.839	2.262	1.174	0.394
47.32	750	—	41.08	13.58	5.517	2.520	1.339	0.449
50.48	800	—	—	15.35	6.237	2.915	1.513	0.507
53.60	850	—	—	17.32	6.999	3.71	1.698	0.569
56.79	900	—	—	19.20	7.801	3.646	1.893	0.635
59.94	950	—	—	21.28	8.645	4.041	2.097	0.703

Flow		Pipe diameter, cm						
Liters/s	GPM	7.5	10	12.5	15	17.5	20	25
63.10	1000	—	—	23.45	9.530	4.454	2.312	0.775
69.49	1100	—	—	28.11	11.42	5.338	2.771	0.929
75.72	1200	—	—	31.75	13.58	6.298	3.269	1.096
82.03	1300	—	—	—	15.69	7.333	3.806	1.277
88.34	1400	—	—	—	18.06	8.441	4.382	1.470
94.65	1500	—	—	—	20.59	9.624	4.996	1.675
101.0	1600	—	—	—	23.28	10.88	5.648	1.894
107.3	1700	—	—	—	26.12	21.21	6.337	2.125
14.0	1800	—	—	—	—	13.61	7.064	2.369
120.0	1900	—	—	—	—	15.08	7.829	2.625
126.0	2000	—	—	—	—	16.62	8.630	2.894

Friction loss (m per 100 m length of lateral lines) of portable aluminum pipe with couplings: Based on Scobey's formula.

Flow, Liters/s	Pipe diameter, cm				
	5.0	7.5	10	12.5	15
	KS = 0.34	KS = 0.33		KS = 0.32	
1.26	—	—	—	—	—
1.89	0.32	—	—	—	—
2.52	2.53	—	—	—	—
3.15	4.40	0.565	0.130	—	—
3.79	6.85	0.858	0.198	—	—
4.42	9.67	1.21	0.280	—	—
5.05	12.9	1.63	0.376	0.122	—
5.68	16.7	2.10	0.484	0.157	—
6.31	20.8	2.63	0.605	0.196	—
7.57	25.4	3.20	0.738	0.240	0.099
8.83	—	4.54	1.04	0.339	0.140
10.10	—	6.09	1.40	0.454	0.188
11.36	—	7.85	1.80	0.590	0.242

Flow, Liters/s	Pipe diameter, cm				
	5.0	7.5	10	12.5	15
	KS = 0.34	KS = 0.33		KS = 0.32	
12.62	—	9.82	2.26	0.733	0.302
13.88	—	12.0	2.76	0.896	0.370
15.14	—	14.4	3.30	1.07	0.443
16.41	—	16.9	3.90	1.26	0.522
17.67	—	19.7	4.54	1.47	0.608
18.93	—	22.8	5.22	1.70	0.700
20.19	—	25.9	5.96	1.93	0.798
21.45	—	29.3	6.74	2.18	0.904
22.72	—	32.8	7.56	2.45	1.02
23.98	—	36.6	8.40	2.74	1.13
25.24	—	40.6	9.36	3.03	1.26
26.50	—	44.7	10.3	3.34	1.38
27.76	—	—	11.3	3.66	1.521
29.03	—	—	12.3	4.00	1.66
30.29	—	—	13.4	4.35	1.80
31.55	—	—	14.6	4.72	1.95
34.70	—	—	15.8	5.10	2.12
37.86	—	—	18.9	6.12	2.52
41.01	—	—	22.2	7.22	2.98
44.17	—	—	25.9	8.40	3.46
47.32	—	—	29.8	9.68	3.99
50.48	—	—	33.8	11.0	4.54
53.63	—	—		12.5	5.15
56.79	—	—		14.0	5.78
59.94	—	—		15.6	6.44
63.10	—	—		17.3	7.14

APPENDIX C PERCENTAGE OF DAILY SUNSHINE HOURS: FOR NORTH AND SOUTH HEMISPHERES

Latitude	Jan	Feb	Mar	Apr	May	Jun	Jul	Aug	Sep	Oct	Nov	Dec
NORTH												
0	8.50	7.66	8.49	8.21	8.50	8.22	8.50	8.49	8.21	8.50	8.22	8.50
5	8.32	7.57	8.47	3.29	8.65	8.41	8.67	8.60	8.23	8.42	8.07	8.30
10	8.13	7.47	8.45	8.37	8.81	8.60	8.86	8.71	8.25	8.34	7.91	8.10
15	7.94	7.36	8.43	8.44	8.98	8.80	9.05	8.83	8.28	8.20	7.75	7.88
20	7.74	7.25	8.41	8.52	9.15	9.00	9.25	8.96	8.30	8.18	7.58	7.66
25	7.53	7.14	8.39	8.61	9.33	9.23	9.45	9.09	8.32	8.09	7.40	7.52
30	7.30	7.03	8.38	8.71	9.53	9.49	9.67	9.22	8.33	7.99	7.19	7.15
32	7.20	6.97	8.37	8.76	9.62	9.59	9.77	9.27	8.34	7.95	7.11	7.05
34	7.10	6.91	8.36	8.80	9.72	9.70	9.88	9.33	8.36	7.90	7.02	6.92
36	6.99	6.85	8.35	8.85	9.82	9.82	9.99	9.40	8.37	7.85	6.92	6.79
38	6.87	6.79	8.34	8.90	9.92	9.95	10.1	9.47	3.38	7.80	6.82	6.66
40	6.76	6.72	8.33	8.95	10.0	10.1	10.2	9.54	8.39	7.75	6.72	7.52
42	6.63	6.65	8.31	9.00	10.1	10.2	10.4	9.62	8.40	7.69	6.62	6.37
44	6.49	6.58	8.30	9.06	10.3	10.4	10.5	9.70	8.41	7.63	6.49	6.21
46	6.34	6.50	8.29	9.12	10.4	10.5	10.6	9.79	8.42	7.57	6.36	6.04
48	6.17	6.41	8.27	9.18	10.5	10.7	10.8	9.89	8.44	7.51	6.23	5.86
50	5.98	6.30	8.24	9.24	10.7	10.9	11.0	10.0	8.35	7.45	6.10	5.64
52	5.77	6.19	8.21	9.29	10.9	11.1	11.2	10.1	8.49	7.39	5.93	5.43
54	5.55	6.08	8.18	9.36	11.0	11.4	11.4	10.3	8.51	7.20	5.74	5.18

Latitude	Jan	Feb	Mar	Apr	May	Jun	Jul	Aug	Sep	Oct	Nov	Dec
NORTH												
56	5.30	5.95	8.15	9.45	11.2	11.7	11.6	10.4	8.53	7.21	5.54	4.89
58	5.01	5.81	8.12	9.55	11.5	12.0	12.0	10.6	8.55	7.10	4.31	4.56
60	4.67	5.65	8.08	9.65	11.7	12.4	12.3	10.7	8.57	6.98	5.04	4.22
SOUTH												
0	8.50	7.66	8.49	8.21	8.50	8.22	8.50	8.49	8.21	8.50	8.22	8.50
5	8.68	7.76	8.51	8.15	8.34	8.05	8.33	8.38	8.19	8.56	8.37	8.68
10	8.86	7.87	8.53	8.09	8.18	7.86	8.14	8.27	8.17	8.62	8.53	8.88
15	9.05	7.98	8.55	8.02	8.02	7.65	7.95	8.15	8.15	8.68	8.70	9.10
20	9.24	8.09	8.57	7.94	7.85	7.43	7.76	8.03	8.13	8.76	8.87	9.33
25	9.46	8.21	8.60	7.74	7.66	7.20	7.54	7.90	8.11	8.86	9.04	9.58
30	9.70	8.33	8.62	7.73	7.45	6.96	7.31	7.76	8.07	8.97	9.24	9.85
32	9.81	8.39	8.63	7.69	7.36	6.85	7.21	7.70	8.06	9.01	9.33	9.96
34	9.92	8.45	8.64	7.64	7.27	6.74	7.10	7.63	8.05	9.06	9.42	10.1
36	10.0	8.51	8.65	7.59	7.18	6.62	6.99	7.56	8.04	9.11	9.35	10.2
38	10.2	8.57	8.66	7.54	7.08	6.50	6.87	7.49	8.03	9.16	9.61	10.3
40	10.3	8.63	8.67	7.49	6.97	6.37	6.76	7.41	8.02	9.21	9.71	10.5
42	10.4	8.70	8.68	7.44	6.85	6.23	6.64	7.33	8.01	9.26	9.8	10.6
44	10.5	8.78	8.69	7.38	6.73	6.08	6.51	7.25	7.99	9.31	9.94	10.8
46	10.7	8.86	8.90	7.32	6.61	5.92	6.37	7.16	7.96	9.37	10.1	11.0

Mean daily maximum duration of bright sunshine hours (n) for different months and latitudes.

North South	Jan.-July	Feb.-Aug	Mar-Sept.	April-Oct.	May-Nov.	June-Dec.	July-Jan.	Aug.-Feb.	Sept.-Mar	Oct.-April	Nov.-May	Dec.-June
50	8.5	10.1	11.8	13.8	15.4	16.3	15.9	14.5	12.7	10.8	9.1	8.1
48	8.8	10.2	11.8	13.6	15.2	16.0	15.6	14.3	12.6	10.9	9.3	8.3
46	9.1	10.4	11.9	13.5	14.9	15.7	15.4	14.2	12.6	10.9	9.5	8.7
44	9.3	10.5	11.9	13.4	14.7	15.4	15.2	14.0	12.6	11.0	9.7	8.9
42	9.4	10.6	11.9	13.4	14.6	15.2	14.9	13.9	12.6	11.1	9.8	9.1
40	9.6	10.7	11.9	13.3	14.4	15.0	14.7	13.7	12.5	11.2	10.0	9.3
35	10.1	11.0	11.9	13.1	14.0	14.5	14.3	13.5	12.4	11.3	10.3	9.8
30	10.4	11.1	12.0	12.9	13.6	14.0	13.9	13.2	12.4	11.5	10.6	10.2
25	10.7	11.3	12.0	12.7	13.3	13.7	13.5	13.0	12.3	11.6	10.9	10.6
20	11.0	11.5	12.0	12.6	13.1	13.3	13.2	12.8	12.3	11.7	11.2	10.9
15	11.3	11.6	12.0	12.5	12.8	13.0	12.9	12.6	12.2	11.8	11.4	11.2
10	11.6	11.8	12.0	12.3	12.6	12.7	12.6	12.4	12.1	11.8	11.6	11.5
5	11.8	11.9	12.0	12.2	12.3	12.4	12.3	12.3	12.1	12.0	11.9	11.8
0	12.1	12.1	12.1	12.1	12.1	12.1	12.1	12.1	12.1	12.1	12.1	12.1

Mean daily percentage (P) of annual daytime hours for different latitudes.

Latitude North South	Jan.–July	Feb.–Aug	March–Sept.	April–Oct.	May–Nov.	June–Dec.	July–Jan.	Aug.–Feb.	Sept.–March	Oct.–April	Nov.–May	Dec.–June
60°	0.15	0.20	0.26	0.32	0.38	0.41	0.40	0.34	0.28	0.22	0.17	0.13
58°	0.16	0.21	0.26	0.32	0.37	0.40	0.39	0.34	0.28	0.23	0.18	0.15
56°	0.17	0.21	0.26	0.32	0.36	0.39	0.38	0.33	0.28	0.23	0.18	0.16
54°	0.18	0.22	0.26	0.31	0.36	0.38	0.37	0.33	0.28	0.23	0.19	0.17
52°	0.19	0.22	0.27	0.31	0.35	0.37	0.36	0.33	0.28	0.24	0.20	0.17
50°	0.19	0.23	0.27	0.31	0.34	0.36	0.35	0.32	0.28	0.24	0.20	0.18
48°	0.20	0.23	0.27	0.31	0.34	0.36	0.35	0.32	0.28	0.24	0.21	0.19
46°	0.20	0.23	0.27	0.30	0.34	0.35	0.34	0.32	0.28	0.24	0.21	0.20
44°	0.21	0.24	0.27	0.30	0.33	0.35	0.34	0.31	0.28	0.25	0.22	0.20
42°	0.21	0.24	0.27	0.30	0.33	0.34	0.33	0.31	0.28	0.25	0.22	0.21
40°	0.22	0.24	0.27	0.30	0.32	0.34	0.33	0.31	0.28	0.25	0.22	0.21
35°	0.23	0.25	0.27	0.29	0.31	0.32	0.32	0.30	0.28	0.25	0.23	0.22
30°	0.24	0.25	0.27	0.29	0.31	0.32	0.31	0.30	0.28	0.26	0.24	0.23*
25°	0.24	0.26	0.27	0.29	0.30	0.31	0.31	0.29	0.28	0.26	0.25	0.24
20°	0.25	0.26	0.27	0.28	0.29	0.30	0.30	0.29	0.28	0.26	0.25	0.25
15°	0.26	0.26	0.27	0.28	0.29	0.29	0.29	0.28	0.28	0.27	0.26	0.25
10°	0.26	0.27	0.27	0.28	0.28	0.29	0.29	0.28	0.28	0.27	0.26	0.26
5°	0.27	0.27	0.27	0.28	0.28	0.28	0.28	0.28	0.28	0.27	0.27	0.27
0°	0.27	0.27	0.27	0.27	0.27	0.27	0.27	0.27	0.27	0.27	0.27	0.27

APPENDIX D   PSYCHOMETRIC CONSTANT (γ) FOR DIFFERENT ALTITUDES (Z)

$$\gamma = 10^{-3}\,[(C_p.P) \div (\varepsilon.\lambda)] = (0.00163) \times [P \div \lambda]$$

γ, psychrometric constant [kPa C^{-1}]

c_p, specific heat of moist air = 1.013

[kJ kg^{-10}C^{-1}]

P, atmospheric pressure [kPa].

ε, ratio molecular weight of water

vapor/dry air = 0.622

λ, latent heat of vaporization [MJ kg^{-1}]

= 2.45 MJ kg^{-1} at 20°C.

Z (m)	γ kPa/°C	z (m)	γ kPa/°C	z (m)	γ kPa/°C	z (m)	γ kPa/°C
0	0.067	1000	0.060	2000	0.053	3000	0.047
100	0.067	1100	0.059	2100	0.052	3100	0.046
200	0.066	1200	0.058	2200	0.052	3200	0.046
300	0.065	1300	0.058	2300	0.051	3300	0.045
400	0.064	1400	0.057	2400	0.051	3400	0.045
500	0.064	1500	0.056	2500	0.050	3500	0.044
600	0.063	1600	0.056	2600	0.049	3600	0.043
700	0.062	1700	0.055	2700	0.049	3700	0.043
800	0.061	1800	0.054	2800	0.048	3800	0.042
900	0.061	1900	0.054	2900	0.047	3900	0.042
1000	0.060	2000	0.053	3000	0.047	4000	0.041

APPENDIX E   SATURATION VAPOR PRESSURE [e_s] FOR DIFFERENT TEMPERATURES (T)

Vapor pressure function = $e_s = [0.6108]*\exp\{[17.27*T]/[T + 237.3]\}$							
T °C	e_s kPa	T °C	e_s kPa	T °C	e_s kPa	T °C	e_s kPa
1.0	0.657	13.0	1.498	25.0	3.168	37.0	6.275
1.5	0.681	13.5	1.547	25.5	3.263	37.5	6.448
2.0	0.706	14.0	1.599	26.0	3.361	38.0	6.625
2.5	0.731	14.5	1.651	26.5	3.462	38.5	6.806
3.0	0.758	15.0	1.705	27.0	3.565	39.0	6.991
3.5	0.785	15.5	1.761	27.5	3.671	39.5	7.181
4.0	0.813	16.0	1.818	28.0	3.780	40.0	7.376
4.5	0.842	16.5	1.877	28.5	3.891	40.5	7.574
5.0	0.872	17.0	1.938	29.0	4.006	41.0	7.778

Vapor pressure function = e_s = [0.6108]*exp{[17.27*T]/[T + 237.3]}							
T °C	e_s kPa	T °C	e_s kPa	T °C	e_s kPa	T °C	e_s kPa
5.5	0.903	17.5	2.000	29.5	4.123	41.5	7.986
6.0	0.935	18.0	2.064	30.0	4.243	42.0	8.199
6.5	0.968	18.5	2.130	30.5	4.366	42.5	8.417
7.0	1.002	19.0	2.197	31.0	4.493	43.0	8.640
7.5	1.037	19.5	2.267	31.5	4.622	43.5	8.867
8.0	1.073	20.0	2.338	32.0	4.755	44.0	9.101
8.5	1.110	20.5	2.412	32.5	4.891	44.5	9.339
9.0	1.148	21.0	2.487	33.0	5.030	45.0	9.582
9.5	1.187	21.5	2.564	33.5	5.173	45.5	9.832
10.0	1.228	22.0	2.644	34.0	5.319	46.0	10.086
10.5	1.270	22.5	2.726	34.5	5.469	46.5	10.347
11.0	1.313	23.0	2.809	35.0	5.623	47.0	10.613
11.5	1.357	23.5	2.896	35.5	5.780	47.5	10.885
12.0	1.403	24.0	2.984	36.0	5.941	48.0	11.163
12.5	1.449	24.5	3.075	36.5	6.106	48.5	11.447

APPENDIX F SLOPE OF VAPOR PRESSURE CURVE (Δ) FOR DIFFERENT TEMPERATURES (T)

$$\Delta = [4098.\ e^0(T)] \div [T + 237.3]^2$$
$$= 2504\{exp[(17.27T) \div (T + 237.2)]\} \div [T + 237.3]^2$$

T °C	Δ kPa/°C	T °C	Δ kPa/°C	T °C	Δ kPa/°C	T °C	Δ kPa/°C
1.0	0.047	13.0	0.098	25.0	0.189	37.0	0.342
1.5	0.049	13.5	0.101	25.5	0.194	37.5	0.350
2.0	0.050	14.0	0.104	26.0	0.199	38.0	0.358
2.5	0.052	14.5	0.107	26.5	0.204	38.5	0.367
3.0	0.054	15.0	0.110	27.0	0.209	39.0	0.375
3.5	0.055	15.5	0.113	27.5	0.215	39.5	0.384
4.0	0.057	16.0	0.116	28.0	0.220	40.0	0.393
4.5	0.059	16.5	0.119	28.5	0.226	40.5	0.402
5.0	0.061	17.0	0.123	29.0	0.231	41.0	0.412
5.5	0.063	17.5	0.126	29.5	0.237	41.5	0.421

T °C	Δ kPa/°C	T °C	Δ kPa/°C	T °C	Δ kPa/°C	T °C	Δ kPa/°C
6.0	0.065	18.0	0.130	30.0	0.243	42.0	0.431
6.5	0.067	18.5	0.133	30.5	0.249	42.5	0.441
7.0	0.069	19.0	0.137	31.0	0.256	43.0	0.451
7.5	0.071	19.5	0.141	31.5	0.262	43.5	0.461
8.0	0.073	20.0	0.145	32.0	0.269	44.0	0.471
8.5	0.075	20.5	0.149	32.5	0.275	44.5	0.482
9.0	0.078	21.0	0.153	33.0	0.282	45.0	0.493
9.5	0.080	21.5	0.157	33.5	0.289	45.5	0.504
10.0	0.082	22.0	0.161	34.0	0.296	46.0	0.515
10.5	0.085	22.5	0.165	34.5	0.303	46.5	0.526
11.0	0.087	23.0	0.170	35.0	0.311	47.0	0.538
11.5	0.090	23.5	0.174	35.5	0.318	47.5	0.550
12.0	0.092	24.0	0.179	36.0	0.326	48.0	0.562
12.5	0.095	24.5	0.184	36.5	0.334	48.5	0.574

APPENDIX G   NUMBER OF THE DAY IN THE YEAR (JULIAN DAY)

Day	Jan	Feb	Mar	Apr	May	Jun	Jul	Aug	Sep	Oct	Nov	Dec
1	1	32	60	91	121	152	182	213	244	274	305	335
2	2	33	61	92	122	153	183	214	245	275	306	336
3	3	34	62	93	123	154	184	215	246	276	307	337
4	4	35	63	94	124	155	185	216	247	277	308	338
5	5	36	64	95	125	156	186	217	248	278	309	339
6	6	37	65	96	126	157	187	218	249	279	310	340
7	7	38	66	97	127	158	188	219	250	280	311	341
8	8	39	67	98	128	159	189	220	251	281	312	342
9	9	40	68	99	129	160	190	221	252	282	313	343
10	10	41	69	100	130	161	191	222	253	283	314	344
11	11	42	70	101	131	162	192	223	254	284	315	345
12	12	43	71	102	132	163	193	224	255	285	316	346
13	13	44	72	103	133	164	194	225	256	286	317	347
14	14	45	73	104	134	165	195	226	257	287	318	348
15	15	46	74	105	135	166	196	227	258	288	319	349
16	16	47	75	106	136	167	197	228	259	289	320	350
17	17	48	76	107	137	168	198	229	260	290	321	351

Day	Jan	Feb	Mar	Apr	May	Jun	Jul	Aug	Sep	Oct	Nov	Dec
18	18	49	77	108	138	169	199	230	261	291	322	352
19	19	50	78	109	139	170	200	231	262	292	323	353
20	20	51	79	110	140	171	201	232	263	293	324	354
21	21	52	80	111	141	172	202	233	264	294	325	355
22	22	53	81	112	142	173	203	234	265	295	326	356
23	23	54	82	113	143	174	204	235	266	296	327	357
24	24	55	83	114	144	175	205	236	267	297	328	358
25	25	56	84	115	145	176	206	237	268	298	329	359
26	26	57	85	116	146	177	207	238	269	299	330	360
27	27	58	86	117	147	178	208	239	270	300	331	361
28	28	59	87	118	148	179	209	240	271	301	332	362
29	29	(60)	88	119	149	180	210	241	272	302	333	363
30	30	—	89	120	150	181	211	242	273	303	334	364
31	31	—	90	—	151	—	212	243	—	304	—	365

APPENDIX H   STEFAN-BOLTZMANN LAW AT DIFFERENT TEMPERATURES (T):

$$[\sigma^*(T_K)^4] = [4.903 \times 10^{-9}], \text{ MJ K}^{-4} \text{ m}^{-2} \text{ day}^{-1}$$
$$\text{where: } T_K = \{T[°C] + 273.16\}$$

T	$\sigma^*(T_K)^4$	T	$\sigma^*(T_K)^4$	T	$\sigma^*(T_K)^4$
			Units		
°C	MJ m^{-2} d^{-1}	°C	MJ m^{-2} d^{-1}	°C	MJ m^{-2} d^{-1}
1.0	27.70	17.0	34.75	33.0	43.08
1.5	27.90	17.5	34.99	33.5	43.36
2.0	28.11	18.0	35.24	34.0	43.64
2.5	28.31	18.5	35.48	34.5	43.93
3.0	28.52	19.0	35.72	35.0	44.21
3.5	28.72	19.5	35.97	35.5	44.50
4.0	28.93	20.0	36.21	36.0	44.79
4.5	29.14	20.5	36.46	36.5	45.08
5.0	29.35	21.0	36.71	37.0	45.37
5.5	29.56	21.5	36.96	37.5	45.67
6.0	29.78	22.0	37.21	38.0	45.96

T	$\sigma^*(T_K)^4$	T	$\sigma^*(T_K)^4$	T	$\sigma^*(T_K)^4$
			Units		
°C	MJ m^{-2} d^{-1}	°C	MJ m^{-2} d^{-1}	°C	MJ m^{-2} d^{-1}
6.5	29.99	22.5	37.47	38.5	46.26
7.0	30.21	23.0	37.72	39.0	46.56
7.5	30.42	23.5	37.98	39.5	46.85
8.0	30.64	24.0	38.23	40.0	47.15
8.5	30.86	24.5	38.49	40.5	47.46
9.0	31.08	25.0	38.75	41.0	47.76
9.5	31.30	25.5	39.01	41.5	48.06
10.0	31.52	26.0	39.27	42.0	48.37
10.5	31.74	26.5	39.53	42.5	48.68
11.0	31.97	27.0	39.80	43.0	48.99
11.5	32.19	27.5	40.06	43.5	49.30
12.0	32.42	28.0	40.33	44.0	49.61
12.5	32.65	28.5	40.60	44.5	49.92
13.0	32.88	29.0	40.87	45.0	50.24
13.5	33.11	29.5	41.14	45.5	50.56
14.0	33.34	30.0	41.41	46.0	50.87
14.5	33.57	30.5	41.69	46.5	51.19
15.0	33.81	31.0	41.96	47.0	51.51
15.5	34.04	31.5	42.24	47.5	51.84
16.0	34.28	32.0	42.52	48.0	52.16
16.5	34,52	32.5	42.80	48.5	52.49

APPENDIX I   THERMODYNAMIC PROPERTIES OF AIR AND WATER

1. Latent Heat of Vaporization (λ)

$$\lambda = [2.501 - (2.361 \times 10^{-3})\ T]$$

where: λ = latent heat of vaporization [MJ kg^{-1}]; and T = air temperature [°C].

The value of the latent heat varies only slightly over normal temperature ranges. A single value may be taken (for ambient temperature = 20°C): λ = 2.45 MJ kg^{-1}.

2. Atmospheric Pressure (P)

$$P = P_o [\{T_{Ko} - \alpha(Z - Z_o)\} \div \{T_{Ko}\}]^{(g/(\alpha.R))}$$

Where: P, atmospheric pressure at elevation z [kPa]

P_o, atmospheric pressure at sea level = 101.3 [kPa]

z, elevation [m]

z_o, elevation at reference level [m]

g, gravitational acceleration = 9.807 [m s^{-2}]

R, specific gas constant = 287 [J kg^{-1} K^{-1}]

α, constant lapse rate for moist air = 0.0065 [K m^{-1}]

T_{Ko}, reference temperature [K] at elevation z_o = 273.16 + T

T, means air temperature for the time period of calculation [°C]

When assuming P_o = 101.3 [kPa] at z_o = 0, and T_{Ko} = 293 [K] for T = 20 [°C], above equation reduces to:

$$P = 101.3[(293 - 0.0065Z)\,(293)]^{5.26}$$

3. Atmospheric Density (ρ)

$$\rho = [1000P] \div [T_{Kv}\,R] = [3.486P] \div [T_{Kv}], \text{ and } T_{Kv} = T_K[1 - 0.378(e_a)/P]^{-1}$$

where: ρ, atmospheric density [kg m^{-3}]

R, specific gas constant = 287 [J kg^{-1} K^{-1}]

T_{Kv}, virtual temperature [K]

T_K, absolute temperature [K]: T_K = 273.16 + T [°C]

e_a, actual vapor pressure [kPa]

T, mean daily temperature for 24-hour calculation time steps.

For average conditions (e_a in the range 1–5 kPa and P between 80–100 kPa), T_{Kv} can be substituted by: $T_{Kv} \approx 1.01\,(T + 273)$

4. Saturation Vapor Pressure function (e_s)

$$e_s = [0.6108]*\exp\{[17.27*T]/[T + 237.3]\}$$

where: e_s, saturation vapor pressure function [kPa]

T, air temperature [°C]

5. Slope Vapor Pressure Curve (Δ)

$$\Delta = [4098. \, e°(T)] \div [T + 237.3]^2$$

$$= 2504\{\exp[(17.27T) \div (T + 237.2)]\} \div [T + 237.3]^2$$

where: Δ, slope vapor pressure curve [kPa C^{-1}]

T, air temperature [°C]

$e^0(T)$, saturation vapor pressure at temperature T [kPa]

In 24-hour calculations, Δ is calculated using mean daily air temperature. In hourly calculations T refers to the hourly mean, T_{hr}.

6. Psychrometric Constant (γ)

$$\gamma = 10^{-3} \, [(C_p.P) \div (\varepsilon.\lambda)] = (0.00163) \times [P \div \lambda]$$

where: γ, psychrometric constant [kPa C^{-1}]

c_p, specific heat of moist air = 1.013 [kJ kg^{-10}C^{-1}]

P, atmospheric pressure [kPa]: equations 2 or 4

ε, ratio molecular weight of water vapor/dry air = 0.622

λ, latent heat of vaporization [MJ kg^{-1}]

7. Dew Point Temperature (T_{dew})

When data is not available, T_{dew} can be computed from e_a by:

$$T_{dew} = [\{116.91 + 237.3 \mathrm{Log}_e(e_a)\} \div \{16.78 - \mathrm{Log}_e(e_a)\}]$$

where: T_{dew}, dew point temperature [°C]

e_a, actual vapor pressure [kPa]

For the case of measurements with the Assmann psychrometer, T_{dew} can be calculated from:

$$T_{dew} = (112 + 0.9T_{wet})[e_a \div (e^0 \, T_{wet})]^{0.125} - [112 - 0.1T_{wet}]$$

8. Short Wave Radiation on a Clear-Sky Day (R_{so})

The calculation of R_{so} is required for computing net long wave radiation and for checking calibration of pyranometers and integrity of R_{so} data. A good approximation for R_{so} for daily and hourly periods is:

$$R_{so} = (0.75 + 2 \times 10^{-5} z)R_a$$

where: z, station elevation [m]

R_a, extraterrestrial radiation [MJ m^{-2} d^{-1}].

Equation is valid for station elevations less than 6000 m having low air turbidity. The equation was developed by linearizing Beer's radiation extinction law as a function of station elevation and assuming that the average angle of the sun above the horizon is about 50°.

For areas of high turbidity caused by pollution or airborne dust or for regions where the sun angle is significantly less than 50° so that the path length of radiation through the atmosphere is increased, an adoption of Beer's law can be employed where P is used to represent atmospheric mass:

$$R_{so} = (R_a) \exp[(-0.0018P) \div (K_t \sin(\Phi))]$$

where: K_t, turbidity coefficient, $0 < K_t < 1.0$ where $K_t = 1.0$ for clean air and

$K_t = 1.0$ for extremely turbid, dusty or polluted air.

P, atmospheric pressure [kPa]

Φ, angle of the sun above the horizon [rad]

R_a, extraterrestrial radiation [MJ m^{-2} d^{-1}]

For hourly or shorter periods, Φ is calculated as:

$$\sin \Phi = \sin \varphi \sin \delta + \cos \varphi \cos \delta \cos \omega$$

where: φ, latitude [rad]

δ, solar declination [rad] (Eq. (24) in Chapter 3)

ω, solar time angle at midpoint of hourly or shorter period [rad]

For 24-hour periods, the mean daily sun angle, weighted according to R_a, can be approximated as:

$$\sin(\Phi_{24}) = \sin[0.85 + 0.3 \varphi \sin\{(2\pi J/365)-1.39\}-0.42 \varphi^2]$$

where: Φ_{24}, average Φ during the daylight period, weighted according to R_a [rad]

φ, latitude [rad]

J, day in the year.

The Φ_{24} variable is used to represent the average sun angle during daylight hours and has been weighted to represent integrated 24-hour transmission effects on 24-hour R_{so} by the atmosphere. Φ_{24} should be limited to > 0. In some situations, the estimation for R_{so} can be improved by modifying to consider the effects of water vapor on short wave absorption, so that: $R_{so} = (K_B + K_D) R_a$ where:

$$K_B = 0.98\exp[\{(-0.00146P) \div (K_t \sin \Phi)\} - 0.091\{w/\sin \Phi\}^{0.25}]$$

where: K_B, the clearness index for direct beam radiation

K_D, the corresponding index for diffuse beam radiation

$K_D = 0.35 - 0.33 K_B$ for $K_B > 0.15$

$K_D = 0.18 + 0.82 K_B$ for $K_B < 0.15$

R_a, extraterrestrial radiation [MJ m^{-2} d^{-1}]

K_t, turbidity coefficient, $0 < K_t < 1.0$ where $K_t = 1.0$ for clean air and $K_t = 1.0$ for extremely turbid, dusty or polluted air.

P, atmospheric pressure [kPa]

Φ, angle of the sun above the horizon [rad]

W, perceptible water in the atmosphere [mm] $= 0.14 e_a P + 2.1$

e_a, actual vapor pressure [kPa]

P, atmospheric pressure [kPa]

APPENDIX J   PSYCHROMETRIC CHART AT SEA LEVEL.

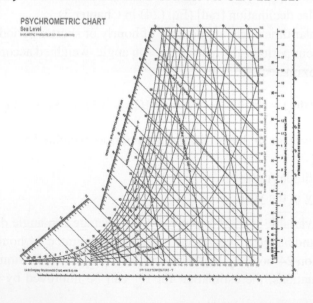

APPENDIX K

[<http://www.fao.org/docrep/T0551E/t0551e07.htm#5.5%20field%20 management%20practices%20in%20wastewater%20irrigation>]

1. **Relationship between applied water salinity and soil water salinity at different leaching fractions (FAO 1985)**

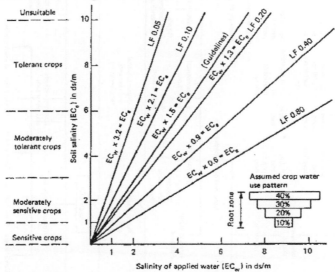

2. **Schematic representations of salt accumulation, planting positions, ridge shapes and watering patterns.**

3. **Main components of general planning guidelines for wastewater reuse (Cobham and Johnson 1988)**

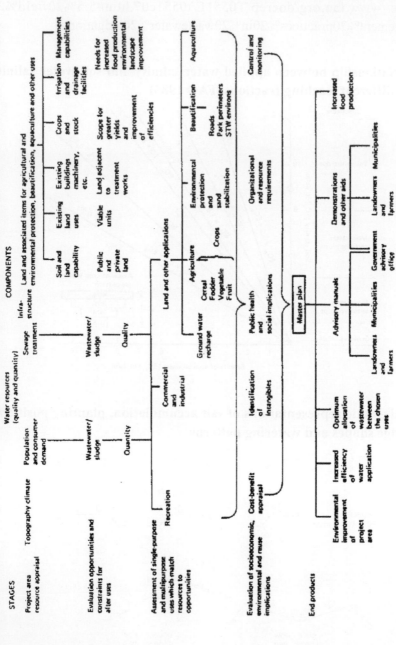

APPENDIX L   VALUES OF *Kc* FOR FIELD AND VEGETABLE CROPS FOR DIFFERENT CROP GROWTH STAGES AND PREVAILING CLIMATIC CONDITIONS.

Crop	Relative humidity				
		RHmin > 70%		RHmin < 20%	
	Crop stage	Wind speed, m/sec			
	Initial 1	0–5	5–8	0–5	5–8
	Crop development 2	Values of K_c			
	Mid-season 3	0.95	0.95	1.0	1.05
	Late season/maturity 4	0.9	0.9	0.95	1.0
Barley	3	1.05	1.1	1.15	1.2
	4	0.25	0.25	0.2	0.2
Beans (green)	3	0.95	0.95	1.0	1.05
	4	0.85	0.85	0.9	0.9
Beans, dry / pulses	3	1.05	1.1	1.15	1.2
	4	0.3	0.3	0.25	0.25
Beets	3	1.0	1.0	1.05	1.1
	4	0.9	0.9	0.95	1.0
Carrots	3	1.0	1.05	1.1	1.15
	4	0.7	0.75	0.8	0.85
Sweet corn (maize)	3	1.05	1.1	1.15	1.2
	4	0.95	1.0	1.05	1.1
Cotton	3	1.05	1.15	1.2	1.25
	4	0.65	0.65	0.65	0.7
Crucifers (cabbage, cauliflower, broccoli)	3	0.95	1.0	1.05	1.1
	4	0.80	0.85	0.9	0.95
Cucumber	3	0.9	0.9	0.95	1.0
	4	0.7	0.7	0.75	0.8
Lentil	3	1.05	1.1	1.15	1.2
	4	0.3	0.3	0.25	0.25
Melons	3	0.95	0.95	1.0	1.05
	4	0.65	0.65	0.75	0.75
Millet	3	1.0	1.05	1.1	1.15
	4	0.3	0.3	0.25	0.25
Oats	3	1.05	1.1	1.15	1.2
	4	0.25	0.25	0.2	0.2
Onion (dry)	3	0.95	0.95	1.05	1.1
	4	0.75	0.75	0.8	0.85

Crop	Relative humidity				
		RHmin > 70%		RHmin < 20%	
	Crop stage		Wind speed, m/sec		
	Initial 1	0–5	5–8	0–5	5–8
	Crop development 2		Values of K_c		
	Mid-season 3	0.95	0.95	1.0	1.05
	Late season/maturity 4	0.9	0.9	0.95	1.0
Onion	3	0.95	0.95	1.0	1.05
(green)	4	0.95	0.95	1.0	1.05
Peanuts	Mid-season 3	0.95	1.0	1.05	1.1
(Groundnut)	Late season/maturity 4	0.55	0.55	0.6	0.6
Peas	3	1.05	1.1	1.15	1.2
	4	0.95	1.0	1.05	1.1
Potato	3	1.05	1.1	1.15	1.2
	4	0.7	0.7	0.75	0.75
Radish	3	0.8	0.8	0.85	0.9
	4	0.75	0.75	0.8	0.85
Safflower	3	1.05	1.1	1.15	1.2
	4	0.25	0.25	0.2	0.2
Sorghum	3	1.0	1.05	1.1	1.15
	4	0.5	0.5	0.55	0.55
Soybeans	3	1.0	1.05	1.1	1.15
	4	0.45	0.45	0.45	0.45
Spinach	3	0.95	0.95	1.0	1.05
	4	0.9	0.9	0.95	1.0
Sugarbeet	3	1.05	1.1	1.15	1.2
	4	0.9	0.95	1.0	1.0
Sunflower	3	1.05	1.1	1.15	1.2
	4	0.4	0.4	0.35	0.35
Tomato	3	1.05	1.1	1.2	1.25
	4	0.6	0.6	0.65	0.65
Wheat	3	1.05	1.1	1.15	1.2
	4	0.25	0.25	0.2	0.2

Note: Values of Kc in this table are for field and vegetable crops; values of Kc for other crops are reported by Doorenbos and Pruitt (1977).

APPENDIX M CROP TOLERANCE AND YIELD POTENTIAL OF CROPS AFFECTED BY IRRIGATION WATER SALINITY (EC_w) OR SOIL SALINITY (EC_e)

Field crops	100%		90%		75%		50%		0% Maximum	
	EC_e	EC_w	EC_e	EC_w	EC_e	EC_w	EC_e	EC_w	EC_e	EC_w
Barley (*Hordeum vulgare*)	8.0	5.3	10	6.7	13	8.7	18	12	28	19
Cotton (*Gossypium hirsutum*)	7.7	5.1	9.6	6.4	13	8.4	17	12	27	18
Sugarbeet (*Beta vulgaris*)	7.0	4.7	8.7	5.8	11	7.5	15	10	24	16
Sorghum (*Sorghum bicolor*)	6.8	4.5	7.4	5.0	8.4	5.6	9.9	6.7	13	8.7
Wheat (*Triticum aestivum*)	6.0	4.0	7.4	4.9	9.5	6.3	13	8.7	20	13
Wheat, durum (*Triticum turgidum*)	5.7	3.8	7.6	5.0	10	6.9	15	10	24	16
Soybean (*Glycine max*)	5.0	3.3	5.5	3.7	6.3	4.2	7.5	5.0	10	6.7
Cowpea (*Vigna unguiculata*)	4.9	3.3	5.7	3.8	7.0	4.7	9.1	6.0	13	8.8
Groundnut (Peanut) (*Arachis hypogaea*)	3.2	2.1	3.5	2.4	4.1	2.7	4.9	3.3	6.6	4.4
Rice (paddy) (*Oriza sativa*)	3.0	2.0	3.8	2.6	5.1	3.4	7.2	4.8	11	7.6
Sugarcane (*Saccharum officinarum*)	1.7	1.1	3.4	2.3	5.9	4.0	10	6.8	19	12
Corn (maize) (*Zea mays*)	1.7	1.1	2.5	1.7	3.8	2.5	5.9	3.9	10	6.7
Flax (*Linum usitatissimum*)	1.7	1.1	2.5	1.7	3.8	2.5	5.9	3.9	10	6.7
Broadbean (*Vicia faba*)	1.5	1.1	2.6	1.8	4.2	2.0	6.8	4.5	12	8.0
Bean (*Phaseolus vulgaris*)	1.0	0.7	1.5	1.0	2.3	1.5	3.6	2.4	6.3	4.2

Field crops	100%		90%		75%		50%		0% Maximum	
	EC_e	EC_w	EC_e	EC_w	EC_e	EC_w	EC_e	EC_w	EC_e	EC_w
Vegetables										
Squash, zucchini (courgette) (*Cucurbita pepo melopepo*)	4.7	3.1	5.8	3.8	7.4	4.9	10	6.7	15	10
Beet, red (*Beta vulgaris*)	4.0	2.7	5.1	3.4	6.8	4.5	9.6	6.4	15	10
Squash, scallop (*Cucurbita pepo melopepo*)	3.2	2.1	3.8	2.6	4.8	3.2	6.3	4.2	9.4	6.3
Broccoli (*Brassica oleracea botrytis*)	2.8	1.9	3.9	2.6	5.5	3.7	8.2	5.5	14	9.1
Tomato (*Lycopersicon esculentum*)	2.5	1.7	3.5	2.3	5.0	3.4	7.6	5.0	13	8.4
Cucumber (*Cucumis sativus*)	2.5	1.7	3.3	2.2	4.4	2.9	6.3	4.2	10	6.8
Spinach (*Spinacia oleracea*)	2.0	1.3	3.3	2.2	5.3	3.5	8.6	5.7	15	10
Celery (*Apium graveolens*)	1.8	1.2	3.4	2.3	5.8	3.9	9.9	6.6	18	12
Cabbage (*Brassica oleracea capitata*)	1.8	1.2	2.8	1.9	4.4	2.9	7.0	4.6	12	8.1
Potato (*Solanum tuberosum*)	1.7	1.1	2.5	1.7	3.8	2.5	5.9	3.9	10	6.7
Corn, sweet (maize) (*Zea mays*)	1.7	1.1	2.5	1.7	3.8	2.5	5.9	3.9	10	6.7
Sweet potato (*Ipomoea batatas*)	1.5	1.0	2.4	1.6	3.8	2.5	6.0	4.0	11	7.1
Pepper (*Capsicum annuum*)	1.5	1.0	2.2	1.5	3.3	2.2	5.1	3.4	8.6	5.8
Lettuce (*Lactuca sativa*)	1.3	0.9	2.1	1.4	3.2	2.1	5.1	3.4	9.0	6.0

Field crops	100%		90%		75%		50%		0% Maximum	
	EC_e	EC_w	EC_e	EC_w	EC_e	EC_w	EC_e	EC_w	EC_e	EC_w
Radish (*Raphanus saivus*)	1.2	0.8	2.0	1.3	3.1	2.1	5.0	3.4	8.9	5.9
Onion (*Allium cepa*)	1.2	0.8	1.8	1.2	2.8	1.8	4.3	2.9	7.4	5.0
Carrot (*Daucus caroka*)	1.0	0.7	1.7	1.1	2.8	1.9	4.6	3.0	8.1	5.4
Bean (*Phaseolus vulgaris*)	1.0	0.7	1.5	1.0	2.3	1.5	3.6	2.4	6.3	4.2
Turnip (*Brassica rapa*)	0.9	0.6	2.0	1.3	3.7	2.5	6.5	4.3	12	8.0
Wheatgrass, tall (*Agropyron elongatum*)	7.5	5.0	9.9	6.6	13	9.0	19	13	31	21
Wheatgrass, fairway crested (*Agropyron cristatum*)	7.5	5.0	9.0	6.0	11	7.4	15	9.8	22	15
Bermuda grass (*Cynodon dactylon*)	6.9	4.6	8.5	5.6	11	7.2	15	9.8	23	15
Barley (forage) (*Hordeum vulgare*)	6.0	4.0	7.4	4.9	9.5	6.4	13	8.7	20	13
Ryegrass, perennial (*Lolium perenne*)	5.6	3.7	6.9	4.6	8.9	5.9	12	8.1	19	13
Trefoil, narrow leaf birds foot[8] (*Lotus corniculatae tenuifolium*)	5.0	3.3	6.0	4.0	7.5	5.0	10	6.7	15	10
Harding grass (*Phalaris tuberosa*)	4.6	3.1	5.9	3.9	7.9	5.3	11	7.4	18	12
Fescue, tall (*Festuca elatior*)	3.9	2.6	5.5	3.6	7.8	5.2	12	7.8	20	13
Wheatgrass, standard crested (*Agropyron sibiricum*)	3.5	2.3	6.0	4.0	9.8	6.5	16	11	28	19
Vetch, common (*Vicia angustifolia*)	3.0	2.0	3.9	2.6	5.3	3.5	7.6	5.0	12	8.1
Sudan grass (*Sorghum sudanense*)	2.8	1.9	5.1	3.4	8.6	5.7	14	9.6	26	17

Field crops	100%		90%		75%		50%		0% Maximum	
	EC_e	EC_w	EC_e	EC_w	EC_e	EC_w	EC_e	EC_w	EC_e	EC_w
Wildrye, beardless (*Elymus triticoides*)	2.7	1.8	4.4	2.9	6.9	4.6	11	7.4	19	13
Cowpea (forage) (*Vigna unguiculata*)	2.5	1.7	3.4	2.3	4.8	3.2	7.1	4.8	12	7.8
Trefoil, big (*Lotus uliginosus*)	2.3	1.5	2.8	1.9	3.6	2.4	4.9	3.3	7.6	5.0
Sesbania (*Sesbania exaltata*)	2.3	1.5	3.7	2.5	5.9	3.9	9.4	6.3	17	11
Sphaerophysa (*Sphaerophysa salsula*)	2.2	1.5	3.6	2.4	5.8	3.8	9.3	6.2	16	11
Alfalfa (*Medicago sativa*)	2.0	1.3	3.4	2.2	5.4	3.6	8.8	5.9	16	10
Lovegrass (*Eragrostis sp.*)	2.0	1.3	3.2	2.1	5.0	3.3	8.0	5.3	14	9.3
Corn (forage) (maize) (*Zea mays*)	1.8	1.2	3.2	2.1	5.2	3.5	8.6	5.7	15	10
Clover, berseem (*Trifolium alexandrinum*)	1.5	1.0	3.2	2.2	5.9	3.9	10	6.8	19	13
Orchard grass (*Dactylis glomerata*)	1.5	1.0	3.1	2.1	5.5	3.7	9.6	6.4	18	12
Foxtail, meadow (*Alopecurus pratensis*)	1.5	1.0	2.5	1.7	4.1	2.7	6.7	4.5	12	7.9
Clover, red (*Trifolium pratense*)	1.5	1.0	2.3	1.6	3.6	2.4	5.7	3.8	9.8	6.6
Clover, alsike (*Trifolium hybridum*)	1.5	1.0	2.3	1.6	3.6	2.4	5.7	3.8	9.8	6.6
Clover, ladino (*Trifolium repens*)	1.5	1.0	2.3	1.6	3.6	2.4	5.7	3.8	9.8	6.6
Clover, strawberry (*Trifolium fragiferum*)	1.5	1.0	2.3	1.6	3.6	2.4	5.7	3.8	9.8	6.6

Field crops	100%		90%		75%		50%		0% Maximum	
	EC_e	EC_w	EC_e	EC_w	EC_e	EC_w	EC_e	EC_w	EC_e	EC_w
Fruit crops										
Date palm (*phoenix dactylifera*)	4.0	2.7	6.8	4.5	11	7.3	18	12	32	21
Grapefruit (*Citrus paradisi*)	1.8	1.2	2.4	1.6	3.4	2.2	4.9	3.3	8.0	5.4
Orange (*Citrus sinensis*)	1.7	1.1	2.3	1.6	3.3	2.2	4.8	3.2	8.0	5.3
Peach (*Prunus persica*)	1.7	1.1	2.2	1.5	2.9	1.9	4.1	2.7	6.5	4.3
Apricot (*Prunus armeniaca*)	1.6	1.1	2.0	1.3	2.6	1.8	3.7	2.5	5.8	3.8
Grape (*Vitus* sp.)	1.5	1.0	2.5	1.7	4.1	2.7	6.7	4.5	12	7.9
Almond (*Prunus dulcis*)	1.5	1.0	2.0	1.4	2.8	1.9	4.1	2.8	6.8	4.5
Plum, prune (*Prunus domestica*)	1.5	1.0	2.1	1.4	2.9	1.9	4.3	2.9	7.1	4.7
Blackberry (*Rubus* sp.)	1.5	1.0	2.0	1.3	2.6	1.8	3.8	2.5	6.0	4.0
Boysenberry (*Rubus ursinus*)	1.5	1.0	2.0	1.3	2.6	1.8	3.8	2.5	6.0	4.0
Strawberry (*Fragaria* sp.)	1.0	0.7	1.3	0.9	1.8	1.2	2.5	1.7	4	2.7

Source: Maas and Hoffman (1977) and Maas (1984).

INDEX

Printed and bound by CPI Group (UK) Ltd, Croydon, CR0 4YY

23/10/2024

01777701-0019